Biology for GCSE

Pauline Alderson and Martin Rowland

The Authors

Pauline Alderson was formerly O-level chief examiner in Biology for The Associated Examining Board. She is now moderator in CSE Biology and Human Biology for the Southern Regional Examinations Board, moderator in Joint 16+ and O-level Biology and Human Biology for the University of London School Examinations Board, and A-level chief examiner in Biology for The Associated Examining Board.

Martin Rowland is Senior Lecturer and Deputy Head of the Department of Science, Mathematics and Computing at East Herts College and O-level chief examiner in Human Biology for the University of London School Examinations Board. He was formerly an assistant examiner for The Associated Examining Board in Biology and Human Biology at O-level and in Human Biology at A-level.

Macmillan Education

© Pauline Alderson and Martin Rowland 1985

All rights reserved. No reproduction, copy or transmission of this publication may be made without written permission.

No paragraph of this publication may be reproduced, copied or transmitted save with written permission or in accordance with the provisions of the Copyright Act 1956 (as amended).

Any person who does any unauthorised act in relation to this publication may be liable to criminal prosecution and civil claims for damages.

First published 1985
Reprinted 1987

Published by
MACMILLAN EDUCATION LTD
Houndmills, Basingstoke, Hampshire RG21 2XS
and London
Companies and representatives
throughout the world

Printed in Hong Kong

ISBN 0-333-36535-6

Acknowledgements

The authors and publishers wish to acknowledge the following photograph sources:

Pete Addis p. 55; Heather Angel pp. 2 top, 4, 5, 6 bottom right; Donald P. Bennett p. 8; Biophoto Associates pp. 78, 102; British Museum (Natural History) p. 94; Cage & Aviary Birds/Photo Dennis Avon p. 1 bottom right; Camera Press p. 54; Bruce Coleman Ltd p. 1 top right; Gene Cox pp. 1 left, 44, 64, 65, 74, 120, 123, 175 left, 190, 196; Family Planning Assoc. p. 213; W.J. Garnett pp. 59, 156; Griffin & George p. 179; Philip Harris p. 60; Popperfoto p. 6 left; Martin Rowland pp. 18, 19, 51, 85, 86, 105, 180, 223; Jim Turner pp. 175 right, 203; Water Research Centre p. 57; Wellcome Institute for the History of Medicine p. 96.

The publishers have made every effort to trace the copyright holders, but if they have inadvertently overlooked any, they will be pleased to make the necessary arrangements at the earliest opportunity.

Contents

Experiments		iv
Introduction for teachers		vi

Unit 1	Variety of organisms	1
Unit 2	Simple organisms; increase in complexity; names; identification; keys	6
Unit 3	Study of a habitat	13
Unit 4	Measurement of environmental factors	18
Unit 5	Energy: release and use	22
Unit 6	Energy: transfer	29
Unit 7	Energy: capture and loss	32
Unit 8	Energy: flow through an ecosystem	37
Unit 9	Cycles in the ecosystem	41
Unit 10	Decomposition; bacteria and fungi; saprophytes; parasites; mutualism	47
Unit 11	Population size: colonisation, succession, competition and population control	50
Unit 12	Human influences: use of land; non-renewable resources; pollution; pest and weed control; recycling	54
Unit 13	The cell	59
Unit 14	Cell specialisation	64
Unit 15	Diffusion and active transport	67
Unit 16	Diffusion of water: osmosis	71
Unit 17	Enzymes	77
Unit 18	Enzyme activity	82
Unit 19	Plant nutrition 1: photosynthesis	86
Unit 20	Plant nutrition 2	90
Unit 21	Human diet	94
Unit 22	Digestion and absorption	100
Unit 23	Digestion and transport in plants	106
Unit 24	Transport in humans	111
Unit 25	Blood and its functions	119
Unit 26	Antibodies; allergy; blood groups of the ABO system	123
Unit 27	Gas exchange in flowering plants and humans	128
Unit 28	Gas exchange in humans; exercise; smoking and pollution	134
Unit 29	Excretion in flowering plants and humans	141
Unit 30	Kidney: structure and function; kidneys and health	146
Unit 31	Sensitivity and response 1	151
Unit 32	Sensitivity and response 2	155
Unit 33	The brain and drugs	161
Unit 34	Hormones; human hormones	164
Unit 35	Temperature regulation; the liver; homeostasis	167
Unit 36	Hormones in plants; artificial use of hormones	171
Unit 37	Living on land: support in flowering plants and arthropods; surface area and volume	174
Unit 38	Support and size in mammals; movement of bones	179
Unit 39	Growth	184
Unit 40	Cell division and mitosis; asexual reproduction	189
Unit 41	Life cycles of humans and flowering plants; sexual reproduction	195
Unit 42	Life cycle of the French bean	201
Unit 43	Sexual reproduction in humans 1	208
Unit 44	Sexual reproduction in humans 2; sexual reproduction in trout	215
Unit 45	Inheritance and genetics	218
Unit 46	Human inheritance	223
Unit 47	Genetic variation	229
Unit 48	Variation due to the environment; natural and artificial selection	234
Appendix A	Mathematical, physical and chemical background	237
Appendix B	How to draw; how to answer examination questions; how to revise	242
Appendix C	for teachers Practical work	245
Index		248

Experiments

3.1 To investigate the distribution of one plant species using quadrats at regular intervals along a line transect

3.2 To investigate the distribution of algae on a tree trunk using a line transect and to measure one environmental variable which might influence it

3.3 To determine the density of an aquatic population using a volume-sampling technique

5.1 To measure the heat energy released during the combustion of a peanut

5.2 To investigate the products of food combustion

5.3 To test whether peas use oxygen in respiration

5.4 To investigate the products of anaerobic respiration

5.5 To measure the rate of anaerobic respiration in a yeast-dough mixture

5.6 To measure the rate of respiration of French-bean seeds using a simple respirometer

7.1 To investigate the hypothesis that light energy is needed for the production of sugars by plant leaves

7.2 To determine the volume of oxygen produced by a submerged aquatic plant during photosynthesis

9.1 To investigate the influence of nitrogen-fixing bacteria on the growth rate of the French bean

10.1 To investigate the influence of temperature on the rate of decomposition

13.1 To investigate the structure of human cheek cells

13.2 To investigate the structure of onion epidermis cells

14.1 To investigate the structure of cells in a moss leaf

15.1 To investigate the rate of diffusion of ammonia

15.2 To investigate the diffusion of methylene blue in gelatin

16.1 To investigate the diffusion of water (osmosis) through a selectively permeable membrane

16.2 To investigate the effect of diffusion of water (osmosis) in dried sultanas

17.1 To investigate the importance of the enzyme amylase in the breakdown (hydrolysis) of starch to reducing sugar

17.2 To investigate the importance of the enzyme glucose (starch) phosphorylase in the synthesis of starch from reducing sugar

18.1 To investigate the effect of different temperatures on the activity of amylase

18.2 To investigate the effect of pH on the activity of amylase

20.1 To investigate the effects of nitrate and iron deficiencies on the growth of plant seedlings

21.1 To carry out qualitative tests for biologically important chemicals

22.1 To investigate the hypothesis that saliva contains an amylase enzyme

22.2 To investigate the hypothesis that fungi secrete digestive enzymes on to their substratum

23.1 To investigate the ability of a potato tuber and a germinating French-bean seed to break down starch

23.2 To investigate the movement of water through a plant leaf stalk or stem

24.1 To dissect a mammalian heart

25.1 To investigate the structure of blood in a human blood smear

26.1 To determine a human ABO blood group

27.1 To investigate the relation between the production of carbon dioxide and its absorption by plants

27.2 To investigate the structure of the respiratory system of a mammal

28.1 To compare the carbon-dioxide content of inhaled (atmospheric) air and exhaled air

28.2 To compare the oxygen content of atmospheric air and exhaled air

28.3 To measure the vital capacity of human lungs

28.4 To investigate the effect of exercise on the pulse rate and on the breathing rate

28.5 To investigate whether there is tar in cigarette smoke

29.1 To dissect a mammalian kidney

30.1 To investigate the microscopic structure of a mammalian kidney
30.2 To test samples of human urine for protein and reducing sugar
31.1 To investigate the effect of unidirectional light on the growth of plant shoots
31.2 To investigate the variety of stimuli to which the human skin responds
32.1 To investigate the human skin's ability to distinguish close simultaneous stimuli
32.2 To investigate the distribution of tissues in human skin
32.3 To demonstrate human reflexes
35.1 To investigate the skin and muscle changes that occur during temperature regulation
36.1 To investigate the effect of rooting powder on the production of roots by stem cuttings
37.1 To investigate the distribution of supporting tissues in a plant stem and a mammalian limb bone
38.1 To investigate the role of calcium salts and protein in producing strength in bones
39.1 To investigate the growth of an insect
39.2 To investigate the growth of a plant stem
40.1 To demonstrate a number of plant propagation methods
40.2 To demonstrate mitosis with the use of bead models
41.1 To investigate the structure of a flower
42.1 To determine the conditions needed for the germination of French-bean seeds
42.2 To determine whether a flower is capable of self-pollination
45.1 To investigate inheritance in maize cobs
46.1 To demonstrate reduction division with the use of bead models
47.1 To investigate the influence of light on the growth of genetically identical plants
48.1 To investigate the effect of competition for space on the germination of cress seeds

Introduction for teachers

This book covers the Southern Group Syllabus for the General Certificate of Secondary Education Biology examination, including the practical work. The syllabus differs from traditional ones in requiring less knowledge of descriptive detail and technical terms and more understanding of biological processes and of techniques for studying them. By comparison with traditional textbooks, this one presents less to learn, more to understand and more to do.

Written for the whole GCSE ability range, it contains something for everyone. So far as possible, each unit begins with simple but essential matter and becomes more difficult as it goes on. In working through a unit students should first master the early explanations and then feel free to concentrate on parts they find easy and interesting.

The book contains 48 units. If it is studied as a two-year course, a week can be devoted to each unit and there will still be spare weeks for more visits to a habitat, extra practical work and revision. To make the book suitable also for a one-year course, the units have been paired to the extent that a topic is begun in an odd-numbered unit and concluded in the next unit. Because availability of practical materials depends on the season of the year, it will not often be possible for a class to study the units in order from 1 to 48. The pairs of units have therefore been made largely self-sufficient (which involves some repetition), but the two units in each pair must be studied in their correct order.

Guidance on drawing graphs, calculating means and the like is given in Appendix A, guidance on drawing biological specimens, answering examination questions and revising for the GCSE examination in Appendix B.

Appendix C lists the apparatus and reagents needed for the practical work. Before beginning an experiment, students should read right through the sections on Materials and Method. Practical work is not divided equally among the units because some topics require more than others. Alternatives have been suggested in units in which there is little or no practical work.

The number of questions and the work they involve also vary from unit to unit. The purpose of the questions is not to provide the same amount of homework each week but to give students whatever work is necessary to consolidate their study of each topic and to test the same abilities as will be tested in the GCSE examination. Questions can be answered either as homework or in class. Questions 'left over' should be answered when time becomes available during units studied later. Students should answer all the questions in the book at one time or another. Indeed they should answer other questions as well, in particular those in recent examination papers.

The biological terms in the syllabus are used consistently in the book and each is printed in bold type at least once, usually where it is explained. Established alternative terms are also mentioned. Unless otherwise stated, *plants* means *flowering plants*. *Protoplasm* is used in its restricted modern sense of cytoplasm plus nuclei.

The syllabus requires detailed study of only two species, humans and French beans. Humans were chosen by the Southern Group because they are the animals of most interest to students and because much is known of their structure and physiology. French beans were chosen because they are readily obtainable, because they can be grown indoors at any time of year and because much is known about them.

In addition to studying humans and French beans in detail, students are required by the syllabus to be aware of the diversity of living organisms. The syllabus refrains from naming any other species so that students, under the guidance of their teachers, may study a variety of local organisms in their local environments. The early units describe principles and techniques which may be illustrated by whatever organisms are familiar and available to the students.

Examples of biotechnology, the use of biological agents in the production of goods and the provision of services, occur throughout the book. The examples range from traditional biotechnology – fermentation, bread-making, sewage treatment and immunisation – to cloning by tissue culture of valuable plants, biological control of pests and genetic engineering.

Nothing in this book requires students or teachers to kill animals. Animals brought into the classroom for study should be returned to their habitats. Hearts, lungs, kidneys and other mammalian parts should be obtained from abattoirs or butchers.

UNIT 1

Variety of organisms

We share our homes, not only with our families and pets, but with thousands of smaller living organisms. Almost all of them are too small for us to see without a microscope. The most numerous group we cannot see are the **bacteria**. Though some are on the floors and the furniture and in the air, most are in and on our bodies, the bodies of our pets and the food we have in the house. They live especially where there is constant moisture. Even when enlarged thousands of times by a microscope as in Figure 1.1, bacteria are only small shapes with little of their internal structure visible.

Fig. 1.2 Dog

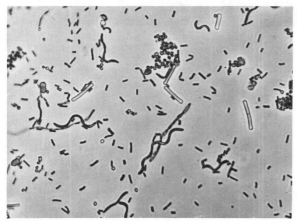

Fig. 1.1 Bacteria: different forms

Mammals

Dogs and cats, which many of us have as pets, belong to the same group of animals as we do, the **mammals**. Mammals are different from all other living organisms because: they have **hair**, which on cats and dogs we call **fur**; they have **external ears**, which means a part of their ears sticks out from their bodies; they have different kinds of teeth in their mouths, whereas other animals have only one kind; female mammals have **mammary glands** on their chests or abdomens, ending in **nipples** from which the young drink milk (and male mammals have functionless nipples). You can see fur, external ears and teeth in the photograph of the dog in Figure 1.2.

Birds

Some of us have a **bird**, probably a budgerigar or canary, as a pet. Birds are different from all other living organisms because: they have **feathers**; they have a **beak** instead of teeth; they have **wings** formed from **forelimbs**. These three features are shown in Figure 1.3.

Fig. 1.3 Canary

Fig. 1.4 Goldfish

Fish

Some of us keep **fish**. They have **streamlined bodies** with powerful **muscular tails** which help them swim; they have two pairs of **fins**, as well as other fins along the middle of their bodies; they have a special sense organ, called a **lateral line**, just visible on each side from head to tail, by which they can detect movement of water. They breathe by passing water in through their mouths and out over their **gills**. The muscular tail, fins and lateral line can be seen in Figure 1.4.

Since mammals, birds and fish all have backbones, they are **vertebrates**. The animals without backbones are **invertebrates**.

Spiders, mites and ticks

Spiders, **mites** and **ticks** are invertebrates in our homes. They have eight legs and at the front they have strong sense organs of touch called **pedipalps**. But spiders belong to one group, mites and ticks to another. A spider is different from a mite or tick because its body is in two parts divided by a waist. Its eight legs are all on the front part. In those spiders that spin webs the back part, the **abdomen**, has a pair of **spinnerets**. A spider's abdomen, unlike an insect's, is not divided into segments. Cats and dogs and people too, particularly if they live in the country, sometimes get ticks which suck their blood

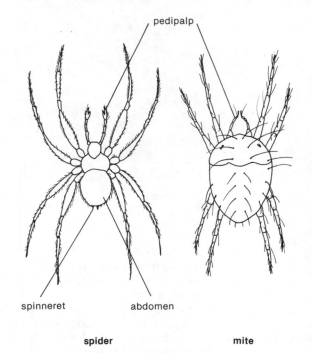

Fig. 1.5 Spider and mite: from above

and then drop off. Mites, as their name suggests, are very small. 'Red spiders' are really mites: if you examine one, you will see that it has no waist. Figure 1.5 shows the bodies of a spider and a mite.

Insects

Insects are probably the most numerous group of animals in our homes which are large enough for us to see. There are more than a quarter of a million different insects in the world. In most of our homes there are **flies**, **moths**, **beetles** and **silverfish**, and some of us are plagued by **ants**. Insects' bodies are divided into three parts: a **head**, a **thorax** (the chest region) and an **abdomen**. Adult insects have six legs, all on the thorax. If they have wings, they are attached to the thorax. The abdomen is divided into obvious sections or **segments**. Sticking out from the head are a pair of **antennae**, organs sensitive not only to touch but to chemicals in the air. Flies have one pair of large wings. Moths have two pairs of wings and often have huge featherlike antennae. Beetles also have two pairs of wings: one pair, which are delicate, are used for flight; the other pair, which are thick and hard, protect the flight wings and stretch right over the abdomen. Ants, which live in colonies, have narrow waists: reproductive ants have two pairs of wings; other ants have no wings at all. Silverfish are very primitive insects without wings: often found in kitchens and bathrooms, they seem to glide away when they are disturbed. Figure 1.6 shows the features of these five insects.

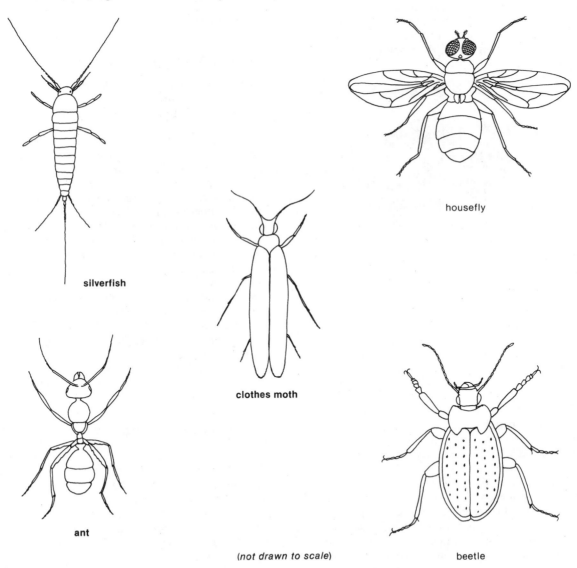

Fig. 1.6 Insects: from above

A number of wingless insects in our homes have probably descended from winged insects. These are the ones that live and feed on us and our mammal pets: lice, bugs and fleas. As they no longer have to move about much for their food, they can live without wings. Even without wings, fleas are very mobile: by means of their extra-long hind legs they can jump two hundred times their own height.

Woodlice

If you live in a damp part of the country, **woodlice** will come into your home. Woodlice have seven pairs of legs and completely segmented bodies with no obvious division into thorax and abdomen. Though they are land animals, they are **crustaceans**, most of whom, for example shrimps and lobsters, live solely or mainly in water. A woodlouse is shown in Figure 1.7.

Fig. 1.7 Woodlouse

Spiders, insects and woodlice are all **arthropods**, a group which have mastered the problems of living on land even though some of their members live in water.

Fungi

Moulds or **fungi** are a group of organisms which thrive only in damp conditions. If you keep a piece of damp bread or fruit under a glass, sooner or later there will appear on it a cotton-wool-like growth which is the threadlike body of a fungus. Black, pink or green **spores**, which will reproduce the fungus, appear above its threadlike body. In damp conditions a fungus called, somewhat misleadingly, **dry rot** may turn wooden beams to powder and bring down a whole building. Mushrooms we eat are a common example of fungi in our homes. If you place the top of a mushroom gill-downwards on a sheet of white paper, you will get a brown spore print after a few hours.

Plants

Animals, bacteria and fungi are not the only organisms in our homes. We also have plants, and not only those we grow in pots as decoration. Much of the food in our homes is in the form of plants and plant products. Some of the vegetables we buy are live plants which would grow if we planted them in moist warm soil: potatoes, carrots, onions and garlic are examples. The seeds of the fruits we buy would also grow if planted. Flour, sugar, tea and coffee are all plant products.

The major groups

In our homes we have examples of the three major groups of living organisms: animals; bacteria and fungi; plants. These three groups are distinct because of the ways they feed. Animals eat other organisms or the products of other organisms: they are **consumers**. Bacteria and fungi also feed on other organisms or the products of other organisms, but they cannot eat them as animals do. Their importance is that they rot the dead remains of organisms: they are **decomposers**. Plants feed on simple substances such as water and carbon-dioxide gas and ions dissolved in water. We call them **producers** because the food they make is used not only by themselves but by other organisms.

You know some of the organisms that are living in most homes. Quite different organisms live in gardens, in soil, in compost heaps, in woods, in freshwater ponds and rivers, on the sea-shore and in marshes and bogs. Places where organisms live are **habitats**. For reasons described in Unit 8, all self-supporting habitats have consumers, producers and decomposers.

History of living organisms

Life on earth probably began in bubbling rock pools about 4000 million years ago. There is evidence that simple organisms like bacteria lived three and a half thousand million years ago. By 500 million years ago simple plants such as seaweeds and the major invertebrates such as worms, arthropods and molluscs (shelled animals) lived in the sea. Fish appeared about 400 million years ago, land-living animals and plants about 300 million years ago. Among the earliest land animals were spiders and mites, wingless insects and animals like woodlice. They were able to live on land because they had developed a hard outer skeleton which not only supported them in air but stopped them drying out. Mammals appeared about 200 million years ago, birds about 150 million years ago, and flowering plants not until about 100 million years ago. To sum up, bacteria, or bacteria-like organisms, came first and were followed in order by seaweeds and simple animals, complex marine invertebrates, fish, land-living arthropods, mammals, birds and, youngest of all, flowering plants.

Questions

Q 1.1 How do you know the animal in Figure 1.8 is a mammal?

Q 1.2 Is the animal in Figure 1.9 a spider, a mite, an insect or a crustacean? Give your reasons.

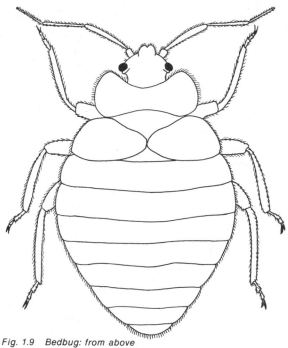

Fig. 1.9 Bedbug: from above

Fig. 1.8 Mouse

PRACTICAL WORK

Note for teachers
Examples of living organisms from widely different groups and of different body structure should be studied. They should range from micro-organisms to mammals and flowering plants. Familiar local organisms should be chosen.

UNIT 2

Simple organisms; increase in complexity; names; identification; keys

Simple organisms

Simple animals

Not many simple animals live on land. The reason is that simple animals are little more than blobs of jelly, called **protoplasm**, and, unless they have a fairly waterproof covering, they dry out in air. Among small animals only one group, the arthropods, have developed such a covering and members of this group are able to live successfully on land.

Amoeba is a simple animal and lives in water, usually fresh water. Figure 2.1 shows an amoeba with food granules inside its single blob of protoplasm called a **cell**. The scientific name for an organism is begun with a capital letter and is printed in italic (sloping letters) or underlined if handwritten or typed. The common name is written like any other word. *Amoeba* is an organism, one of many, for which the scientific and common names are the same.

If you look at pond water under a microscope, you can expect to see a number of different single-celled animals, and even single-celled plants, swimming through the water. Single-celled animals are **protozoans**. Their scientific name is **Protozoa**. The scientific name of a group is begun with a capital letter but, unlike the scientific name of an organism, is printed in ordinary letters, not in italic.

Hydra is a slightly larger animal found in fresh water. Consisting of a few thousand cells, it is just visible to the naked eye. Organisms with more than one cell are **multicellular**. A hydra is an unusual animal because it remains more or less in one position and catches food with its long tentacles: Figure 2.2 shows a hydra. *Hydra* belongs to a group called the **coelenterates**.

In freshwater streams, particularly under stones, there are **flatworms** less than 1 cm long. Although they consist of millions of cells, flatworms are simple animals which cannot survive on land. The flatworms form a group called the **platyhelminths**. Figure 2.3 shows a flatworm.

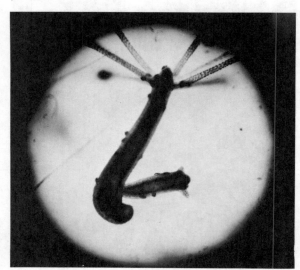

Fig. 2.2 Hydra

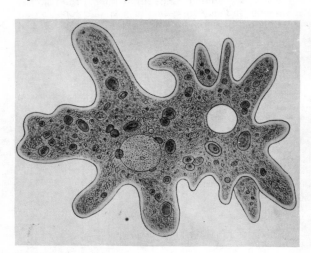

Fig. 2.1 Amoeba

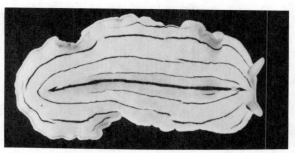

Fig. 2.3 Flatworm

Other worms living in water are **roundworms (nematodes)** and **segmented roundworms (annelids)**. The best-known segmented roundworms are **earthworms**, which are able to live in moist soil where they do not dry out quickly. For animals smaller and simpler than earthworms, moist soil is not very different from fresh water as a habitat. Moist soil contains Protozoa so small that a drop of water in soil is the equivalent to them of a small pond to a fish. A square metre of woodland soil can contain a thousand different animals: ten million nematode worms and Protozoa; half a million mites and **springtails** (wingless insects); ten thousand other invertebrates. A cubic centimetre of such soil can contain between six and ten million bacteria and between one and two kilometres of fungal threads.

Bacteria and fungi

Most bacteria and fungi live on land and in damp places. They survive and spread in dry air as thick-walled **spores** or reproductive cells. Bacteria and fungi live also in water: where water is well oxygenated they thrive.

Simple plants

A number of single-celled plants live in water. Many of them, unlike most plants, can move by swimming. Simple plants, without leaves and flowers, can grow quite large in water, especially the seaweeds on rocky sea-shores. All these simple plants are members of a group called the **algae**. There are some algae of just one cell, but others are large and many-celled, even several metres long.

Simple plants survive in air more easily than simple animals because all their cells are surrounded by a firm **cell wall**. *Pleurococcus*, a common alga, consists of only one or two cells. It forms a green powder on many surfaces such as tree trunks, branches, walls and roofs. An unusually thick cell wall enables it to survive in air without difficulty.

Increase in complexity

Organisms range from the microscopic to the huge. There are many more groups of common plants and animals than the dozen or so mentioned in this unit and in Unit 1. Most of those mentioned have been animals: Figure 2.4 summarises them and gives some idea, as you read from left to right, of their increasingly complex body structure.

Names

Organisms must have names. The most useful names are those that are the same in any language and can be recognised in any country in the world. A universal language for naming organisms has been in use for the last two hundred years, since the Swedish botanist Carl Linnaeus (1707–78) introduced the **binomial system**. Binomial means that the system gives all organisms two names. Our name is *Homo sapiens*. The first name, the **genus**, starts with a capital letter; the second name, the **species**, starts with a small letter. The names are usually derived from Latin or Greek. Both names are printed in italic or are underlined if handwritten or typed. A species is always a subdivision of a genus. In *Homo sapiens* the genus *Homo* means man and the species *sapiens* means intelligent. There are no longer any other *Homo* species, but we have probably evolved from *Homo erectus*, another *Homo* species that lived in Africa and Asia less than a million years ago.

A generic name in the binomial system may be used on its own: *Amoeba* and *Hydra* are generic names. A specific name in the binomial system is never used on its own. But, when you are writing about a species and have given the genus name in full once, you can usually refer to it again by its initial capital letter followed by a full stop. For example, *Homo sapiens* becomes *H. sapiens*. Only if you are discussing two or more genera beginning with the same letter will you not be able to abbreviate them.

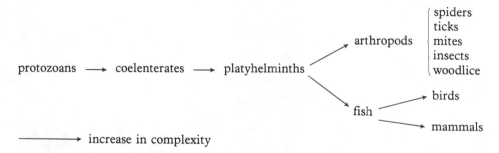

Fig. 2.4 Animals: different groups

Every species belongs to a genus. Among organisms that reproduce sexually, those that mate with one another in natural conditions (i.e. are **interfertile**) are classed as members of the same species. In captivity different species do sometimes mate and have offspring. An ass and a mare belong to different species: if they mate, the offspring will be a mule. That it is unnatural for them to mate is shown by the fact that mules themselves are **infertile**, i.e. cannot have offspring. Under natural conditions members of the same species usually look alike. But members of the same species do not *have* to look alike. Lions and lionesses look different; so do cocks and hens and many other males and females of the same species. Great Danes and dachshunds are both dogs and belong to the species *Canis familiaris*.

There are no hard and fast rules about what forms a genus. Closely related species are members of a single genus and will have a great deal in common. The lion, tiger and leopard are different species because they do not breed naturally with one another, but they are so similar in body form that they are put in one genus, *Panthera*. The lion is *Panthera leo*, the tiger *Panthera tigris* and the leopard *Panthera pardus*.

Identification

No one knows the names of all the animals and plants in this country, let alone in the world. Even an insect expert (an entomologist) would not be able to identify every insect. There are books on all the large groups of organisms, such as algae, flowering plants, fungi, Protozoa, insects, fish, birds and mammals. If you are trying to identify an organism and you know or can find out the large group to which it belongs, you can look it up in such a book. If you know where an organism lives, you may also be able to look it up in a book on a habitat such as ponds or the sea-shore. An illustrated book may enable you to identify an organism just by looking at the pictures. It is more likely that you will have to study the organism's features and go through the text looking for them. Some books have keys to help with identification.

Keys

To use a key you start at paragraph 1 and read the alternatives. Keys with only two alternatives are **dichotomous**. In a good key the alternatives are mutually exclusive: one and only one fits your organism. For example, if you are trying to identify a fern at the bottom of a hedge and you consult a book on plants, the first paragraph may read:

1 Plant free-floating on or below the surface of water   2
   Land plant or rooted at bottom of water   3

You have to decide which of the alternatives is your plant. Having decided, you look along the line to the right. Usually you will find a reference number that tells you which paragraph to read next. Only when you are at the end of the trail will you find the name of your organism. Since your fern is at the bottom of a hedge, you must read paragraph 3. Paragraph 3 may read:

3 Plant reproducing by spores; flowers 0; always herbs   4
   Plant reproducing by seeds; flowers; often woody   9

You need to know that 'flowers 0' means 'no flowers' and that herbs are not woody but soft. Now you must look at your fern. Does it have spores or seeds? Does it have flowers or not? Is it soft or is it woody?

Your fern will satisfy none of the conditions of the second line: it will not have seeds or flowers and it will not be woody. It will definitely satisfy two of the conditions of the first line: it will have no flowers and it will be soft. **Spores** of ferns are single cells developed in reproductive organs under the large leaves: you will need a microscope to see any on your fern. Even if you do not find spores, you can be fairly confident that your plant is the first alternative. Next, therefore, you must read paragraph 4.

These are examples of a good key. The two alternative lines in fact contain three alternatives, which makes it easier to choose right. To use keys you need to learn some of the scientific language in which organisms are described.

Figure 2.5 is a skull of a deer. Try to identify the kind of deer from the key. Three technical terms are

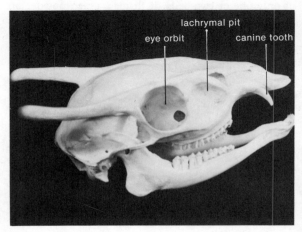

Fig. 2.5 Deer skull

used in the key: the **orbit** is the eye orbit, the bony socket of the eye with a hole at the back where the eye nerve passes to the brain; a **lachrymal pit** is a depression in the skull, in front of the eye orbit; **canines** are the fang-like teeth so obvious in cats and dogs.

1 Lachrymal pit in front of each orbit — 2
  No lachrymal pit — Roe deer
2 Canines 0 — Fallow deer
  Canines present — 3
3 Lachrymal pit round, more than half the diameter of the orbit; straight ridge above the lachrymal pit and orbit — Muntjac deer
  Lachrymal pit elongated, less than half the diameter of the orbit; no ridge above the lachrymal pit and orbit — 4
4 Lachrymal pit deep and slit-like — Water deer
  Lachrymal pit shallow — 5
5 Lachrymal pit small — Sika deer
  Lachrymal pit large — Red deer

The answer is at the end of this unit.

Try to identify the five buttercup fruits shown in Figure 2.6 from the next key. The style is the projection at the top of the fruits.

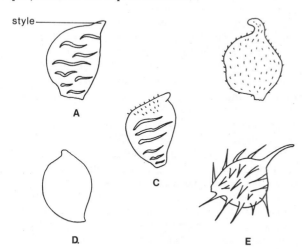

Fig. 2.6 Buttercup fruits

1 Ridges on fruit wall — 2
  No ridges on fruit wall — 3
2 Hairs — *Ranunculus sphaerospermus*
  No hairs — *R. fluitans*
3 Short style; smooth surface — *R. bulbosus*
  Long style; surface not smooth — 4
4 Hairy — *R. auricomus*
  Spiny — *R. arvensis*

The answers are at the end of this unit.

Making a key

You should be able to make up keys of your own. You need to have a good look at the organisms or parts of the organisms first. It is much easier when you can handle the organisms, turn them over and look at them from all angles, but you can also work from diagrams and photographs. Size and colour are not good features to use in a key: size varies with the age and development of the organism; colour is difficult to describe and is useful only if the colours are completely different. If you want to use a measurement, you should make it comparative:

1 Antennae longer than body — 2
  Antennae the same length as or shorter than body — 3

You have to assume that the antennae and body will always be in the same proportion to one another.

Figure 2.7 shows five animals from different groups: four are arthropods and the snail is a mollusc, a shelled animal. Features you are likely to be unfamiliar with are labelled for you. How can you make a dichotomous key from which someone unfamiliar with the five animals can identify them?

Take a good look at the five animals and write down ways in which some are obviously different from others. For example, only four have legs, only two have antennae visible and only two have waists. Arrange them in sets according to these distinguishing features:

legs – only A, B, D and E
eight legs – only A and D
six legs – only B
28 legs – only E
antennae visible – only B and E (*antennae* is the plural of *antenna*)
waist – only A and B
pedipalps – only A
wings – only B
shell – only C
tentacle – only C
claws – only D
rami – only E (*rami* is the plural of *ramus*)

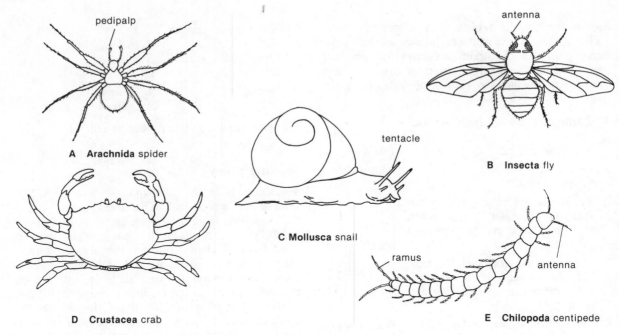

Fig. 2.7 Five animals

There are many different ways you could make a dichotomous key for the five organisms in Figure 2.7 with these features. You may like to try to make your own before reading on. Remember, the more alternatives the better. The crab has antennae but they are small and can be seen only from underneath. For the purpose of your key 'no antennae visible' is characteristic of both an animal with no antennae and of a crab.

If only one animal of the five has a certain feature, it can be identified immediately from it and the reader need not be referred to another paragraph. If only one animal of the five does *not* have a certain feature, it can be identified from its absence. Such negative features are not as reliable as positive ones: an animal with tentacles may have withdrawn them or have lost them in an accident. Even positive features are not entirely reliable: an apparently six-legged animal may have begun with eight and lost two in an accident. This is one reason for having as many features as possible in a key: 'Only C has a shell; only C has tentacles; only C has no legs'. Any one of these features is enough to identify C as a snail, but no legs on its own would be unreliable. Giving all three features enables someone who recognises only one of them to know the animal is a snail; it also helps if one of the features proves unreliable. The alternative line in the same paragraph must be 'Shell 0; tentacles 0; legs present'. Since no animal can be identified from this line, it must be given a reference to paragraph 2.

1 Shell present; tentacles present; legs 0   snail
   Shell 0; tentacles 0; legs present   2

You can now forget about shells and tentacles, since none of the other animals has them, and can concentrate on legs. Of the four animals remaining, two, A and D, have eight legs and two, B and E, do not. In the first line of paragraph 2 you can write 'Legs 8' and in the second line you can write 'Legs fewer or more than 8'. Look to see if there are any other features that distinguish A and D from B and E. There is just one: A and D have no antennae visible; B and E have antennae visible. Each line of paragraph 2 will therefore have two features. You can give the first line a reference to paragraph 3. You will start a good habit if you do not yet number the reference to the second line. If paragraph 3 did not succeed in identifying all the animals with eight legs, you would want to carry on with eight-legged animals in paragraph 4. While you are making your key, you may find it helps to put 'A and D' in brackets at the end of the first line and 'B and E' in brackets at the end of the second line. Remember to omit these notes when you write your key out neatly.

1 Shell present; tentacles present; legs 0   snail
   Shell 0; tentacles 0; legs present   2
2 Legs 8; no antennae visible   3 (A and D)
   Legs fewer or more than 8; antennae
   visible   (B and E)

To distinguish between A and D in paragraph 3 you will have to use some feature or features other than legs and antennae. There are three features, any one of which makes possible an immediate identification of either animal: A, the spider, has a waist and pedipalps but no claws; D, the crab, has no waist or pedipalps but does have claws. Having disposed of A and D, you now know that the second line of paragraph 2 must give a reference to paragraph 4.

1	Shell present; tentacles present; legs 0	snail
	Shell 0; tentacles 0; legs present	2
2	Legs 8; no antennae visible	3 (A and D)
	Legs fewer or more than 8; antennae visible	4 (B and E)
3	Waist present; pedipalps present; claws 0	spider
	Waist 0; pedipalps 0; claws present	crab

In paragraph 4 you know that B, the fly, has 6 legs while E, the centipede, has 28 legs. Though this is enough to make possible an immediate identification, you should add any other distinguishing features you can find. There are two more: B, the fly, has wings but no rami; E, the centipede, has no wings but does have rami. You have now completed a key and it is a good one containing several alternatives.

1	Shell present; tentacles present; legs 0	snail
	Shell 0; tentacles 0; legs present	2
2	Legs 8; no antennae visible	3
	Legs fewer or more than 8; antennae visible	4
3	Waist present; pedipalps present; claws 0	spider
	Waist 0; pedipalps 0; claws present	crab
4	Legs 6; wings present; rami 0	fly
	Legs 28; wings 0; rami present	centipede

Look back at the distinguishing features to see if there are any you have not used. On this occasion there is none. Sometimes, even though you have completed a key, you will find distinguishing features that you have described and that can be added to it. Other times you will find unused features but will not be able to use them because they do not distinguish between the groups of organisms in the key you have made. Someone else's key may group the organisms differently and make use of the features you have discarded while discarding features you have used.

Look at the five British mammals in Figure 2.8 and try to devise a dichotomous key to enable someone who has never seen them before to identify them. The phrase 'not drawn to scale' means that you cannot tell the sizes of the animals from the drawings. Although these animals look about the same size in the diagram, they may really be very different sizes. Do not expect your key to be the same as everyone else's. If yours identifies the five mammals, it is a good key. If it uses all the possible contrasting features, it is a very good key.

Make a note of the features that are different in these mammals: fur and whiskers will not be of any use because they seem much the same in all five. It looks as though the shape of the ears, nose, tail and forelimbs will be useful. Make a list of the different features and note in which animals they occur. Try to construct a key using as many pairs of alternatives in one paragraph as you can.

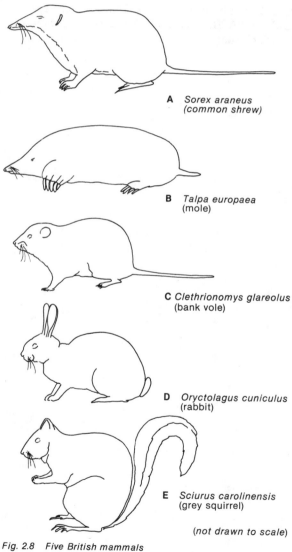

A *Sorex araneus* (common shrew)

B *Talpa europaea* (mole)

C *Clethrionomys glareolus* (bank vole)

D *Oryctolagus cuniculus* (rabbit)

E *Sciurus carolinensis* (grey squirrel)

(not drawn to scale)

Fig. 2.8 Five British mammals

Questions

Q 2.1 Name any two animals. Using their external features only, list (a) as many *differences* between them as you can and (b) as many *similarities* between them as you can.

Q 2.2 Find any two leaves. Name them if you can. Draw them accurately, being particularly careful to get the outline and veining drawn exactly. If the leaves are large, do not try to draw all the veins: draw a small representative portion accurately. Before you draw the leaves, read Appendix B on 'How to draw'.

Q 2.3 Find any five leaves from different species. Call them A, B, C, D and E. Make a dichotomous key by which someone else could identify them.

PRACTICAL WORK

Note for teachers
Using local material, get students both to identify five organisms from a dichotomous key and to construct a dichotomous key for five organisms.

Problem answers

The deer skull in Figure 2.5 is of a Muntjac deer. The buttercup *Ranunculus* fruits are: A. *Ranunculus fluitans*; B. *R. auricomus*; C. *R. sphaerospermus*; D. *R. bulbosus*; E. *R. arvensis*.

UNIT 3 Study of a habitat

If you enjoy walking in the countryside, you may have noticed that different plants grow in different places: for example, bluebells grow in woods and rushes in marshlands. Walking in a park, you can see that grass grows better in the open than under trees. Plants and animals are not equally distributed throughout the world. Each species lives in a **habitat** where the conditions suit it. Banana trees grow in warm climates and do not survive well in Britain in natural conditions. Even in Britain they will grow and produce fruit if kept in a heated greenhouse. But it is not always so easy to decide why a plant or animal grows where it does.

If you are in doubt, you need more precise information. You need to know, not just that a plant or animal lives in a wood or in marshland, but in which part of the wood or marshland it lives. You need to know what other plants and animals live in its habitat and how many of them there are. If you are inquiring about an animal, you need to know what it feeds on throughout the year. You need to know about the soil in the habitat. You need to know about the temperature and humidity and about changes in the temperature and humidity throughout the year. Does it freeze and snow in winter? When you have such information, you can form a **hypothesis** (untested theory) to explain why the plant or animal lives in its particular habitat. You can then conduct experiments to test whether your hypothesis is true.

This unit describes ways of finding the distribution of plants and animals in different habitats. Unit 4 describes how conditions such as temperature and humidity can be measured.

Plant and animal distribution

It is easier to study the distribution of plants than of animals. This is because plants do not normally move around and because their dependence on sunlight means that they live where they can be seen easily.

Mapping

If the study area is small enough, a grid can be made across it by fixing string with pegs at regular intervals. Figure 3.1 shows a string grid. You can use a piece of graph paper to represent the grid. Using the intersections of the string as reference points, you can draw a map showing where each species lives.

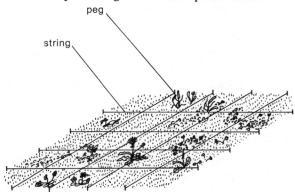

Fig. 3.1 Mapping using a string grid

Sampling

Suppose you want to know about the plants and animals in a large area. To map it all would take too long and it would be impossibe to search carefully for every plant and animal. It is better to study carefully a number of small parts, or **samples**, of the area. Studying one sample is not enough because it may not be representative of the area. You should choose at least ten samples from different parts of the area, and you should ensure that they are representative of the whole.

Quadrats

You can use a square frame made of metal, wood or plastic to mark out sample areas called **quadrats**. You can count the number of different animals within the quadrat. You can also try counting the number of different plants, but it may be difficult to decide what is a separate plant and it is usually more informative to estimate the percentage of the quadrat covered by each species of plant. This is the **percentage cover**. You will find it easier to estimate the percentage cover if you use a wooden comb which lies across the quadrat frame with ten equally spaced nails in it. Figure 3.2 shows a quadrat frame and comb. You will use them in Experiment 3.1.

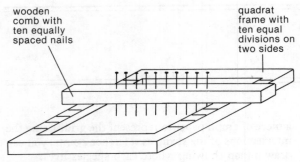

Fig. 3.2 Quadrat frame and comb

Line transect

A line transect can be used to position the quadrat frame for each sample. This means making a line across the study area with string or rope and placing the quadrat frame against the line at regular intervals. Line transects are used in Experiments 3.1 and 3.2.

Random sampling

Ideally you should take your samples at random. But random sampling is not as easy as it sounds. It is not good enough to throw the quadrat frame and take a sample where it falls, because you choose where you stand and in which direction you throw. You should make use of random-number tables which are provided in books of statistical tables. You can make a rope grid across the area, number the grid squares and choose which of them to sample by using the random-number tables.

Point frame

Another way of sampling is to use a point frame (Figure 3.3). This is a comb that stands on its own. Its ten nails are loose, so that they can be lowered to the ground and raised again. The point frame is put down in different places within the area being studied. In each place each of the nails is lowered to the ground and a record is made of each different plant species it touches. The frequency of a species can be calculated as a percentage of the total number of nail lowerings. This is the percentage cover of the species:

$$\frac{\text{number of times species A is touched by nail}}{\text{number of nail lowerings}} \times 100.$$

Volume sampling

Quadrats can be used only when the organisms being investigated live on a surface such as that of a field, a tree or a wall. If the organisms live in water or in soil or leaf litter, **volume sampling** must be used. The distribution of earthworms can be estimated by placing quadrat frames at random and digging up the soil beneath them to a certain depth. The volume of earth examined in each sample is the area of the quadrat multiplied by the depth of the soil. The results are expressed as the number of earthworms per unit volume. The same method can be used to find the distribution of other animals in soil, in leaf litter or in sand at the sea-shore. But sifting through these substances by hand is time-consuming and it is easy to miss an animal.

A **Tullgren funnel** can be used to extract small active animals from a sample of soil, leaf litter or sand. A known volume is put in the funnel with a light bulb above it, as shown in Figure 3.4. The light and heat from the bulb make the animals in the sample move downwards until they fall out of the funnel into the collecting vessel.

You can catch small organisms that live in water by using a net with a collecting tube fastened at one end. But this does not tell you the density of the organisms, because you do not know what volume of water has passed through the net while you have been collecting them. To discover their density you must remove measured volumes of water and count the different species in them. Your samples of water should be chosen so that they are representative of the

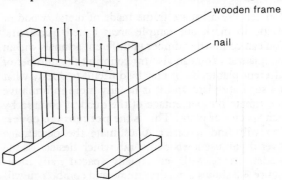

Fig. 3.3 Point frame

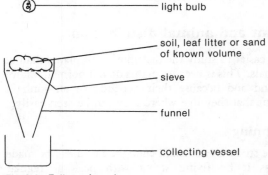

Fig. 3.4 Tullgren funnel

pond or rockpool or whatever it is you are investigating. If you are interested in both small and large organisms, you will need to take both small and large samples of water. Experiment 3.3 describes a volume-sampling technique.

Animal traps

Animals that move quickly or hide well cannot be investigated by any of the techniques described so far. One way to find their distribution is to leave traps over the study area and count the animals caught in them over a period of time. **Longworth traps** contain bait to capture small mammals alive without harming them. Though they are equipped with food and bedding, you must visit them regularly in order to release trapped animals. A **pitfall trap** is used to catch insects and other small animals crawling or running on the ground. It is a small jar or cup set in the ground so that its mouth is level with a pathway used by animals. The trap should be set where the pathway is at a high level so that it does not fill with rainwater. The mouth must be covered by wood or stones so that rain and larger animals do not fall into the trap. Of course the wood or stones must be raised a little above the mouth to leave room for animals to go under them. The trap must be emptied frequently in case some of the animals caught in it are eaten by others. A **light trap** attracts night-flying insects and birds. A simple light trap is a lamp shining on a sheet at night. Flying animals attracted to the light settle on the sheet, from which they can be removed and counted. A mercury-vapour lamp, standing on the ground surrounded by egg cartons, is a good way to catch moths: they fall and settle on the egg cartons. A mercury-vapour lamp is more successful than an ordinary tungsten lamp because it gives out ultra-violet light as well as visible light.

A **capture-recapture** technique can be combined with any method of catching animals. Each animal caught is marked with a ring, a tag or a spot of paint. The marked animals are counted and released where they were captured and given time to mix with the rest of their population. A second sample is taken in the same place and counted. The recaptured animals in it are also counted. The total population can then be estimated:

$$\text{population} = \frac{\text{number of marked animals released} \times \text{number in second sample}}{\text{number of marked animals in second sample}}$$

Because animals move so much, none of the methods of investigating their distribution is as good as the methods of investigating the distribution of plants.

Questions

Q 3.1 Figure 3.5 shows a pitfall trap to catch small animals.

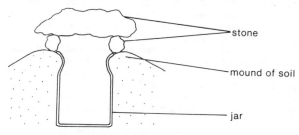

Fig. 3.5 Pitfall trap

Suggest (a) why the stone is placed above the trap, (b) why the soil at the sides of the trap slopes away from the mouth of the jar, (c) why the contents of the jar may not show the variety of the population.

Q 3.2 Quadrat frames with sides of 50 cm were used to sample the distribution of daisies in a field. The numbers found in ten quadrats were: 4, 9, 3, 2, 1, 5, 3, 1, 4, 2. Estimate the number of daisies per square metre in the field. Show your working.

Q 3.3 Suggest why it is often more informative to find the percentage cover of a plant rather than the number per unit area.

Q 3.4 Thirty marked snails were released in the field in which they had been caught. Random samples taken from the same field two days later contained a total of 60 snails, of which nine had been marked. Estimate the number of snails in the field. Show your working.

PRACTICAL WORK

Note for teachers
While studying the density of a single population, students observe, and may collect, specimens of other species. If an aquatic habitat is sampled in Experiment 3.3, collecting specimens of other species is unavoidable. Any such specimens collected from a single habitat can be used in the practical work in Unit 6.

Experiment 3.1 To investigate the distribution of one plant species using quadrats at regular intervals along a line transect

Note for teachers
Any terrestrial habitat is suitable for this experiment and it is assumed that most centres have playing fields,

lawns, parks or wasteland nearby. Teachers with no such land may be able to do Experiment 3.2 instead. Land open to roads should be avoided and areas should be chosen so that the class can easily be supervised. Where appropriate the teacher should seek permission to use the land.

The syllabus requires the study of only one plant species in this experiment, but more may be studied if the students are willing to do more work. The choice of plant (or plants) is left to individual teachers and will depend on the nature of the area chosen. Some time before the class carry out their work, the teacher should visit the area to select a plant which is common, easily identified and permanent to the site: dandelions and plantains are often suitable. Specimens of the chosen plant should be taken into the laboratory so that students can learn to recognise them before they begin work outside.

This experiment needs twenty samples. The experimental description prescribes a 21 m length of line with 1 m intervals. This provides for twenty samples at 1 m intervals with space over at each end to secure the washing-line with pegs. If the transect is to be longer, provide the class with a longer line with twenty indelible marks at equal intervals and space over at each end.

Warn the students some days before this experiment that they will be required to work outdoors and should bring suitable clothing and footwear.

Materials required by each group of four students
1 metal, wooden or plastic quadrat frame with inside edges 50 cm × 50 cm and with ten 5 cm intervals marked along opposite sides
1 wooden comb 55 cm long with ten nails protruding at 5 cm intervals
2 short wooden or metal pegs
21 metres of plastic-coated washing-line with indelible marks at 1 m intervals
1 polythene bag big enough for the note-taker to write inside if it rains
1 notebook and pencil

Method
1 Appoint one person in your group to act as note-taker and to begin by making in the group notebook a table like the one illustrated.

Distance along ground (m)	Number of times a nail touches your chosen species position										Percentage cover of plant species
	1	2	3	4	5	6	7	8	9	10	
0											
1											
⋮											
20											

2 Lay the washing-line along the ground in a place indicated by the teacher and fasten the ends down using the pegs.
3 Put the quadrat frame on the ground so that one of the sides with regular marks on it touches the right-hand side of the line while the back of the frame is level with the first mark on the line.
4 Place the comb across the quadrat frame as in Figure 3.6: the back of the comb is level with the

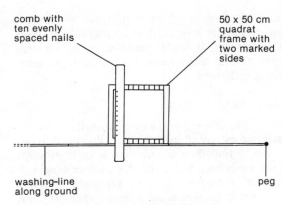

Fig. 3.6 Use of a quadrat frame and line

inside edge of the frame; the end nails of the comb are equidistant (2.5 cm) from the two sides of the frame. Record the number of times a nail touches your chosen species in column 1.
5 Place the comb across the quadrat frame in nine more positions, each time with its back in line with the marks on the side of the frame. Record in columns 2 to 10 the number of times a nail touches your chosen species in each of the nine positions.
6 To find the percentage cover in the quadrat, simply add together the figures in columns 1 to 10. Because there are a hundred nail positions in the quadrat, your total figure is a fraction of a hundred, i.e. the percentage.
7 Repeat instructions 4 to 6 at each of the 1 metre intervals marked on the line.

Interpretation of results
1 Act now individually, no longer as a group, and copy the table in your notebook.
2 Draw a bar chart of the data with distance on the x-axis and percentage cover on the y-axis. Appendix A shows you what a bar chart looks like.
3 Was the plant species evenly distributed along the transect? If not, suggest what might have caused the uneven distribution.

Experiment 3.2 To investigate the distribution of algae on a tree trunk using a line transect and to measure one environmental variable which might influence it

Note for teachers
This experiment is similar to Experiment 3.1 and can be used as a substitute for it if the class has no access to open land. Isolated trees may be used, even those at the sides of roads, provided that attention is paid to students' safety.

The plastic quadrat frames used in this experiment can be made by ruling a grid of a hundred squares of side 1 cm on a 10 cm × 10 cm plastic sheet with a spirit pen, or can be adapted from the 12 cm × 12 cm grids available from biological suppliers by cutting off 2 cm on two adjacent sides.

Warn the students some days before this experiment that they will be required to work outdoors and should bring suitable clothing and footwear.

Materials required by each group of four students
1 100 cm³ measuring cylinder
1 plastic quadrat frame 10 cm × 10 cm divided into a hundred squares 1 cm × 1 cm
1 compass
2 2 m lengths of string with indelible marks at 25 cm intervals
4 plastic cups
1 polythene bag big enough for the note-taker to write inside if it rains
1 roll of waterproof adhesive tape
1 notebook and pencil

Method
1 Appoint one person in your group to act as note-taker and to begin by making in the group notebook two tables like the one illustrated.

Side of tree facing ....

Height above ground (cm)	Percentage cover of algae
0	
25	
50	
:	
200	

2 Use adhesive tape to stick the two lengths of string on the north and south sides of the tree. Use the compass to find north. Stick the lengths of string so that one end touches the ground and the other reaches 2 m up the tree.
3 Put the plastic quadrat frame over the bottom of the string on the north side so that half of it lies on each side of the string.
4 Use the hundred small squares to estimate the percentage area under the quadrat which is covered by green dust-like algae and record it in the table opposite height 0 cm.
5 In the same way, estimate the percentage cover of algae at each 25 cm mark on the string until you reach the top of the string or until you can reach no higher. Record all the estimates of percentage cover of algae in the table.
6 Repeat instructions 3 to 5 on the south side of the tree.
7 Use adhesive tape to stick two of the plastic cups on the north side of the tree, one where there are many algae and one where there are few algae. Do the same on the south side of the tree.
8 Empty each cup daily for the next few weeks. Using the measuring cylinder, measure the volume of water in each cup each day and record the data in a table of your own design.

Interpretation of results
1 Act now individually, no longer as a group, and copy the two tables in your notebook.
2 Plot two bar charts, one for the north side and one for the south side of the tree, with height above the ground on the x-axis and percentage cover of algae on the y-axis. Appendix A shows you what a bar chart looks like.
3 Were the algae distributed evenly up the tree on (a) the north side and (b) the south side?
4 Were the algae equally common on the north and south sides of the tree?
5 Calculate the mean (average) daily volume of water collected at the four sites on the tree trunk. Appendix A tells you how to calculate a mean. Where does this water come from?
6 Is there any relation between the amount of water collected in each cup and the density of algae on the part of the tree near it? Suggest an explanation for your answer.

Experiment 3.3 To determine the density of an aquatic population using a volume-sampling technique

Note for teachers
Once this experiment has been performed in the laboratory it can be repeated with natural populations.
Crustacea provided by biological suppliers are accompanied by instructions for their culture.

Materials required by each student
1 5 cm³ pipette fitted with a pipette filler
1 dropper pipette
1 test-tube rack with 10 test-tubes
1 petri dish
1 spirit marker
1 beaker or jar containing a culture of water fleas (*Daphnia*) or brine shrimp (*Artemia*)

Method
1 Use the pipette to put 5 cm³ of tap water in each tube. Use the spirit marker to mark the levels of the water and then empty the tubes.
2 Put the rubber teat on the sharp end of the dropper pipette.
3 Use the inverted dropper pipette to transfer water from the culture to the level of the mark on each tube.
4 Pour the contents of one tube into the petri dish and remove the crustacea with the inverted dropper pipette. Record the number of crustacea.
5 Tip the contents of the petri dish back into the culture and repeat instruction 4 with the other tubes.

Interpretation of results
1 Calculate the mean number of crustacea per 5 cm³. Appendix A tells you how to calculate a mean.
2 Use this value to estimate the number of crustacea per dm³ of culture.
3 Compare your estimate of the density of crustacea with others in your class. What do the differences suggest about the reliability of this method? How may the reliability be improved?

UNIT 4 Measurement of environmental factors

Scientists are not content just to describe something: they like to explain it as well. Unit 3 tells how to study the distribution of animals and plants within a habitat. To *explain* their distribution we need to know what factors affect the ability of individuals to survive in the habitat. For example, if there are more individuals where it is sunny or wet or windy, we shall obviously ask ourselves if light or rain or wind help individuals to survive. Before we can ask ourselves such questions, we must be able to measure light, rain, wind and other factors in the **environment**, i.e. the surroundings.

Data about environmental factors, notably temperature, are given in local newspapers and on local radio. But the temperature given for your district cannot apply to every part of it. Shady places are cooler than sunny ones and windy corners are cooler than sheltered nooks. Temperature and other environmental factors can change within centimetres, with profound effects on living organisms. The term **microclimate** is used of the prevailing environmental conditions in a very small locality. In a wood, for example, the microclimate of the herbaceous plants at ground level is quite different from that of the canopy of leaves in the trees.

Sophisticated equipment is often needed to measure the differences between microclimates – differences in, for example, temperature, light, humidity, rainfall, air movement and soil. You should know a little about all these factors and should study at least one of them thoroughly.

Even within a microclimate environmental factors are not constant. Temperature in the canopy of leaves in a wood, for example, may change from minute to minute. Just as it is important to take samples when studying the distribution of animals and plants, as described in Unit 3, so it is important to take samples when studying an environmental factor. From your various measurements of a single environmental factor you can calculate the mean.

Temperature measurement

If the habitat is large, its temperature can usually be measured with an ordinary **liquid-in-glass thermometer**. The most common type is the mercury thermometer you will use in later units. It is suitable for measuring the temperature of the air, of water and of soil.

Figure 4.1 shows a **maximum–minimum thermometer**, which records both the highest and lowest temperatures during a period of time. Thus a maximum–minimum thermometer records the temperature range. It is really two thermometers in one and contains wire markers which are pushed by the liquid inside the two thermometers to the highest and lowest temperatures reached.

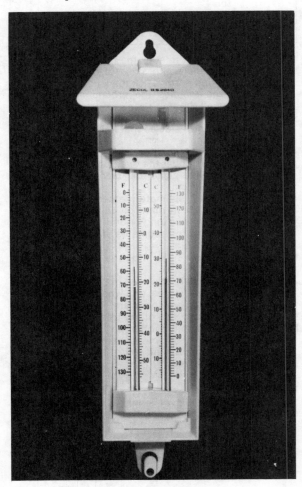

Fig. 4.1 Maximum-minimum thermometer

A liquid-in-glass thermometer is too big to use in very small habitats and is not accurate enough when fine measurements of temperature are needed. Some of the heat is conducted away from its bulb up the stem, and it absorbs any radiant heat in the environment. An electrical thermometer, which uses a temperature-sensitive device called a **thermistor**, is both smaller and more accurate. Provided that it is waterproofed, an electrical thermometer can be used to measure water and soil temperature as well as air temperature.

Light measurement

Total solar radiation can be measured by comparing the temperatures of two thermometers, one in the shade and the other in the sun, separated by a metal foil reflector.

Most instruments for measuring **light intensity** incorporate an electrical device called a **photoelectric cell**. (*Photo* means *light*.) Photoelectric cells can be used for measuring the light intensity in any environment, provided that they are waterproofed.

An ordinary **photographic light meter** can be used to measure light intensity. It should be pointed directly at the part of the environment whose light intensity is to be measured. **Light duration** can be measured as the number of hours each day when the light intensity in a habitat is above a predetermined value. The **light spectrum** can be measured with different coloured filters and photoelectric cells.

Humidity measurement

Humidity is the mass of water vapour per unit mass of dry air. It is difficult to measure this directly. Instead we measure **relative humidity**, the ratio of the pressure of the water vapour in the air to the pressure of the water vapour in the air when it is saturated at the same temperature.

Relative humidity (R.H.) is most easily measured with a **wet-and-dry-bulb hygrometer**, an instrument like a football fan's rattle. A wet-and-dry-bulb hygrometer is shown in Figure 4.2. Inside the hygrometer are two thermometers, one wetted with water and the other dry. The hygrometer is whirled around until each thermometer shows a constant temperature. The two temperatures enable the relative humidity to be read off from a set of tables.

In very small habitats a wet-and-dry-bulb hygrometer is too big to use. Instead, tests can be conducted in which the change in colour of paper dipped into a suitable chemical gives an indication of the humidity. Cobalt chloride turns from blue to pink in damp conditions; other chemicals such as cobalt

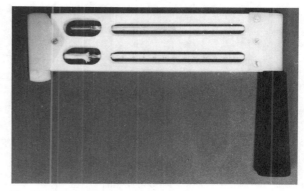

Fig. 4.2 Wet-and-dry-bulb hygrometer

thiocyanate show similar changes. A paper is exposed for a certain length of time and its colour compared with a sheet of coloured standards to obtain the relative humidity. The results of such tests are accurate to about two per cent. Since very little paper is needed, the tests can be used in the smallest habitats.

Precipitation measurement

A **rain gauge** collects rainwater in a funnel which empties into a collecting vessel, as shown in Figure 4.3. A calculation involving the height of water collected in the vessel and the ratio of the squares of the diameter of the funnel rim and the diameter of the collecting vessel gives the rainfall in centimetres during the time it was collected. The collection time should be shorter in very wet weather in case the collecting vessel overflows.

Fig. 4.3 Rain gauge

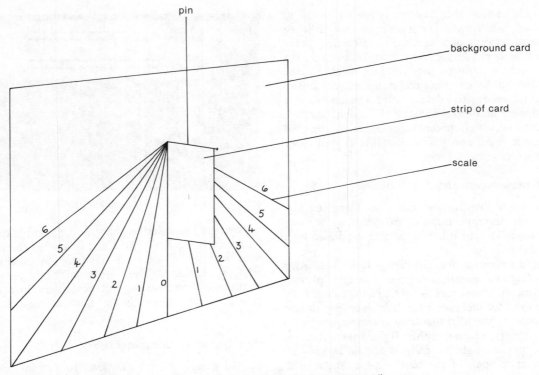

Fig. 4.4 Simple wind gauge (a stronger version can be used to measure water movement)

Air-movement measurement

Wind-vanes and **cup anemometers** show the speed and direction of wind in the environment, but are of little use for measuring air movement in microclimates.

A stopwatch can be used to determine the rate of movement of soap bubbles when blown by the wind.

If a small strip of card or paper is mounted on a simple scale, the extent to which the wind blows the strip away from the vertical can be measured. Figure 4.4 shows such a wind gauge, often called a **deflection flow meter**. A stronger version of it is used to measure water movement.

Devices involving thermistors, the temperature-sensitive devices in electrical thermometers, have been used to measure the speed of slow winds. In the sensitive part of the device a heating element is cooled by any wind to which it is exposed. The wind speed is found by measuring the difference between the temperature of the heating element and the temperature of the surrounding air.

Soil factors

The **pH** and the **concentration of inorganic ions** in soil samples can be found by fairly simple chemical tests purchased in kit form. All the tests rely on chemical reactions between the soil and the reagents to produce colour changes which can be compared with colour charts. Each kit contains the reagents, apparatus and colour chart needed for the test and instructions for carrying it out.

Questions

Q 4.1 The data show the relative humidity (R.H.) and the most common species of harvestman (a spider) at different layers in Wytham Woods (Berkshire).

(a) How could the relative humidity have been measured?
(b) Suggest why the relative humidity is different at each layer.
(c) How may organisms living at each level of the wood be affected by its relative humidity?
(d) To what extent do you think that differences in relative humidity determine which organisms live at each layer?

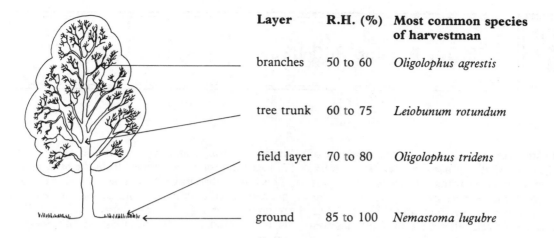

Layer	R.H. (%)	Most common species of harvestman
branches	50 to 60	*Oligolophus agrestis*
tree trunk	60 to 75	*Leiobunum rotundum*
field layer	70 to 80	*Oligolophus tridens*
ground	85 to 100	*Nemastoma lugubre*

Q 4.2 *Sphagnum* is a species of moss which forms dense carpets in peat bogs. The table shows temperatures during one day at the surface of a *Sphagnum* carpet and at a depth of 10 cm below its surface.

Time of day (h)	Temperature (°C)	
	at surface	10 cm below surface
08.00	27.0	21.0
10.00	31.0	21.0
12.00	34.0	22.0
14.00	36.0	23.0
16.00	34.5	24.0
18.00	29.0	24.0
20.00	25.0	23.0
22.00	23.0	22.0
24.00	20.0	22.0
02.00	20.0	21.0
04.00	18.0	21.0
06.00	19.0	21.0
08.00	26.0	21.0

(a) How could these data have been obtained?
(b) Using a single set of axes, draw and label graphs of the surface temperatures and of the temperatures 10 cm below the surface throughout the day.
(c) Compare the two graphs in respect of (i) the maximum temperature reached, (ii) the time at which the maximum temperature was reached, (iii) the minimum temperature reached, (iv) the fluctuation in temperature during the 24-hour period.

PRACTICAL WORK

Note for teachers
At least one environmental factor should be measured in the habitat studied in Experiment 3.1. Ideally a factor should be studied which might account for differences in the distribution of the chosen species.

UNIT 5

Energy: release and use

An organism lives only as long as it has a supply of energy. Just to stay alive it needs energy. To do work, such as moving or growing, it needs even more energy. The same is true of each cell within an organism. Without energy an organism or a cell dies.

No organism or cell relies on a constant and continuous supply of energy from outside. Every organism and every cell stores a source of energy within itself. Just as the Electricity Boards have stores of coal, oil and gas which they can convert into electricity, so organisms and cells have stores of **reserve food** which they can convert into energy.

If you set light to a piece of food such as a peanut, it will flame and burn. As it does so, it will give off light and heat – both forms of energy. To burn, a substance needs the gas oxygen, which is in air. Because it uses oxygen, burning is a process of **oxidation**. In our bodies we have oxygen that we have breathed in. Energy is released in our bodies by a process of oxidation similar to burning but not as sudden and uncontrolled. This process, called **respiration**, releases energy in controlled stages.

A human weighing 70 kilograms contains about one kilogram of reserve food in the form of **glycogen** and twelve kilograms in the form of **fat**. During fasting all the glycogen is used up within the first three days and two kilograms of fat are used up by the end of the week. During the first week half a kilogram of protein is also used up. Protein is not reserve food but is actually body tissues (protoplasm) which the organism needs. It is easy to understand that fasting for longer than a week is not good for you.

Respiration: releasing energy

What happens to all the glycogen, fat and protein that the body uses as food? These complex substances cannot just disappear. In the first stage of their breakdown all these substances form simpler compounds such as sugar, which can be used directly in respiration. Glycogen and fat contain only the elements carbon, hydrogen and oxygen: just rearranging them to form compounds that can be used in respiration releases some energy. Protein contains not only carbon, hydrogen and oxygen but also nitrogen. The nitrogen in protein is removed from the body as urea. This is described in Unit 29. The remaining elements are rearranged to form compounds that can be used in respiration, and again some energy is released. But it is in respiration itself that most of the energy is released.

Aerobic respiration

Most plants store food as starch. In both animals and plants the compound most commonly used in respiration is **sugar**. Glycogen in animals and starch in plants are first broken down to sugar. Most of the energy in glycogen and starch is still present in the sugar. It is this energy that is released by respiration. Respiration in which oxygen is used is **aerobic**. Aerobic respiration is summed up in the equation:

sugar + oxygen → carbon dioxide + water + energy

For students who understand chemical formulae, the balanced equation is:

$C_6H_{12}O_6$ + $6O_2$ → $6CO_2$ + $6H_2O$ + energy
(glucose, (oxygen)   (carbon   (water)
a sugar)   dioxide)

Aerobic respiration occurs in human beings and other animals, in plants, and in bacteria and fungi. All these organisms convert their food stores to simpler compounds, such as sugar, and use oxygen to respire them to release energy. Carbon dioxide and water are waste products of the process of respiration: they contain no usable energy.

Anaerobic respiration

Oxygen is not always available to the cells of living organisms. All cells are able, at least for a short time, to respire and release energy without oxygen.

If we run fast, our muscles need a great deal of energy very quickly. Some oxygen is stored in muscles, and blood brings fresh supplies all the time. For a while there is enough oxygen. Glycogen is broken down in our muscles first to sugar and then to carbon dioxide and water. But, if we run fast for long enough, there is not enough oxygen. Sugar is then broken down to release energy without oxygen. Without oxygen it cannot form carbon dioxide and water. Instead it forms an intermediate product in the muscles called **lactic acid**. Respiration without oxygen is **anaerobic**. Anaerobic respiration is much less efficient than aerobic and releases much less energy. It is summed up in the equation:

sugar → lactic acid + a little energy

The accumulation of lactic acid in our muscles can give us the pain of **cramp** and make it impossible for us to go on running. The muscles have become **fatigued** and we have to stop to let them recover. Blood continues to supply oxygen to the muscles and the lactic acid is removed. With oxygen, some of the lactic acid is respired further to carbon dioxide and water. The energy released is used to convert the rest of the lactic acid back to glycogen.

Where respiration without sufficient oxygen takes place, an **oxygen debt** builds up. This is what happens in the muscles as lactic acid is formed. After exercise that produces lactic acid, oxygen is used by the muscles above the normal rate for some time until the debt is paid off, i.e. until all the lactic acid has been either converted to carbon dioxide and water or reconverted to glycogen. This explains why, after running a race, we still breathe faster and deeper than normally: it gets more oxygen into our blood and more oxygen to our muscles to pay off the oxygen debt.

Many animal cells produce lactic acid in anaerobic respiration. But some organisms produce different products. **Yeast**, a small single-celled fungus, produces **alcohol** in anaerobic respiration. This is summed up in the fermentation equation:

sugar → alcohol + carbon dioxide + a little energy

This reaction is the basis of making beer and wine: yeast is added to malt to make beer; natural yeasts on the skins of grapes make wine. Yeast, like muscle cells, can respire anaerobically or aerobically. If plenty of oxygen is present, yeast breaks sugar down fully to carbon dioxide and water. When yeast is used in bread-making, carbon dioxide cannot escape from dough: it forms bubbles of gas and the dough 'rises'.

Plant seeds can respire anaerobically like yeast, releasing a little energy from sugar by breaking it down to alcohol and carbon dioxide. It is rare for plants themselves to respire anaerobically, because they usually have plenty of oxygen. The reason for this is explained in Unit 7.

Using energy

Unlike burning, respiration releases energy in stages. Even so, about three-quarters of the energy released is in the form of heat and is not available to the organism for other uses. Three-quarters seems a lot of energy to lose, but most of the engines designed by humans lose even more energy than is lost in respiration.

What do organisms do with the non-heat energy released in respiration? In the first instance they keep it in ready-to-use chemical bonds of a substance called **adenosine triphosphate** or **ATP** for short. From one molecule of sugar, aerobic respiration can make thirty-eight molecules of ATP, but anaerobic respiration makes only two molecules. This emphasises how much more energy is released by aerobic respiration. Both processes are summed up in Figure 5.1.

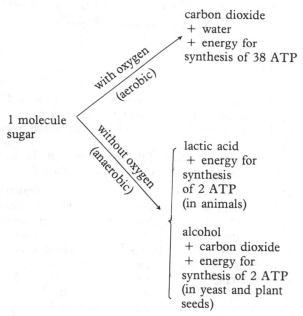

Fig. 5.1 Formation of ATP

The energy in ATP is available in the living cells of all organisms – animals, plants, bacteria and fungi – for work of different kinds. It is used to keep cells alive and to enable them to grow. It is used to make the thousands of different compounds that cells need. In animals (and some small plants) a great deal of the energy in ATP is used to produce movement. Moreover, in birds and mammals, which are warm-blooded, the heat inevitably released in respiration is not entirely wasted: it helps keep the body's temperature above that of its surroundings.

Questions

Q 5.1 Germinating French-bean seeds were kept in one insulated flask and boiled French-bean seeds in another insulated flask as shown in Figure 5.2.

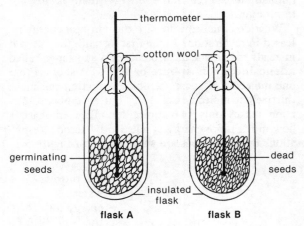

Fig. 5.2

After two days the temperature in flask A was several degrees higher than the temperature in flask B.
(a) Explain the difference in temperature in the two flasks.
(b) Suggest what other differences there might be between the air in flask A and the air in flask B.

Q 5.2 During home wine-making, a solution containing sugar is inoculated with yeast (i.e. has yeast added to it) and is left to ferment in a large flask with a cotton-wool plug. After 48 hours the cotton-wool plug is replaced by an air lock which allows gases to escape from but not to enter the flask.
(a) What kind of respiration will occur during the first 48 hours? What is the advantage of this?
(b) Why is an air lock fitted after 48 hours?

PRACTICAL WORK

Note for teachers
There are six experiments involving respiration. They are grouped at the end of this unit but can be undertaken whenever there is time available during the course. There are other units that have little or no practical work: Unit 8 is one during which time is likely to be available.

Experiment 5.1 To measure the heat energy released during the combustion of a peanut
Materials required by each student
1 test-tube rack with 1 test-tube
1 thermometer
1 Bunsen burner
1 pair of tongs for holding the test-tube
1 length of galvanised or copper wire
1 fresh peanut

Method
1 Pour water in the test-tube to a depth of about 3 cm.
2 Use the thermometer to find the temperature of the water in the tube and record it.
3 Push the wire through the peanut and fold over its end so that the peanut will not fall off.
4 Hold the peanut in the Bunsen flame until it continues to burn when you take it out of the flame.
5 Use the pair of tongs to hold the test-tube in one hand. With the other hand hold the wire so that the burning peanut is directly underneath the tube.
6 When the peanut has fully burned, measure the temperature of the water in the tube and record it.

Interpretation of results
1 What change was there in the temperature of the water after the burning of the peanut?
2 The temperature of the water changed because energy was released by the peanut in the form of heat. In what chemical form was this energy stored inside the peanut?
3 Suggest how the peanut might normally have used this energy store.
4 In what ways does energy release in the burning peanut (a) resemble and (b) differ from energy release in a germinating peanut?

Experiment 5.2 To investigate the products of food combustion
Note for teachers
It is assumed that students are familiar with the effect of water on dry cobalt-chloride paper and of carbon dioxide on clear lime water.
  In this experiment the gases produced by burning food are collected and analysed. It is advisable to avoid fatty foods because they produce an excessively sooty flame. Dry bread and water biscuits are ideal. The spoon on which the burning food is to be held can be made from a length of galvanised or copper wire and a square of aluminium cooking foil: make a loop of 10 to 20 mm diameter in the wire, press the foil into the loop, and twist the other end of the wire to make a handle.

Materials required by each student
2 clean dry jars with tightly fitting lids
1 pair of forceps
1 stopclock
1 Bunsen burner
1 heat-resistant mat
1 aluminium and wire spoon
2 dry strips of cobalt-chloride paper
100 cm^3 of clear lime water
1 piece of dry bread or water biscuit

Method
1 Copy the table.

	Change in colour of cobalt-chloride paper	Appearance of lime water
First jar		
Second jar		

2 Pick up one of the strips of dry cobalt-chloride paper with the forceps, drop it into one of the jars and place the lid on the jar. Note in your table any colour change in the paper after one minute.

3 Briefly remove the lid from the same jar and pour about half the lime water into it. Replace the lid and shake the lime water in the jar. Note the appearance of the lime water in your table.

4 Break the bread or water biscuit and place it in the spoon. Hold the spoon in the Bunsen flame until the food goes on burning when you take it out of the flame.

5 Hold the second jar upside down over the burning food so that any gas produced in the flame will go into it.

6 When the food stops burning, or when the jar gets uncomfortable to hold, put the spoon down on the heat-resistant mat. Put the lid on the jar, turn it the right way up and put it on the bench.

7 Pick up the second piece of dry cobalt-chloride paper with the forceps and put it in this second jar, quickly replacing its lid. Note in your table any colour change in the paper after one minute.

8 Quickly pour the rest of the lime water into the second jar and replace its lid. Shake the lime water in the jar and record its appearance in your table.

9 If you wish to repeat the experiment, replace the square of aluminium foil in the spoon with a clean piece and wash and thoroughly dry both jars.

Interpretation of results

1 Which two gases do your results show have been produced by burning food? Explain your answer.
2 Why were the tests on the first jar needed?
3 Give a brief chemical explanation of how the gases collected were produced.
4 Your observations may indicate other products of burning food. What might these products be? How do your observations indicate that they have been produced?

Experiment 5.3 To test whether peas use oxygen in respiration

Note for teachers
Two practical sessions are needed for this experiment, one to set it up and one some days later to determine the results. It is assumed that students know that oxygen is the gas present in air which allows burning to occur.

Candle holders for this experiment and for Experiment 28.2 should be made with lengths of wire and lids which are too large for the jars or beakers which are to be used. Make two holes about 3 cm apart in each lid and thread a length of wire through them both as shown in Figure 5.3. Bend the longer end and impale a night-light candle on it.

About four days before the experiment soak enough peas for the class in water for 24 hours and then allow

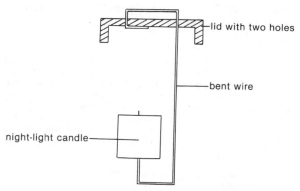

Fig. 5.3 Candle holder for Experiments 5.3 and 28.2

them to germinate. Two hours or so before the experiment, boil half the peas, cool them and soak them for five minutes in 1% sodium-hypochlorite solution. Rinse them in sterile water.

Materials required by each student
2 jars with tightly fitting lids
1 stopclock
1 spirit marker
1 candle holder with attached candle
2 pieces of absorbent cotton wool
25 germinating peas
25 dead peas

Method
1 Label one jar A and the other B. Add your initials and the date to each jar.
2 Put one piece of moist cotton wool in the bottom of each jar.
3 Carefully count 25 live peas into jar A and 25 dead peas into jar B.
4 Put the lid on each jar and screw it down as tightly as you can. Leave the jars in a safe place in the laboratory until your next practical session.
5 In your next practical session, use the candle holder and stopclock to find how long the candle will burn in each jar by carefully following instructions 6 to 9.
6 Unscrew the lid on jar A but do not remove it.
7 Light the candle in the candle holder and start the stopclock.
8 Take the lid off jar A and lower the burning candle into it until the lid of the candle holder is pressed tightly against the rim of the jar. Use the stopclock to find the time in seconds for which the candle burns and make a note of it.
9 Repeat instructions 6 to 8 with jar B.

Interpretation of results

1 Name the gases which were present in the air in the jars at the start of the experiment. Which of these is needed for things to burn?
2 In which jar did the candle burn for the longer time? What does this tell you about the final composition of the gases in the two jars?
3 Explain how the activity of the peas caused the composition of the gas in the two jars to differ.
4 Before you used them, the dead peas had been soaked in a disinfectant. Suggest why this was necessary in this experiment.

Experiment 5.4 To investigate the products of anaerobic respiration

Note for teachers
Prepare for this experiment well in advance by cutting two 25 cm lengths of 5-6 mm glass delivery tubing and flame-polishing both ends. Make two right-angle bends in each tube, one 5 cm and the other 10 cm from one end. Fit the shorter end through a one-hole rubber bung which will then fit into a 25 mm × 150 mm boiling tube. It is a good idea to make a number of spare tubes to allow for class breakages. Allowing about 1 g per student, kill a batch of dried baker's yeast by heating it in an oven at 150°C for fifteen minutes.

Materials required by each student
1 500 cm^3 beaker
1 10 cm^3 graduated pipette fitted with a pipette filler
1 dropper pipette
1 boiling-tube rack with 2 boiling tubes and 2 test-tubes
1 weighing bottle
2 one-hole rubber bungs with fitted bent delivery tubes
1 spatula
1 spirit marker
5 cm^3 cooking oil
30 cm^3 5 % glucose solution
20 cm^3 clear lime water
1 g dried baker's yeast
1 g heat-killed dried baker's yeast

Materials required by the class
balances to weigh 1 g (1 per 4 students)
1 thermometer
hot water

Method
1. Label one boiling tube A and the other B.
2. Use the spatula, weighing bottle and balance to weigh 1 g of dried baker's yeast and add it to tube A.
3. Use the spatula, weighing bottle and balance to weigh 1 g of heat-killed dried baker's yeast and add it to tube B.
4. Pipette 10 cm^3 of glucose solution into each of the two labelled tubes by gently running the solution down their sides. Swirl gently to mix the yeast and the glucose solution but *do not shake*.
5. Use the dropper pipette to cover the mixture in both tubes with a thin layer of oil. Stand both tubes in the boiling-tube rack.
6. Before continuing, study Figure 5.4 showing what the apparatus will look like when you have assembled it. If you do not understand the diagram, or think you lack some apparatus, consult your teacher.
7. Half fill the beaker with water at a temperature of about 40°C.
8. Pipette 10 cm^3 of lime water into each of the two test-tubes.
9. Put the boiling tubes containing the yeast mixture in the beaker of warm water. Fit rubber bungs securely in both boiling tubes so that the delivery tubes dip into the lime water in the test-tubes.
10. Leave the apparatus until towards the end of your practical session and then record the appearance of the lime water in the two test tubes.
11. Remove the bungs from the boiling tubes and smell the contents.

Interpretation of results
1. List the steps which you took in this experiment to ensure that anaerobic (and not aerobic) respiration would occur.
2. Suggest ways in which oxygen might still be present in the yeast mixture.
3. What changes did you observe in the lime water in the test-tubes? What do they indicate about one product of anaerobic respiration?
4. What change in the yeast-glucose mixtures may be associated with the observations you noted in 3?
5. Identify the chemical that produces the characteristic smell in tube A. This is a second product of anaerobic respiration.
6. Explain the importance of (a) the glucose solution and (b) the heat-killed yeast in this experiment.

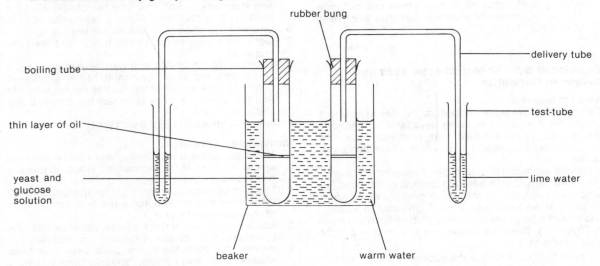

Fig. 5.4 Apparatus used in Experiment 5.4

Experiment 5.5 To measure the rate of anaerobic respiration in a yeast-dough mixture

Note for teachers
In this experiment the measuring cylinders full of risen dough will prove difficult to clean unless they are rinsed out immediately. Prepare the yeast-glucose mixture before the experiment by stirring 8 g of glucose and 6 g of dried baker's yeast into 120 cm^3 of water at 45°C. Leave it to stand in a warm place for fifteen minutes. Allow 10 cm^3 of this mixture per student.

Materials required by each student
1 100 cm^3 measuring cylinder
1 250 cm^3 beaker
1 10 cm^3 graduated pipette fitted with a pipette filler
1 weighing bottle
1 glass rod
1 stopclock
1 spatula
10 g plain flour
10 cm^3 yeast-glucose mixture

Materials required by the class
balances to weigh 10 g (1 per 4 students)

Method
1 Copy the table.

Time (min)	Volume of mixture (cm^3)
0	
5	
10	
15	
⋮	⋮
60	

2 Weigh 10 g of plain flour and pour it into the beaker.
3 Pipette 10 cm^3 of the yeast-glucose mixture into the beaker and stir with the glass rod to produce a smooth dough mixture.
4 Pour the dough mixture into the measuring cylinder, taking care not to get any of it on the sides. Remove any dough mixture from the sides of the cylinder if you make a mistake. It does not matter if you do not pour all the dough mixture from the beaker into the cylinder.
5 As soon as the surface of the dough mixture in the measuring cylinder is level, record the volume it occupies in your table opposite Time 0. Start the stopclock.
6 Leave the measuring cylinder on the bench at room temperature for five minutes. Then record the volume which the dough mixture occupies at Time 5 minutes in your table.
7 Continue to read and record the volume of the dough mixture every five minutes for one hour or until the end of your class, whichever is sooner.

Interpretation of results
1 Assuming that the yeast cells in the dough mixture were respiring, explain (a) why glucose was needed in the original yeast-glucose mixture, (b) why the volume of dough mixture kept increasing, (c) why it is likely that respiration was anaerobic.
2 On a sheet of graph paper draw a jagged-line graph of your results, with time on the x-axis and the volume of the dough mixture on the y-axis. Appendix A tells you how to draw a graph.
3 Calculate the rate of anaerobic respiration of the yeast as its volume increase (cm^3) per minute for the first thirty minutes. Appendix A tells you how to calculate a rate.
4 Was the rate of respiration constant in your experiment? Explain how you can tell this from your results and suggest explanations of any changes in the rate.

Experiment 5.6 To measure the rate of respiration of French-bean seeds using a simple respirometer

Note for teachers
Allow twenty French-bean seeds per student and soak them in water for twenty-four hours four days before the experiment. Place wet paper towels on sheets of plastic film and lay the seeds in single rows across each set of towels. Roll each set of wet towels containing seeds and loosely fasten the ends with elastic bands. Stand these rolls in beakers containing about 4 cm of water and leave them for three days in a warm room. On the day of the experiment remove the seeds from each roll. By this time they should have radicles about 10 mm long. Boil half the germinating seeds in water for a few minutes and then cool them in running water.

The glassware for the respirometers must be prepared well in advance. Cut one 12 cm length of 5-6 mm glass delivery tubing, one 6 cm length of 5-6 mm glass delivery tubing and one 30 cm length of 1 mm bore capillary tubing per student and flame-polish all the cut ends. Make two right-angle bends in each 12 cm length of delivery tubing and fit it and one of the straight 6 cm lengths into a two-hole rubber bung to fit 25 mm × 150 mm boiling tubes.

Tap water with enough cooking dye to produce a deep colouration and with a few drops of washing-up liquid makes a suitable fluid for the manometers.

Materials required by each student
1 250 cm^3 beaker
1 100 cm^3 beaker
1 boiling-tube rack with 1 boiling tube
1 weighing bottle
1 two-hole rubber bung with fitted glass delivery tubing
1 30 cm length of 1 mm bore capillary tubing
1 stopclock
1 tripod and gauze
1 spatula
1 spirit marker
2 30 mm lengths of rubber tubing to fit tightly over the glass tubing
1 2 cm^3 syringe
1 plastic three-way tap
enough cotton wool to make a plug for the boiling tube
30 cm^3 manometer fluid
5 g soda lime
10 live germinating French-bean seeds
10 boiled and cooled germinating French-bean seeds

Materials required by the class
balances to weigh 1 g (1 per 4 students)

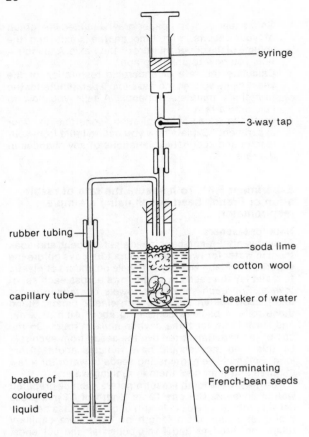

Fig. 5.5 Apparatus used in Experiment 5.6

Method

1. Before starting work, study Figure 5.5 which shows the final appearance of the apparatus.
2. Weigh ten live French-bean seeds and put them in the boiling tube. Record their mass.
3. Make a cotton-wool plug and put it in the tube just above the beans.
4. Weigh 5 g of soda lime and put it on the top of the cotton-wool plug.
5. Stand the large beaker on a tripod and gauze and put the tube in it.
6. Fit two pieces of rubber tubing on the ends of the glass tubing in the rubber bung. Fit a three-way tap to the straight piece and the capillary tubing to the bent piece.
7. Check that the lever of the three-way tap is pointing upwards and push the bung firmly into the boiling tube. Fill the beaker with tap water up to the level of the soda lime in the tube.
8. Pour the coloured fluid into the small beaker and immerse the free end of the capillary tubing in it.
9. Check that the plunger is pushed into the syringe as far as it will go and then fit the syringe in the upper hole of the three-way tap.
10. Turn the lever of the tap to its horizontal position and withdraw the plunger in the syringe. This should make the coloured fluid rise up the capillary tube. Let go of the plunger and watch the liquid. If it moves down the capillary tube, there is a leak in the apparatus and you should make sure that the rubber bung, the rubber connection to the capillary tube and the rubber connection to the tap are all tight. If there is still a leak when you have done so, consult your teacher.
11. When you are sure there is no leak, turn the lever of the tap to point downwards and pull the plunger of the syringe out to the 2 cm³ mark.
12. Move the lever of the tap to point upwards and make a mark on the capillary tube to show the level of the coloured liquid.
13. Start the stopclock and turn the lever of the tap horizontally. The level of liquid in the capillary tube should move slowly upwards. Record how long it takes to move nearly to the top of the capillary tube.
14. Press the plunger on the syringe until the coloured liquid has been pushed down to the mark you made earlier. Record the new position of the plunger.
15. Let the liquid rise up the capillary tube for the same time as before and repeat instruction 14. You now have two readings to indicate the volume change inside the boiling tube over the same length of time.
16. Take the apparatus apart and empty and clean the boiling tube.
17. Repeat instructions 2 to 16 using ten boiled French beans instead of ten live ones.

Interpretation of results

1. Why was the boiling tube kept in a beaker of water?
2. What was the function of the soda lime? Why might the results have been different if no soda lime had been used?
3. What was the mean volume change in the boiling tube containing (a) live beans and (b) boiled beans? Explain why these volume changes occurred.
4. Using your calculated mean volume change, work out the rate of respiration in each boiling tube in cm³ of oxygen per gram of beans per hour. Explain any difference between the two results.

UNIT 6

Energy: transfer

Animals, plants, bacteria and fungi contain thousands of different organic compounds made by themselves or other organisms. Until recently organic compounds could be made only by organisms, but nowadays humans can make them in their laboratories and factories. Organic compounds always contain carbon, usually contain hydrogen, often contain oxygen and sometimes contain nitrogen. Glycogen, starch, sugar, fats and proteins are all examples of organic compounds.

When organic compounds burn, they flame and release energy as heat and light. Living organisms release energy during the similar but slower and controlled process of respiration. Although living organisms lose most of this energy as heat, they can use the remainder to make organic compounds, to make other compounds, to grow and, many of them, to move. Releasing and using energy is described in Unit 5.

Every naturally occurring organic compound is a potential source of energy for some other organism. Otherwise the earth would long ago have been buried under a heap of organic compounds that no other organism could use. Figure 6.1 shows a sycamore tree as an energy source. We know it is an energy source because we burn its wood in fires as a source of heat. In fact every part of it is an energy source: animals such as moth caterpillars eat its leaves; mice eat its seeds; greenfly (aphids) suck its sugary sap; bees drink nectar from its flowers; earthworms, mites and woodlice eat its dead fallen leaves; bacteria and fungi feed on all its dead parts; tar-spot fungi, toadstools and other fungi feed on the living cells in its leaves and in its roots.

By feeding on the tree, all these organisms can use it as a source of energy. We have names for the different categories of 'feeders' on the tree.

Herbivores are animals that eat living plants or products formed by living plants. The caterpillars and bees are herbivores. Other herbivores are rabbits, sheep and cows.

Scavengers are animals that eat dead plants and animals: earthworms, mites and woodlice feeding on dead parts of the sycamore are scavengers. Other scavengers are crows, vultures and the maggots, or larvae, of flies. Scavengers perform a useful function: they get rid of dead bodies of animals and plants.

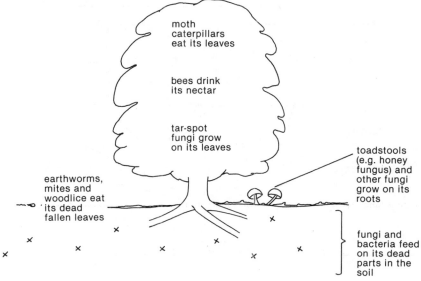

Fig. 6.1 Sycamore tree

Saprophytes are bacteria and fungi that feed on dead plants and animals and on the undigested parts of plants and animals in faeces. Saprophytes are not able to 'eat' their food as scavengers do. They do not have mouths through which they can swallow dead organisms. They feed only on liquid food but they are able to change solid compounds into soluble compounds which will dissolve in water. Saprophytes are described in detail in Unit 10. The bacteria and fungi living on the dead parts of the sycamore in the soil are saprophytes. Saprophytes feed especially on the dead parts that animals are not able to use: these are the parts that pass through the guts of animals as faeces. Saprophytes are **decomposers**.

Parasites are organisms that develop a feeding relationship with other *living* organisms of a different species. Greenfly, tar-spot fungi and the toadstools on the sycamore get their food from its living cells and are parasites. As long as the sycamore stays alive, they get their supply of food from it. Parasites are described in more detail in Unit 10.

Carnivores are animals that catch other living animals for food. Carnivores do not feed directly on the sycamore. Beetles and birds eat caterpillars and greenfly that eat the sycamore leaves and suck its sap. Beetles and birds are carnivores that get their food, and energy, *indirectly* from the sycamore. What about the fox that eats the birds that eat the caterpillars that eat the sycamore leaves? The fox too is a carnivore and it too gets its food, and energy, indirectly from the sycamore.

An animal is a **predator** on other animals that it eats; the animals it eats are its **prey**. Foxes are predators on birds; birds are their prey. Birds are predators on caterpillars; caterpillars are their prey.

Omnivores are animals that eat both living plants and living animals. Badgers are omnivores that can be found living in sets, or burrows, excavated around the roots of sycamores. They eat the seeds and young roots of the tree as well as earthworms, beetles and other animals that have fed directly on the sycamore. Wood mice are also omnivores: they eat both seeds of the sycamore and insects that feed on it. Figure 6.2 is a diagram showing some of the feeding relationships in which the sycamore is involved. A diagram of such relationships is called a **food web**.

Food webs are enormously complicated when all the possible feeding relationships are traced. Figure 6.2 does not begin to show the complexity: dozens of different insects feed on sycamore trees and dozens of different birds feed on these insects. All living organisms are involved in food webs of this kind.

A single line in a food web is a **food chain**. Figure 6.3 shows one food chain from sycamore to fox via caterpillar and blackbird.

sycamore ⟶ moth caterpillar ⟶ blackbird ⟶ fox

Fig. 6.3 One food chain from sycamore

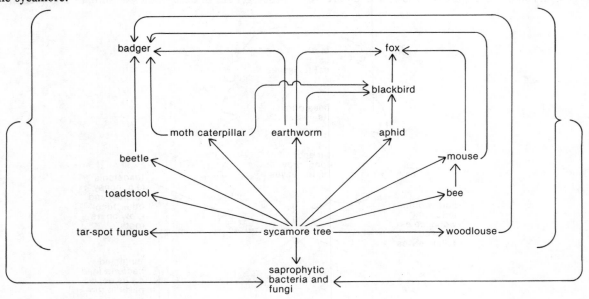

Fig. 6.2 Food web from sycamore

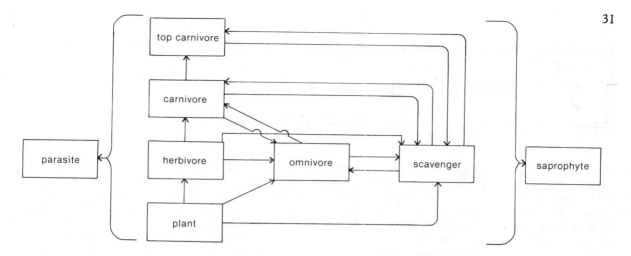

Fig. 6.4 Food web: principles

Arrows in a food web or food chain show the direction of the energy transfers.

Figure 6.4 is a simplified food web in which only the different categories of feeders are described. It shows the principles involved in any food web: herbivores eat the plant; carnivores eat the herbivores, other carnivores, omnivores and scavengers; a carnivore not eaten by other carnivores is a **top carnivore**; omnivores eat herbivores, carnivores and plants; scavengers eat all the dead plants and animals; saprophytes feed on all the dead organisms of the food web; parasites can have a feeding relationship with any of the living organisms.

Questions

Q 6.1 Suppose you have eaten hard-boiled eggs, cheese and tomatoes.
(a) For each of the food items construct a food chain ending with yourself.
(b) In each food chain indicate whether you were feeding as a herbivore or as a carnivore.

Q 6.2 Some bacteria are parasites and others saprophytes. What important difference is there in the feeding of these two kinds of bacteria?

Q 6.3 Use your own knowledge and any reference books to which you have access to classify each of the following animals as a carnivore, a herbivore or an omnivore: cat; mouse; earthworm; song thrush; perch; pike.

PRACTICAL WORK

Note for teachers
While sampling the density of a population in Unit 3 students may have collected a range of organisms from one habitat. If not, further sampling of that habitat, of another convenient habitat or of a laboratory aquarium should be undertaken.

Students should use keys and field guides to identify some of the organisms and should try to construct food chains and simple food webs. Organisms are rarely seen eating one another: feeding relationships will therefore be based largely on information from books. Students should categorise the organisms as plants, herbivores, carnivores, omnivores, scavengers, saprophytes or parasites before constructing food chains and food webs for the habitat.

Teachers must use their discretion in deciding the precision with which specimens are to be identified. A snail may be easily assigned to its species whereas a freshwater unicell may be difficult to place beyond its phylum or class.

UNIT 7

Energy: capture and loss

A carnivore such as a fox gets its energy from eating another animal such as a rabbit. A herbivore such as a rabbit gets its energy from eating a plant such as grass. Where does the grass get its energy from? It does not eat food as an animal does; nor does it feed as saprophytes (bacteria and fungi) do.

Suppose you plant a French-bean seed in soil in a pot, keep it warm and watered, and occasionally add to the water a drop of liquid fertiliser containing a mixture of inorganic ions. It should grow well. After two months the dry mass of the French-bean plant could easily be one hundred times the dry mass of the original seed: it would then have in its organic compounds one hundred times the energy store it had as a seed.

Energy capture

Where has this energy come from? It has certainly not come from the water: if you try to burn water, as you did a peanut in Unit 5, you will not succeed, because there is no food energy to be released from it. Nor will you succeed in burning the inorganic ions in the liquid fertiliser because they have no food energy to release. The only other substance that the plant has been absorbing while it has been growing is a gas, carbon dioxide, which it got from air. This will not burn either, because it has no food energy to release. The only substances a plant takes in are water, inorganic ions and carbon dioxide, yet none of them gives it energy.

The plant's source of energy is light. A plant growing in the open absorbs sunlight as well as water, inorganic ions and carbon dioxide: with these four ingredients it makes all the thousands of organic compounds of which it is composed. (Plants can use energy from artificial light instead of from sunlight.)

You know that sunlight has energy because anything left lying in the sun is warmed. Light energy is easily converted into heat energy. But how is it converted into energy in the compounds of plants?

Photosynthesis

The energy from light is used to make sugar by a process called **photosynthesis**. (*Photo* means *light*; *synthesis* means *building-up*, the opposite of analysis.) Sugar is an organic compound containing only carbon, hydrogen and oxygen: both the carbon and oxygen come from carbon dioxide; the hydrogen comes from water. Since water consists of hydrogen and oxygen, and since only the hydrogen is used in the sugar, the oxygen is a waste product and is given off as a gas. This is summed up in the equation:

carbon dioxide + water + light energy → sugar + oxygen

A simplified balanced equation is:

$$6CO_2 + 6H_2O + \text{light energy} \rightarrow C_6H_{12}O_6 + 6O_2$$

It is not enough to mix carbon dioxide and water and shine light on the mixture. This does not produce sugar. Something else is needed. That something else is part of a green plant and is called a **chloroplast**.

The green leaves of a plant contain cells in which there are small green disc-like bodies: these are chloroplasts. A single cell may have a hundred of them. There is a much enlarged photograph of a chloroplast in Unit 13 (page 60). The dark patches in the chloroplast contain a green pigment called **chlorophyll**. It is chlorophyll that makes plants look green. Chlorophyll is able to capture light energy and to use it to remove hydrogen from water and to form ATP. In the rest of the chloroplast the hydrogen is bound to carbon and oxygen with energy from ATP to form sugar. This is the process of photosynthesis, which captures the energy with which most food is made, food not only for plants themselves but also for animals, which feed on plants directly or indirectly.

Having captured light energy in sugar by photosynthesis, a plant releases some of it through respiration to make the thousands of other organic compounds it needs. Respiration is described in Unit 5. Sugar is made of only carbon, hydrogen and oxygen. To make protein and the other complex organic compounds in a plant, inorganic **ions** supply nitrogen and other elements in the form of, for example, nitrates, ammonium and phosphates. (Appendix A explains ions.) Proteins and the other complex organic compounds contain proportionately more energy than the original sugar.

Because the waste product of photosynthesis is oxygen, plants are rarely short of oxygen and rarely need to respire anaerobically. Photosynthesis in plants provides all the oxygen used by all living organisms in aerobic respiration.

Energy loss

It seems there is an almost inexhaustible supply of energy available to us from sunlight. Why then is there concern about our dwindling energy supplies? Only plants are efficient at capturing light energy, and they capture at most 5 % (a twentieth) of the energy that falls upon them. Another problem is that plants can capture only light that falls directly on their chloroplasts, and plants have already covered all the available surfaces of the earth on which they can grow. Every time we remove a habitat for plants, by building a road or a town, we remove a means of capturing light energy. So far humans have not been able to match a plant's ability to capture light energy.

Loss of light energy

A plant captures 5 % of the light energy that falls on it only in ideal conditions of warmth and plentiful water. These conditions exist in a few places in the world such as tropical rain forests. Most places are either too cold or too dry. In Britain only about 1 % of the light energy falling on a plant is captured in photosynthesis. The other 99 % is reflected, lost as heat or used to evaporate water. This is summed up in Figure 7.1.

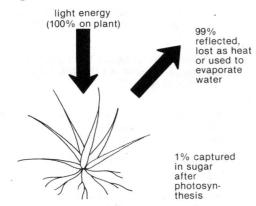

Fig. 7.1 Loss of light energy in Britain

Loss of captured photosynthetic energy

It is in respiration that a plant releases energy trapped in sugar and so is able to make protein and other organic compounds. As explained in Unit 5, about three-quarters of the energy released from sugar in respiration is lost as heat. But not all a plant's sugar is respired. Some of it remains in the plant as sugar and retains all its energy. Only about 20 % of the total energy in a plant's sugar is lost as heat in respiration. About 80 % of the total remains in the plant either in sugar or in other organic compounds made from sugar. This is summed up in Figure 7.2.

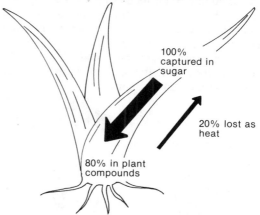

Fig. 7.2 Loss of energy captured by photosynthesis

Loss of energy by herbivores

Plants, which are eaten by herbivores such as rabbits, contain a lot of indigestible fibre. More than a half, about 55 %, of what herbivores eat goes straight through their guts and is lost in faeces. About another 35 % is lost, after food is absorbed through their guts into their bodies, either as heat in respiration or as energy for movement. Only about 10 % of the energy in the original plant compounds ends up as body compounds of the herbivores. This is summed up in Figure 7.3.

Some herbivores are much better than others at converting plant compounds into their own tissues: 10 % is an approximate average.

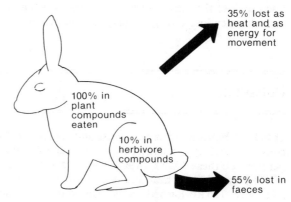

Fig. 7.3 Loss of energy from plant food

Loss of energy by carnivores

Carnivores are more efficient than herbivores at digesting their food. Because their food is animals, like themselves, it contains compounds that are easily converted into their own bodies. Carnivores expel only about 20 % of their food as faeces. Even so, like herbivores, they incorporate only about 10 % of their food into their own tissues. The reason is that they expend much more energy than herbivores in movement in order to get their food: they have to search longer for it and often have to chase it to catch it. This is summed up in Figure 7.4.

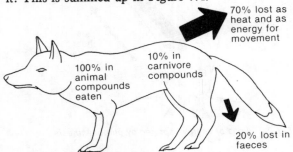

Fig. 7.4 Loss of energy from animal food

Some carnivores are much better than others at converting prey into their own body compounds: for example, an octopus can convert more than 50 %. Warm-blooded herbivores and carnivores are less efficient than others at converting food into their own body compounds: in keeping their body temperature above that of their surroundings they lose much energy as heat.

Loss of energy by omnivores

Like both herbivores and carnivores, omnivores lose about 90 % of the energy in their food. If they eat mainly plants, with a lot of indigestible fibre, they lose much of the energy in faeces. If they eat mainly animals, they lose much of the energy in movement.

Questions

Q 7.1 A bullock grazing on grassland was found to consume 3050 kJ of energy per square metre of grass per year. Of these 3050 kJ, 1900 kJ were lost in faeces, 1025 kJ were lost as heat and energy for movement, and only 125 kJ formed body compounds in the bullock.

(a) Calculate the percentage of energy intake (i) lost in faeces, (ii) lost as heat and energy for movement, (iii) trapped in the bullock's body compounds. Appendix A tells you how to calculate percentages.
(b) Is a bullock more or less efficient at converting its food than the average herbivore shown in Figure 7.3?

Q 7.2 Figure 7.5 is a graph showing the rate of photosynthesis over a range of light intensities.

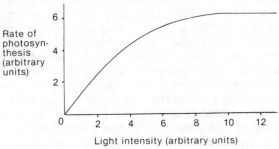

Fig. 7.5

What does this graph show about the effect of light on photosynthesis?

PRACTICAL WORK

Note for teachers
If necessary the practical work in this unit can be completed during Unit 8, for which there is no practical.

Experiment 7.1 To investigate the hypothesis that light energy is needed for the production of sugars by plant leaves

Note for teachers
Students should be familiar with the test for reducing sugars, which is given in Experiment 21.1, and with the relation between simple sugars and starch.

Materials required by each student
1 250 cm^3 beaker
1 filter funnel
1 10 cm^3 graduated pipette fitted with a pipette filler
1 test-tube rack with 2 test-tubes
1 pair of forceps
1 pair of scissors
1 stopclock
1 plain white tile
1 Bunsen burner, tripod and gauze
1 pestle and mortar
1 spirit marker
2 sheets of filter paper
4 cm^3 Benedict's solution
2 g clean fine sand

Materials required by the class
1 cork borer
1 iris plant that has been grown in the light for at least four days before the experiment
1 iris plant that has been grown in the dark for at least four days before the experiment

Method
1. Prepare a water bath by half filling the beaker with water and heating it over the Bunsen burner. When the water is boiling, turn down the heat so that it continues to boil gently.
2. Label one test-tube A and the other B.
3. Cut a piece of leaf about 5 cm long from the iris plant that has been grown in the light. Use the cork borer and the white tile to cut three discs from this piece of leaf and quickly drop them into the water bath.
4. After two minutes use forceps to remove the leaf discs from the water bath and put them into the mortar. Add a little clean sand and grind the discs.
5. Use the graduated pipette to add 10 cm^3 of water to the ground discs and grind the mixture for a few seconds.
6. Filter this mixture into the test-tube labelled A.
7. While the mixture is filtering, wash the pestle and mortar and repeat instructions 3 to 5 using a leaf from the iris plant that has been grown in the dark for the last few days. Filter this mixture into test-tube B.
8. Pipette 2 cm^3 of Benedict's solution into each test-tube and put them both into the water bath for two minutes. Take the tubes from the water bath, and in a table like the one illustrated record the colour change (or lack of it) in each tube.

Iris plants	Result of Benedict's test
Grown in light	
Grown in darkness	

Interpretation of results
1. Why were the leaf discs dropped into boiling water?
2. What do your results show about the presence of sugar in the two plants?
3. Do your results suggest that photosynthesis has taken place in either of the plants? Explain your answer.

Experiment 7.2 To determine the volume of oxygen produced by a submerged aquatic plant during photosynthesis

Note for teachers
The best results are obtained by using *Elodea*. This can be obtained during spring and summer from aquaria suppliers. It should be kept well illuminated just before the experiment.

The microburette can be obtained ready-made from biological suppliers. Teachers with glass-blowing ability can make microburettes using a length of capillary tubing of 1 mm bore, cut and flame-polished at each end. After one end has been softened by heating, a swan neck should be made. After this end has been sealed, a bulb should be blown in it. A small hole should be burned in the end of the bulb. Finally a scale should be stuck to the straight arm of the tube.

After Unit 19 has been studied, this experiment should be adapted to investigate the effect on the rate of oxygen production by photosynthesising plants of (a) light intensity (by varying the distance between the light source and the plant), (b) carbon-dioxide concentration (by adding different concentrations of sodium-hydrogencarbonate solution), (c) temperature (by heating the water in the beaker with a Bunsen burner).

Materials required by each student
1 500 cm^3 beaker
1 microburette
1 scalpel
1 stopclock
1 plain white tile
1 clamp, stand and boss head
1 bench lamp
1 screw clip
1 ruler graduated in cm
1 short length of rubber tubing to fit over the end of the microburette
1 2 cm^3 syringe

Materials required by the class
5% sodium-hydrogencarbonate solution
pieces of *Elodea* of about 10 cm (4 per student)

Method
1. Three-quarters fill the beaker with tap water.
2. Collect two or three pieces of pond weed that you can see are producing bubbles of gas. Make clean cuts across the stems midway between two whorls of leaves to produce three lengths of pond weed of about 10 cm. Put these into the beaker of water and shine a bench lamp on them.
3. Fit the rubber tubing over one end of the microburette and fit the screw clip over this tubing. Put the syringe into the other end of the tubing.
4. Set up the clamp, boss head and stand to hold the microburette vertically with its bulb in the beaker of water as shown in Figure 7.6.
5. Choose the piece of pond weed that is producing most bubbles of gas and push its cut end into the bulb of the microburette so that it is lightly held in place by its own leaves. Discard the other pieces of pond weed.
6. Lower the bulb of the microburette into the water in the beaker.
7. Slowly withdraw the plunger of the syringe to fill the microburette with water and then tighten the screw clip over the rubber tubing.
8. Put the bench lamp 20 cm from the beaker, switch it on and leave the apparatus for a few minutes. Bubbles of gas should stream from the cut end of the pond weed into the bulb of the microburette. If the bubbles stop, add a little sodium-hydrogencarbonate solution to the water.
9. When the pond weed has been producing a regular stream of bubbles for a few minutes, unscrew the clip on the rubber tubing enough to make the gas move from the bulb to the stem of the microburette. Now start the stopclock.
10. After two minutes gently unscrew the clip a little more to draw the gas which has been produced into the stem of the microburette. Carefully read off the length of the bubble formed using the scale which is stuck on to the glass.
11. Withdraw the syringe plunger a little to remove all bubbles from the bulb and stem of the microburette.

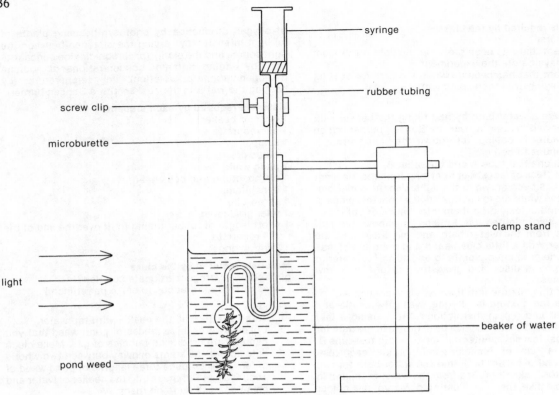

Fig. 7.6 Apparatus used in Experiment 7.2

12 Repeat instructions 9 to 11 to obtain three values for the length of the bubble of gas formed by the pond weed over three two-minute periods.

Interpretation of results

1 Calculate the mean of the lengths of the gas bubbles measured during your investigations.
2 What further information would you need to calculate the mean *volume* of gas produced? Suggest how the apparatus might be used to find the information needed.
3 What gas would you expect to be present in the bubbles produced by the pond weed? Explain how it is produced.
4 Suggest what effect (a) light and (b) sodium-hydrogencarbonate solution may have on gas production by the pond weed.
5 Although the gas bubbles came from the cut end of a stem, it must be assumed that they were produced mainly in the leaves. Suggest how the gas travelled from the leaves to the cut end of the stem.

UNIT 8

Energy: flow through an ecosystem

A **habitat** is a place that is inhabited by organisms and has a recognisable character, such as a wood, a pond, a river, the sea-shore, a marsh and a bog. An **ecosystem** is a habitat plus the community of organisms living in it and interacting with one another and with their environment and climatic conditions. (*Eco* comes from the Greek for *house*.) **Ecology** is the study of the relations between different organisms living in the same habitat and between the organisms and their environment and climatic conditions. Any alteration in an ecosystem may have far-reaching effects. Removing hedges in Eastern England has allowed topsoil to be blown away by wind and, by destroying habitats, has reduced wildlife. Cutting down trees in tropical forests has allowed topsoil to be washed away by heavy rain, while the loss of evaporation from the trees' leaves has reduced rainfall miles away. It is always dangerous to disturb the ecological balance, i.e. the balance of nature.

The kinds of organisms that live in an ecosystem are determined by its physical characteristics and climatic conditions. Plants and animals that live on the sea-shore must be adapted to tides and to sand or rocks. Plants and animals that live in very hot dry places must be able to conserve water and must have an efficient method of keeping cool.

Figure 8.1 shows the energy flow through an ecosystem based on a lake. The units of measurement have been simplified and are called 'eu' for energy units. The figures for energy transfers and losses in this ecosystem are not very different from the typical figures given in Unit 7. Just over 1 % of the light energy falling on the green plants has been captured

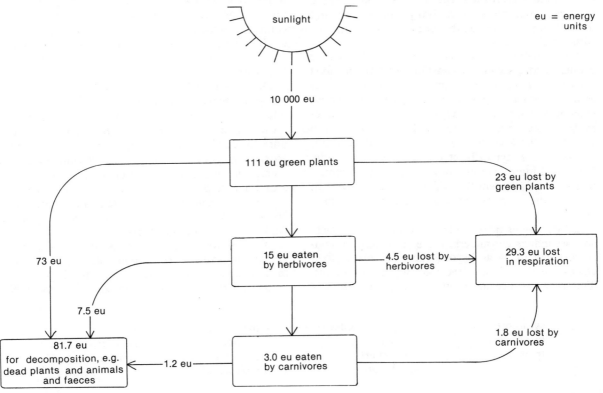

Fig. 8.1 *Energy flow through a lake ecosystem*

by photosynthesis. The plants lose about 20 % of that captured energy in respiration. The herbivores lose about 30 % of their energy in respiration, while the carnivores, which have to move around more to get their food, lose about 60 %. The herbivores' energy content is about 13.5 % of the energy content of the plants they eat, a little more than the typical figure of 10 % given in Unit 7. The carnivores' energy content is 20 % of the energy content of the herbivores they eat, considerably more than the typical figure of 10 % given in Unit 7. Remember the example of the octopus: carnivores often do better than 10 %.

All ecosystems have **producers**, **consumers** and **decomposers**, i.e. plants, animals, and bacteria and fungi. Though the plants are very different in different ecosystems – for example, seaweeds on the sea-shore, mosses and small plants in bogs – they all capture the energy of sunlight and manufacture food. In all ecosystems some or all of the plants form the basis of food chains and food webs, i.e. they are eaten by animals. The animals also vary greatly from one ecosystem to another, but they are all consumers depending, directly or indirectly, on the producers for food. Those eating plants, the herbivores, are **primary consumers**; those eating primary consumers are **secondary consumers**; those eating secondary consumers are **tertiary consumers**; and so on. Figure 8.2 repeats the food chain in Figure 6.3 with technical terms.

It is possible for the same animal to be both a primary and a secondary consumer in different food chains. A wood mouse is an omnivore: when it is eating sycamore seeds, it is a primary consumer; when it is eating moth caterpillars, which themselves eat sycamore leaves, it is a secondary consumer. You would put the wood mouse into the category which it occupied most typically at the time of your study. The wood mouse is more of a carnivore in summer, when insects are plentiful, and more of a herbivore in winter, when insects are scarce and it eats mainly seeds.

All ecosystems have producers, primary consumers and secondary consumers; some have tertiary consumers; not many go beyond this. The reason food chains are short is that the food supply peters out. This is not surprising when you consider how little energy is converted into body compounds every time one organism eats another. A tertiary consumer gets only about 10 per cent of 10 per cent of 10 per cent, i.e. a thousandth, of the energy in a plant. Another word for feeding is **trophic**. The different feeding levels – in an ecosystem those of producer, primary consumer, secondary consumer, tertiary consumer and so on – are called **trophic levels**. Suppose you want to compare the amounts of food provided at different trophic levels. It is no use counting the organisms at each trophic level, because they may vary greatly in size. One tree provides much more food for primary consumers than ten dandelions. It is better to weigh the tree and the dandelions than to count them.

But the **live mass** can itself be misleading because organisms contain very different proportions of water: a lettuce and a heather plant weigh about the same, but the lettuce contains much more water. Since water has no energy value for consumers, the heather is a much more valuable source of food to primary consumers than the lettuce, despite the fact that it weighs the same. To make a reliable comparison between trophic levels you need the **dry mass** of the organisms, i.e. their mass after all the water in them has been evaporated. Even the dry mass is an imperfect measure because it does not contain only food value: it contains also such things as calcium salts in bones and shells which are largely inorganic. But dry mass is a much better guide than live mass.

The same principle applies if you want to compare two ecosystems or two parts of the same ecosystem. The dry mass of all the living organisms in an ecosystem at any one time is the **biomass**. Since it is unrealistic to weigh all the organisms in two ecosystems, you must compare equal sample areas in them. It is usual to measure biomass as mass in a unit area: for example, grams per square metre if you are comparing small areas of a field and tonnes per hectare if you are comparing large areas of forest. Again you want the dry mass.

Of course to measure dry mass you would have to kill the organisms. You would then have to chop them up and dry them in an oven at 105 °C for several hours so that all the water evaporated from them. No

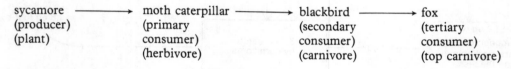

Fig. 8.2 One food chain from sycamore

sycamore ⟶ moth caterpillar ⟶ blackbird ⟶ fox
(producer) (primary consumer) (secondary consumer) (tertiary consumer)
(plant) (herbivore) (carnivore) (top carnivore)

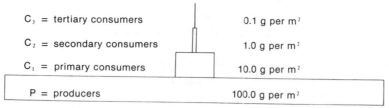

Fig. 8.3 Pyramid of biomass: diagrammatic

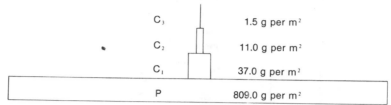

Fig. 8.4 Pyramid of biomass: a river

one expects you to find the dry mass of an ecosystem or even of part of one. If students throughout the land did this every year, there would be wholesale slaughter.

Biomass is a measure of the total food in an ecosystem. Sometimes people are interested, not in the total food, but in certain types of food. A forester, for example, will want to know only the plant biomass, not the animal biomass.

Another way of giving more information is to make **pyramids of biomass**. To do this you first estimate the biomass of all the producers, of all the primary consumers, of all the secondary consumers, and so on. As in Figure 8.3, you then stack blocks, representing the different trophic levels, on top of each other.

Pyramids of biomass are another way of showing how energy is lost at each trophic level in a food web. Figure 8.4 shows a pyramid of biomass of a river. On average, under a square metre of the river surface the dry mass of the producers (plants) is 809 grams, yet the dry mass of the primary consumers (herbivores) is only 37 grams. The dry mass of the primary consumers is only $\frac{37.0 \times 100}{809.0}$, or 4.6 % of that of the producers. This is less than the typical figure of 10 % given in Unit 7. The dry mass of the secondary consumers (carnivores) is $\frac{11.0 \times 100}{37.0}$, or 29.7 % of the dry mass of the primary consumers, much more than the typical figure of 10 %. The dry mass of the tertiary consumers (top carnivores) is $\frac{1.5 \times 100}{11.0}$, or 13.6 % of the dry mass of the secondary consumers, a little more than the typical figure of 10 %.

Pyramids of biomass have serious weaknesses. They take account neither of the productivity of an ecosystem nor of the role of decomposers. An ecosystem in which the producers double themselves every day, as small sea plants do, is far more productive of producer biomass than an oak wood, which may increase its producer biomass by only 1 % a year. If you weigh all the cheese you find in a cheese factory, you still have no idea how much is produced there: you need a figure of cheese produced per day, per week or per year.

The decomposers in an ecosystem are ignored in pyramids of biomass, yet they may have a far greater biomass than the consumers. A typical pine forest has a total consumer biomass of only $1.7 \, g$ per m^2, while the biomass of its fungi is $120 \, g$ per m^2 and of its bacteria is $39 \, g$ per m^2: thus its decomposer biomass is $159 \, g$ per m^2, nearly a hundred times that of all its consumers combined.

Omnivores

Humans are omnivores: we can eat producers (plants) or consumers (animals). It is a waste of energy trapped by producers when we eat consumers: there has been about a 90 % energy loss between the producer and the primary consumer. To gain energy efficiently we need to eat producers (plants), not consumers (animals). But we do not eat only to gain energy. We also need substances such as vitamins and amino acids, some of which are found more often in animals than in plants. The most efficient diet for humans is therefore one that is nearly all plants but includes some animal products. It is important to realise this when one of the world's problems is how to improve the diet of the millions of humans who are undernourished.

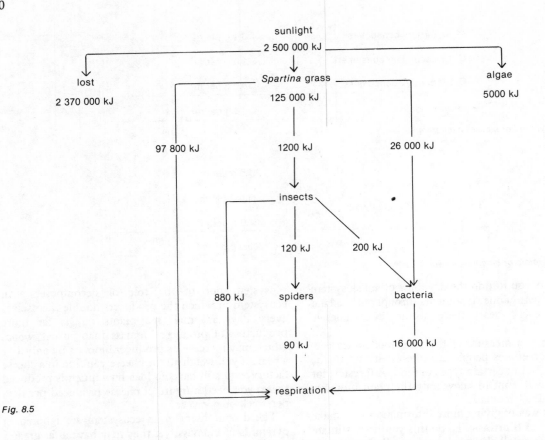

Fig. 8.5

Questions

Q 8.1 Figure 8.5 shows approximate values of the energy flow (measured as kJ per m^2 per year) in a subtropical marsh of *Spartina* grass.
(a) Write out a food chain, shown in the diagram, which includes a producer, a primary consumer and a secondary consumer.
(b) What word is used to describe the levels in a food chain?
(c) Calculate (i) the percentage of sunlight energy captured by *Spartina*, (ii) the percentage of energy taken up by the insects which is available to spiders, (iii) the percentage of their energy intake which spiders lose in respiration.

PRACTICAL WORK

Note for teachers
There is no practical work in this unit. This leaves time to complete the extensive practical work on energy release and capture in Units 5 and 7.

UNIT 9

Cycles in the ecosystem

Every ecosystem has a never-ending supply of energy from outside itself: the sunlight with which plants make sugar. Unlike its energy, its chemical elements, such as carbon, hydrogen, oxygen and nitrogen, are not continuously added to the ecosystem. There is only so much carbon, hydrogen, oxygen and nitrogen available on earth and none enters from outer space. All the chemical elements have to be continuously cycled through every ecosystem and used over and over again.

Carbon

Carbon is in every organic compound and therefore in every organism in thousands of different compounds. Animals get their carbon as organic compounds in their food, whether they eat plants, plant products, other animals or animal products. Ultimately all animals get their carbon in an organic form from plants. Where do plants get their carbon compounds? They do not eat other organisms; nor do they get carbon from the soil. Their only source of carbon is carbon dioxide: land plants get carbon dioxide as gas from the air; plants living under water get dissolved carbon dioxide.

Carbon dioxide is colourless and odourless. Only 0.03 % of air is carbon-dioxide gas. This is enough to provide all the carbon with which plants make sugar by photosynthesis. The plant makes all its other carbon compounds from sugar.

Since there is so little carbon dioxide in the air, why has it not been used up? It is constantly replaced in the air by respiration. Nearly all living organisms — plants, animals, bacteria and fungi — use carbon compounds as a source of energy and most of them give off carbon dioxide to the air.

Figure 9.1 shows the movement of carbon through an ecosystem. Look at carbon dioxide in the cycle: it is being continually removed by one process, photosynthesis, and continually replaced by another process, respiration. Since we end up where we began, the diagram is of a cycle.

Nowadays the natural carbon cycle is complicated by the human process of burning fuel in power stations, factories, motor cars and houses. Burning, though it is rapid and uncontrolled, involves the same chemical process as respiration: it uses oxygen and releases carbon dioxide. Though it is only 0.03 % of air, carbon dioxide is one substance humans are not using up too quickly. With all the burning we do, we

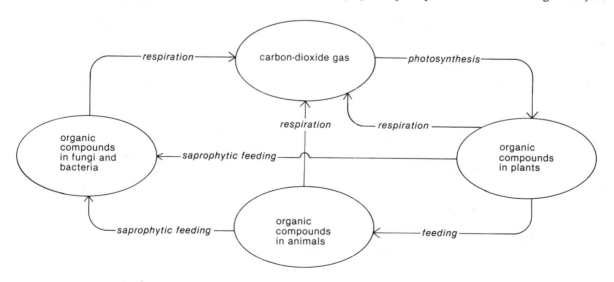

Fig. 9.1 Natural carbon cycle

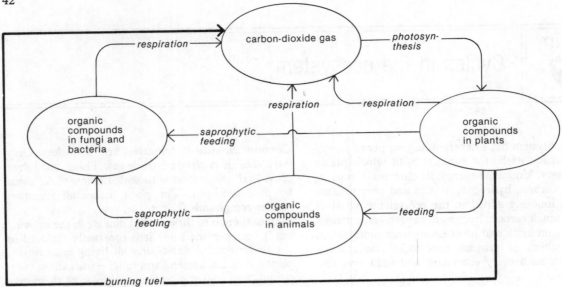

Fig. 9.2 Present-day carbon cycle

may even be producing too much carbon dioxide: a recent prediction is that the carbon-dioxide level of the atmosphere will double by the year 2050. Nuclear energy and some human-made fuels used in rockets are exceptions, but most fuel comes from plants: coal, oil, petrol and natural gas all come from plants that lived millions of years ago, while wood comes from plants that lived in our own time. Figure 9.2 shows the carbon cycle with the human process of burning included.

Oxygen

Oxygen, like carbon dioxide, is a colourless, odourless gas. It occupies about 20 % of the air. While in respiration carbon dioxide is being released into the air, oxygen is being removed from it. Burning also releases carbon dioxide into the air while removing oxygen from it. In photosynthesis the opposite is occurring: carbon dioxide is being removed from the air while oxygen is being released into it. This is all part of the carbon cycle shown in Figures 9.1 and 9.2. Figure 9.3 shows only the carbon dioxide and oxygen in air and explains how photosynthesis compensates for respiration and burning.

Remember that, while respiration occurs in every living organism, photosynthesis, which releases oxygen, occurs only in plants.

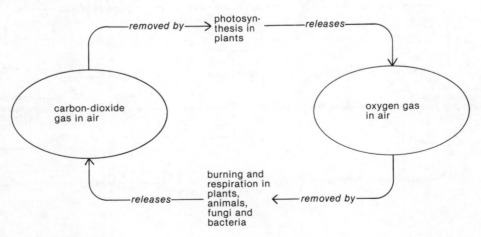

Fig. 9.3 Carbon dioxide and oxygen in air

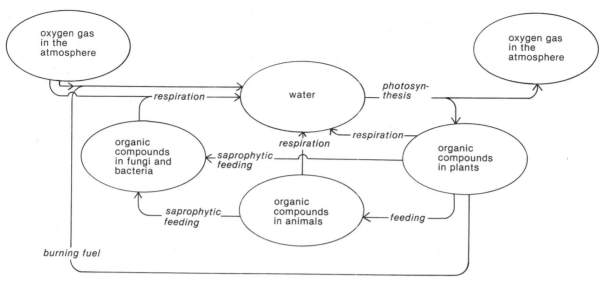

Fig. 9.4 Present-day water cycle

Water

Water, like carbon dioxide, is removed from the environment in photosynthesis and released in respiration. The water cycle is just like the carbon-dioxide cycle. Water contains hydrogen and oxygen. During photosynthesis, water gives up its hydrogen to form sugar and its oxygen is released. Hydrogen is transferred from sugar to all the other organic compounds. During respiration, the hydrogen in the organic compounds combines again with oxygen to form water. Figure 9.4 shows the present-day water cycle. Like respiration, burning the traditional fuels uses oxygen and releases water vapour as well as carbon-dioxide gas. Thus burning influences the carbon, oxygen and water cycles.

Carbon dioxide and water used in photosynthesis enter a plant differently: whereas carbon dioxide enters as a gas by the plant's leaves, water enters as a liquid by its roots. Both are released as gases in respiration – water as water vapour. Water vapour forms clouds and water eventually gets back into the soil when it falls as rain.

Nitrogen

The nitrogen cycle is very different because plants, the producers, cannot use the plentiful supply of the nitrogen gas in the air. Plants get their nitrogen from the soil in the form of nitrate and ammonium ions dissolved in the water absorbed by their roots. The nitrate and ammonium ions are used by the plant to make its organic compounds, particularly protein. The nitrogen stays in the plant until it is either eaten by an animal or decomposed by bacteria and fungi.

Suppose that the plant is eaten by an animal. The nitrogen may stay in the animal in the form of its organic compounds, particularly protein; it may be excreted from the animal as **urea** in **urine** (or as some other nitrogen-waste compound); it may pass straight through the animal undigested and be removed in its **faeces**. Most of the organic nitrogen compounds will sooner or later return to the soil and be decomposed by bacteria and fungi. The nitrogen in them will end up as ammonium ions, ready to be absorbed by the roots of plants. There are even **nitrifying bacteria** which convert ammonium ions into nitrate ions which those plants better able to absorb their nitrogen in the form of nitrates can use. This cycle is shown in Figure 9.5.

The nitrogen cycle would be as simple as the other cycles if all the nitrogen continued going round and round. It does not because nitrogen compounds are constantly lost from the cycle. For example, nitrate and ammonium ions in the soil dissolve in water and are liable to be drained away in rainwater into rivers and the sea where they are no longer available to land plants growing in the soil. Even modern sewage works release large amounts of nitrogen compounds into rivers and the sea.

There are also bacteria which, when they are living in waterlogged soil, do not get enough oxygen dissolved from air. They use nitrate ions as a source of oxygen and give off the nitrogen from the nitrate as a gas to the air where it is no longer available to plants. They are called **denitrifying bacteria**.

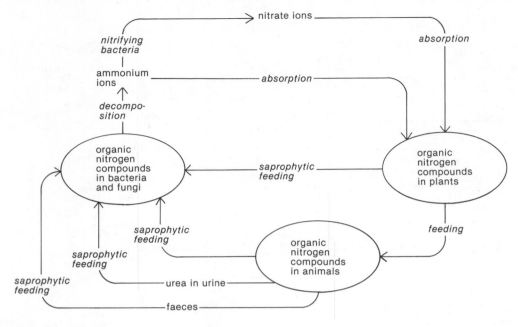

Fig. 9.5 Basic nitrogen cycle

Fortunately there are ways in which nitrogen is added to the cycle to balance these losses. Though plants cannot use the nitrogen in air to make organic compounds, certain bacteria can. They are said to **fix** the nitrogen in air. Some of these **nitrogen-fixing bacteria** live freely in the soil, converting nitrogen gas into their own body compounds. Others live in harmony with certain plants in swellings on their roots called **nodules**, where they not only convert nitrogen gas into their own body compounds but also give excess nitrogen compounds to the plants. In return these bacteria get sugars from the plants. This relationship, in which bacteria and plants benefit one another, is an example of **mutualism**. One of the plants that has this relationship with nitrogen-fixing bacteria is the French bean. Nitrogen-fixing bacteria put nitrogen back into the cycle because, whether they live in plant nodules or freely in the soil, when they and the plant are decomposed by other bacteria, nitrogen is left in the soil in the form of ammonium ions available to plants. Figure 9.6 shows roots of a French-bean plant with nodules.

Another cause of the fixation of nitrogen in the air is lightning. Lightning combines the gases oxygen and nitrogen into compounds which dissolve in rain and end up as nitrate ions in the soil, where they are available to plants.

When we add these processes to the nitrogen cycle in Figure 9.5, the result looks complicated. But the entire cycle makes sense. It will be easier to reproduce it if you build it up in the order in which it is described in this unit. Begin with the basic cycle from soil ions to plants, to animals and bacteria and fungi and back to ions, as shown in Figure 9.5. Next add the ways in which nitrogen is lost from the cycle. Finally add the ways in which nitrogen is regained by the cycle. The whole nitrogen cycle is shown in Figure 9.7.

Fig. 9.6 French bean: root nodules

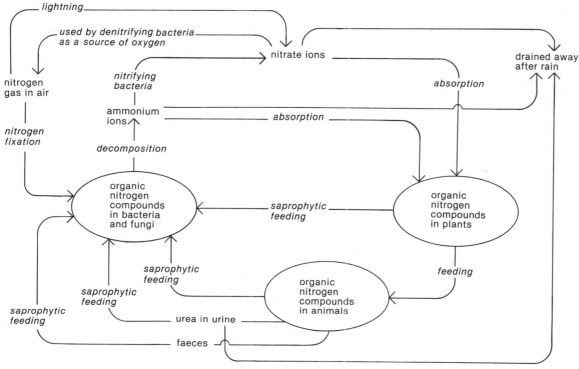

Fig. 9.7 Nitrogen cycle

Questions

Q 9.1 When farmers harvest crops, they break the natural mineral cycle in which plants die and decompose where they grow. As a result the soil becomes deficient in minerals.
(a) Give one mineral which may become deficient in farmed soils.
(b) Explain how the addition of farmyard manure to land will increase the mineral content of the soil.
(c) Every few years a farmer may grow leguminous plants (instead of his usual crop) and plough them into the soil. How does this practice benefit the soil? (Leguminous plants have roots which contain nitrogen-fixing bacteria in nodules.)

Q 9.2 Waterlogged acid soils, such as are found in peat bogs, often contain the undecomposed remains of animals and plants which died many years ago. Suggest why (a) waterlogging and (b) an acid pH help prevent the decomposition of dead organisms.

Q 9.3 Early this century the River Thames became polluted by the discharge of dead organic matter from sewage works, paper mills, flour mills, etc. Suggest:
(a) what effect the organic matter will have had on the number of bacteria involved in decomposition in the River Thames;
(b) what effect the change in the number of bacteria will have had on the oxygen content of the River Thames;
(c) what effect the change in the oxygen content will have had on other organisms in the River Thames;
(d) why salmon have been absent from the River Thames for most of this century.

PRACTICAL WORK

Experiment 9.1 To investigate the influence of nitrogen-fixing bacteria on the growth rate of the French bean

Note for teachers
If samples of nitrogen-fixing bacteria (*Rhizobium leguminosarum*) are obtained from biological suppliers, a *Rhizobium* suspension can be made by washing the surface of the agar slope with sterile water and then decanting the water into a sterile container.

Alternatively, *Rhizobium* can be obtained by rinsing a root nodule of a French-bean plant in tap water and then washing it for three minutes in 70 % ethanol and for a further three minutes in sterile water. The nodule should then be crushed in sterile water and filtered to remove cell debris.

Prepare the seeds by washing them in 1 % sodium-hypochlorite solution for three minutes and rinsing them in sterile water. Dip half the seeds in the *Rhizobium* suspension. These are the inoculated seeds.

Prepare the nitrogen-free agar on which the seeds are to be grown by grinding together the salts listed before dissolving them in 1 dm^3 of distilled water.

Crone's medium:
	KCl	5.0 g
	K_2HPO_4	1.25 g
	$CaSO_4 \cdot 2H_2O$	1.25 g
	$MgSO_4 \cdot 7H_2O$	1.25 g
	$Ca_3(PO_4)_2$	1.25 g
	$Fe_3(PO_4)_2$	0.01 g

Add 8 g of agar and autoclave the mixture. Pour the mixture into sterile boiling tubes to a depth of about 9 cm and plug them with cotton wool. Allow six tubes per student.

Materials required by each student
1 boiling-tube rack with 6 sterile boiling tubes of Crone's medium fitted with cotton-wool plugs
1 pair of forceps
1 spirit marker
3 French-bean seeds inoculated with nitrogen-fixing bacteria
3 uninoculated French-bean seeds

Method
1. Label three of the tubes INOC and the other three O and add your initials to all six.
2. Put one inoculated French-bean seed on the surface of the agar in each of the three tubes labelled INOC and replace the cotton-wool plugs.
3. Put one uninoculated French-bean seed on the surface of the agar in each of the three tubes labelled O and replace the cotton-wool plugs.
4. Leave all the tubes in a warm place away from direct sunlight and inspect them every three or four days. When the plants begin to develop leaves, remove the cotton-wool plugs from the tubes.
5. Five weeks after leaving the tubes, remove the plants from them. Record the lengths of their stems and roots and the numbers of nodules on their roots.

Interpretation of results
1. Calculate the mean stem length, mean root length and mean number of nodules of the two groups of plants (inoculated and uninoculated) and record them in a table.
2. Suggest how the root nodules were formed and explain the difference in the mean number of nodules in the two groups of plants.
3. The nutrient agar contained no nitrates. What use is made of nitrates by plants?
4. What do the mean stem and mean root lengths tell you about the importance of root nodules in French-bean plants?

UNIT 10 — Decomposition; bacteria and fungi; saprophytes; parasites; mutualism

Without the **rotting** or **decomposition** of organic matter, the earth would long ago have been buried under the dead remains of plants and animals. The carbon and nitrogen cycles would have come to a stop because these elements would have been used up. Decomposition, during which organic compounds are broken down, is carried out by bacteria and fungi.

Bacteria and fungi

Bacteria and fungi look different but work in similar ways. Bacteria are shown in the photograph in Figure 1.1: they are tiny single-celled organisms which can hardly be seen under a light microscope and do not show much of their structure even under an electron microscope. Fungi can be large enough to be seen without a microscope at all. Some are bought every day in the shops in the form of **mushrooms**. Most fungi exist in the form of fine living threads called **hyphae**. Mushrooms are made of hyphae bound together in a mass. Part of a hypha is shown in Figure 10.1.

Bacteria and fungi feed in similar ways. They cannot feed like plants by photosynthesis because they do not contain the pigment chlorophyll. (Some bacteria do in fact have chlorophyll and photosynthesise, but they are rare.) Nor can they feed like animals because they have no mouths and cannot take in solid food.

If their food is already dissolved, such as milk, fruit juice and urea in urine, they simply absorb it through their surface walls and it passes into their protoplasm. If their food is solid but dissolves readily, as sugar does, all they need is water to dissolve it: there are various sources of water, such as rain, condensation and the food itself. If their food is solid and does not dissolve, they need both water and **enzymes** that change solid food into a form in which it will dissolve. Enzymes are chemicals produced by living organisms which change the speed of reactions between other chemicals without being used up themselves. In the chemistry laboratory a chemical that does this is called a **catalyst**. An enzyme is a catalyst produced by a living organism. Enzymes are described in Unit 17.

Many fungi and bacteria produce an enzyme that passes out through their walls on to their food and changes starch, which is **insoluble** (unable to dissolve), into sugar, which is **soluble** (able to dissolve). The sugar dissolves in water. The bacteria and fungi absorb the dissolved sugar through their walls. Turning insoluble food into soluble food, whose molecules are small enough to be absorbed, is **digestion**. To feed on insoluble food, bacteria and fungi must digest it so that it will dissolve and they will be able to absorb it. Figure 10.2 shows how a bacterium and a fungus get their food. The bacterium is tiny. The hypha of the fungus, though narrow, is long and may branch. A cubic centimetre of soil can contain between six and ten million bacteria and between one and two kilometres of fungal hyphae.

A substance that is produced inside protoplasm and comes out of it is said to be **secreted**. Bacteria and fungi digest insoluble food by secreting enzymes over it. Because this happens outside their bodies, it is **external digestion**.

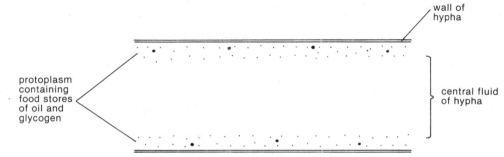

Fig. 10.1 Fungal hypha: section

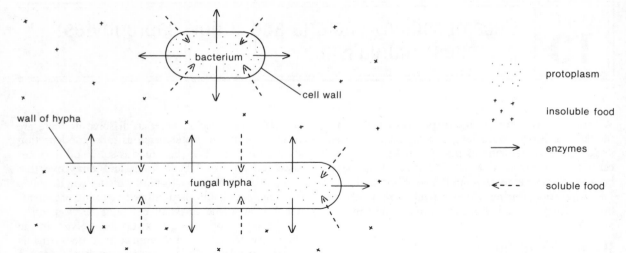

Fig. 10.2 External digestion

Saprophytes

When dead food is digested externally and absorbed in solution, feeding is **saprophytic**. Bacteria and fungi that feed in this way are **saprophytes**. Dead food may be either dead organisms or dead organic products such as wood and faeces.

Parasites

Other bacteria and fungi feed on living organisms. If one organism feeds consistently on a larger one of a different species in a way unlikely to kill it, the organism fed on is a **host**. Obviously the host is harmed by being fed on. If the host also gains some benefit from the organism feeding on it, as do French beans from bacteria in the nodules described in Unit 9, the relationship is one of **mutualism**. If the host is only harmed by the organism feeding on it, i.e. if it does not also gain some benefit from it, the relationship is **parasitic**. Bacteria and fungi that feed on living organisms without benefiting them are **parasites**. Bacteria and fungi that cause infections and diseases in our bodies are parasites feeding on us; we are the hosts. Tar-spot fungi and toadstools on the sycamore tree in Figure 6.1 are parasites; the sycamore tree is the host. An organism feeding on another living organism can be a parasite without feeding in the way that bacteria and fungi do. A tick feeding on the blood of a sheep is a parasite; the sheep is the host. Blackfly (aphids) feeding on the sugary sap of French beans are parasites; the French beans are the hosts.

Bacterial and fungal parasites must have their living food in solution, i.e. dissolved in water, or they cannot absorb it. Their food must therefore be digested and dissolved before they can absorb it. Sometimes they digest it for themselves by secreting enzymes on to it to make it soluble: the fungus that causes athlete's foot does so by secreting enzymes on to the skin. Many internal parasites get their food after their hosts have digested it: an example of such a parasite is the bacterium that can live in our gut and causes bacillary dysentry.

Decomposition

Decomposition usually takes place in soil or water. Both bacteria and fungi are more active when there is plenty of oxygen for respiration. Decomposition is therefore slower in water or in waterlogged soil where there is less oxygen. Foul-smelling intermediate products of decomposition accumulate when bacteria and fungi have insufficient oxygen for aerobic respiration. Fungi are better at decomposing plants: they can make enzymes that digest cellulose and wood, both of which are found in plants. Although some bacteria can also digest cellulose and wood, they are better at digesting protein, the compound that is abundant in animals.

Both fungi and bacteria are more efficient decomposers when the dead plant and animal material has first been broken up by passing through the body of a consumer: the undigested material in faeces is quickly decomposed. Scavengers (animals such as earthworms, woodlice, fly larvae and crows, which eat dead material) are therefore important in the process of decomposition.

Decomposition is an essential part of the carbon, water and nitrogen cycles. Eventually all the dead organic matter is decomposed through the action of saprophytic bacteria and fungi and disappears. When saprophytic bacteria and fungi respire, they give off carbon and oxygen as carbon-dioxide gas; hydrogen and more oxygen as water vapour; nitrogen as ammonium ions. Saprophytic bacteria and fungi eventually die themselves and are digested and absorbed in solution by other saprophytic bacteria and fungi.

Questions

Q 10.1 (a) What is a saprophyte?
(b) What is a parasite?

Q 10.2 Both scavengers and saprophytes feed on dead organisms. What are the major differences between them?

PRACTICAL WORK

Experiment 10.1 To investigate the influence of temperature on the rate of decomposition

Note for teachers
Any soft-skinned fruits (apples, peaches, plums, tomatoes, etc.) are suitable for this experiment. Choose whatever is cheaply available at the time of year.

Materials required by each student
6 glass containers to hold one fruit each
1 thermometer
1 spirit marker
enough plastic cling film to cover the containers
6 soft-skinned fruits

Materials required by the class
1 refrigerator
1 dark cupboard at room temperature

Method
1. Lightly bruise the fruits by tapping them against the bench. Leave them on the bench as long as you can before carrying out instruction 3.
2. Put the date and your initials on the containers. Label three COLD and three WARM.
3. Put one fruit in each container and cover each container with cling film.
4. Put the three containers labelled COLD in the refrigerator and the three labelled WARM in the cupboard.
5. Each day for one week examine the fruits in the containers and record which have become discoloured, which show signs of fungal growth and which have lost their texture. At the same time, record the temperature of the cupboard and of the refrigerator.

Interpretation of results
1. Which fruit decomposed most quickly?
2. Assuming that decomposition occurs owing to the action of bacterial and fungal enzymes, suggest:
   (a) why one group of fruits decomposed more quickly than the other;
   (b) why you bruised the fruits at the beginning of the experiment;
   (c) why you exposed the bruised fruits to the air for as long as possible before enclosing them in the containers.
3. Do the results of your experiment suggest that light is necessary for decomposition to occur?
4. Suggest *three* ways in which the decomposition of fruit could be prevented.

UNIT 11: Population size: colonisation, succession, competition and population control

Individuals of one species living in a certain area are called a **population**. The British are a population of Great Britain; oak trees are a population of an oak wood. Populations of different species living together in one habitat form a **community**. The populations of different species in an oak wood are an oak-wood community.

Colonisation

Suppose you leave a bucket of heat-sterilised soil in your garden. You will not be surprised if after a few weeks some plant seedlings are growing in the soil and if some time after that there are animals in it as well. Since the soil has been sterilised, the plants and animals in it have come from outside the bucket. The soil has been **colonised**.

Which species of plants and animals you find in your bucket will depend on the species living in your garden and nearby, on their mobility and on the physical conditions in your bucket of soil. The mobility of the plants depends on their methods of seed or spore dispersal: the most likely colonisers are plants that produce large numbers of light seeds or spores easily carried by air currents. But they will not take root unless the physical conditions of the soil, such as the size of its particles and the amount of moisture in it, are suitable for them. Nor will animals that find their way into the bucket stay and survive unless the conditions suit them.

Your bucket of soil is an 'empty habitat'. Because you prepared it, it is an artificial empty habitat. Other artificial empty habitats are a newly built stone wall, a valley flooded to form a reservoir and the scorched earth left after a forest fire. These are not such perfect examples as your bucket of sterilised soil, because they will not be completely without forms of life: at the very least they will contain bacteria, and seeds will have survived the forest fire. Empty habitats can occur naturally too. A volcanic island erupting in the sea is an example of a perfect natural empty habitat. A deposit of dung from an animal is another example of a natural empty habitat, though it will contain bacteria and seeds may have survived in it.

Events after colonisation

An empty habitat must first be colonised by a plant (or by one of the rare plant-like bacteria) because they are the only organisms that can make food from sunlight: other organisms need to find their food in the habitat.

The growth of a colonising plant species will change the habitat: plants will provide food for animals where there was none before; dead plants and animals will provide food for saprophytes where there was none before; decomposing plants and animals will retain water in an environment that previously dried out quickly. Food chains will be established.

The first species to colonise an empty habitat may be a simple plant such as an alga of the kind on the tree trunk in Experiment 3.2. Or it may be a special kind of plant called a **lichen**. Lichens are the dry coloured patches on rocks, stone walls, roofs of houses and branches of trees. Strictly lichens are a mixture of algae and fungi in a relationship of mutualism: both benefit from the relationship and indeed they cannot live without each other. The thousands of tiny spores lichens produce are easily carried by wind and are able to colonise bare rock. The fine threads of the fungi penetrate small cracks in the rock, absorbing rainwater and inorganic ions. (They absorb and are killed by mere traces of industrial poisons such as sulphur dioxide.) The algae make sugar by photosynthesis. As a result of lichen growth, rock faces crumble. Crumbled rock and the decomposing remains of dead lichen form a soil-like substance, containing inorganic ions and water, in which seeds can germinate and roots can grow.

Succession

Colonisation by lichens has changed the habitat and enabled other plants to grow there which could not have done so before. In turn, colonisation by these plants will change the habitat further and enable yet other organisms, including animals that feed on the plants, to live and grow there. The later colonising species may destroy or drive out some of the earlier ones. Change over a period of time in the organisms in a single community is called **succession**. In time a community will include many different populations, i.e. individuals of many different species, with complex relationships between them. A stable community, one in which the populations are unlikely to change in the foreseeable future, is a **climax community**. A mature oak wood is an example of a climax community. Figure 11.1 summarises the events that occur during any succession.

```
empty habitat
    ↓
habitat colonised by plants
    ↓
changes occur in the habitat
as a result of the early colonisers
    ↓
other organisms including
animals inhabit the environment
and may replace the early colonisers
    ↓
a climax community develops
```

Fig. 11.1 Events during succession

Look at some rocks, a stone wall round a house, the roof of a house or a blocked drainpipe or gutter and you may be able to see how far colonisation and succession have gone. The speed of colonisation and succession varies greatly. Figure 11.2 shows a gravestone which is nearly a hundred years old and took many years to be colonised by the plants you can see on it. In dung, colonisation and succession are so rapid that small animals are usually living in it within days.

Fig. 11.2 Gravestone succession

Competition

Why do some species, particularly among the early colonisers, disappear from a habitat? Plants growing near to one another make similar demands on the environment. They need: space in which to grow and spread their roots for anchorage; light, water and carbon dioxide for photosynthesis; inorganic ions for protein production; oxygen for aerobic respiration. Inevitably some plants will be better at obtaining limited resources than others: plants whose roots grow faster occupy more space in the soil and absorb more of the water and inorganic ions; plants which reproduce rapidly fill more of the available spaces with their offspring; plants which grow tall absorb more sunlight and shade shorter plants. Similarly some animals will be better than others at obtaining limited space, nesting sites, food, oxygen and mates.

Whenever the organisms in a habitat need more resources than are available, there is **competition** both between members of the same species (**intraspecific competition**) and between members of different species (**interspecific competition**). When we are studying populations and communities, we are interested in competition between species. A species may become less common and die out because it is not as good as others at obtaining the resources it needs or because it is eaten by other species. A species of plant may disappear from a community after the arrival of a herbivore species; in turn the herbivore species may disappear from a community because it is preyed on by a carnivore or omnivore species. Successive new competitors make the habitat less suitable for earlier colonisers. Succession continues until there is no other species which is in reach of the habitat and which can displace one or more of the species already in it. For example, no other species can displace a mature oak tree.

Population control

Any population is capable of producing far more offspring than are needed to replace the parents. Unit 39 explains that yeast increases more than a hundredfold in less than 36 hours and Unit 44 that a female trout lays more than a thousand eggs a year.

In spite of this reproductive capacity, the number of individuals in most populations stays more or less constant. This can be either because individuals fail to reproduce to their full capacity or because most offspring die. The reproductive capacity of parents and the survival of young are affected by a number of environmental factors.

Biological (biotic) factors

The section on competition describes how any one population is affected by other populations in the same habitat. A food shortage, for whatever reason, will reduce reproductive capacity. Consumers, such as herbivores, predators and parasites, may kill parents and offspring.

Physical and chemical (abiotic) factors

Temperature and pH (acidity or alkalinity) of the environment affect the rates of chemical changes in the bodies of organisms (by affecting the rates at which enzymes work). Warm-blooded animals, such as birds and mammals, which can control their body temperatures, have an advantage and are less affected by external temperature changes.

Light intensity and duration affect the ability of plants to photosynthesise and hence to produce food and energy for growth and reproduction. Some plants can grow only in bright sunlight while others can also grow in shade.

Humidity affects the rate of evaporation from an organism's body. Organisms poorly adapted to resist water loss can survive only in a humid environment.

Currents of either air or water move the organisms from one part of a habitat to another which may be less suitable for them.

Edaphic (soil) factors

Different types of soil have different physical and chemical properties. Apart from temperature and pH, which have already been mentioned, those that affect plant growth include concentration of inorganic ions, concentration of oxygen in the soil air and water-holding capacity.

Figure 11.3 summarises natural controls on population size. Human influences on populations are described in Unit 12.

Human population

Human population size is a special case. Figure 11.4 shows how the world population, for centuries more or less constant at 350 million, rose during medieval times and rose dramatically from the seventeenth century onwards until it is now more than four and a half thousand million (4 500 000 000) and threatens to rise to six thousand million by the year 2000.

The Black Death of the fourteenth century is estimated to have killed one third of the population of Europe. As late as the sixteenth and seventeenth centuries great epidemics periodically killed millions throughout the world. Even after the First World War a flu epidemic killed twenty-five millions. Now, however, we have much improved hygiene and immunisation programmes to prevent people from catching diseases, and we have antibiotics and other drugs to cure diseases. Thus there are fewer natural

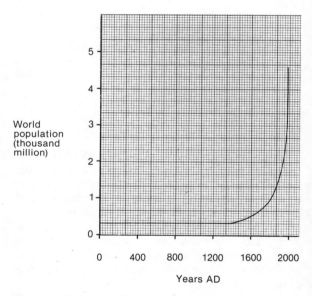

Fig. 11.4 World population

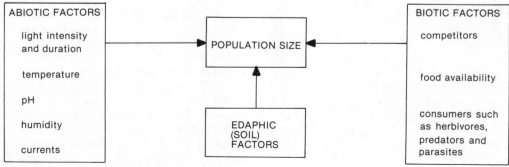

Fig. 11.3 Natural controls on population size

checks on population size. It is true that greater scientific and technological knowledge has enabled us to increase food production and so support a larger world population. Nonetheless there are millions of deaths every year from starvation and disease, and many of those who succumb to serious disease today do so because they are poorly fed.

Both the world population and the world death rate would be higher still were it not for birth control. To realise the importance of birth control in an industrialised country such as Britain, you have only to remember that without it the typical family might contain between ten and twenty children. Modern contraceptive methods are described in Unit 43. These are freely available only in richer countries whose governments allow them, but more and more countries are introducing birth-control programmes.

Questions

Q 11.1 The diagram shows a succession in a terrestrial habitat.

bare rock
↓
lichen-dominated community
↓
moss-and-fern-dominated community
↓
grass-dominated community
↓
climax woodland

(a) Why are plants the initial colonisers of the habitat?
(b) Suggest what changes made to the habitat by lichens, mosses and ferns allow the growth of grasses.
(c) Suggest how the trees in the climax community may affect the previous dominant species in the habitat.
(d) Suggest what might happen in this habitat if an accidental fire cleared part of the wood.

Q 11.2 Clay is obtained for industrial processes by washing it from suitable soil. The rest of the washed soil is a waste product and is often piled into large tips that look like small mountains. These tips are eventually colonised and there is succession from an algae-dominated community to a community dominated by leguminous plants.
(a) Suggest two reasons why most plants will not grow in the washed waste soil.
(b) Suggest why the succession is dominated by leguminous plants.

Q 11.3 Some plants, such as bluebells and dog's mercury, are found in woodlands but not in grasslands. Such plants flower in the early spring before the leaves have grown on the trees.
(a) Suggest why bluebells and dog's mercury can grow in woodlands where other small plants cannot.
(b) Suggest why the woodland plants are unable to grow in grasslands where other small plants thrive.

Q 11.4 The table shows the average number of young which were successfully reared by breeding pairs of robins (*Erithacus rubecula*) during one breeding season in various parts of the world.

Country	Average number of young reared per breeding pair
North Africa	4.2
Spain	4.9
England	5.1
Scotland	5.5
Sweden	6.3

(a) Use an atlas to find the position of the countries listed in the table. What is the relation between the reproductive capacity of robins and latitude (distance from the equator)?
(b) In summer the length of day increases as you move from the Equator to the North Pole. Suggest how this might explain the data in the table.

Q 11.5 A small stream ran from a mountain side into a lake. Both the stream and the lake were searched for flatworms which live under stones and feed on crustacea and small worms. In the stream the only species found was *Polycelis felina*. In the lake the only species found was *Polycelis tenuis*.
(a) Suggest why *Polycelis tenuis* was unable to live in the stream.
(b) Suggest one physical or chemical and one biotic factor which may have prevented *Polycelis felina* from living in the lake.

PRACTICAL WORK

Note for teachers
Colonisation and succession should be studied in a habitat such as freshly dug soil, an old bonfire site, fresh dung or a wall.

UNIT 12

Human influences: use of land; non-renewable resources; pollution; pest and weed control; recycling

A few thousand years ago humans had little more effect on their environment than other animals have today. They killed a few animals for food and clothing; they cleared small areas of vegetation in order to build their homes, grow their food and feed their domestic animals; they used plant products as building materials and tools. Even a few hundred years ago, though much more land was cultivated and there were large towns and cities, the effect of humans was not dramatic. Since then, however, industrialisation and mechanised agriculture have transformed much of the earth's surface, and the use of guns and harpoons has endangered animal species.

Use of land

Deforestation

Deforestation is the cutting down of forests, which are the climax vegetation in most parts of the world. Forests are cut down in order to support increasing human populations by growing food and by building towns and roads. The trees from the cut-down forests are used for building, for paper-making and for fuel. Deforestation has had unforeseen effects. Soil exposed by removing the trees has been washed away by rainfall (**rain erosion**) and blown away by wind (**wind erosion**), sometimes to such an extent that deserts and wastelands have been created. Loss of water evaporating from the leaves of the cut-down trees has reduced rainfall elsewhere. Loss of photosynthesis in the cut-down trees has reduced the oxygen recycled to the atmosphere and has allowed carbon dioxide to accumulate in it.

Afforestation

Replanting of forests, **afforestation**, has been carried out in Britain on a small scale. But natural climax vegetation of oak and beech trees, which provide hardwood, has been replaced by quicker-growing softwoods such as pines and other conifers. Many of the new trees planted are not even native to Britain. Fallen leaves and branches of the newly planted pines and conifers decay more slowly than those of the oaks and beeches they have replaced. The new forests do not provide suitable food for the organisms that lived in the old ones. The habitats have changed so drastically that most species of animals and plants have disappeared from them. Although some new species have appeared in them, there is not the variety that there was in the older forests. The Forestry Commission is now replanting some mixed woodlands in which conifers and pines grow alongside oaks and beeches.

Monoculture

With the increasing use of large and expensive agricultural machinery it has seemed efficient to devote vast areas to a single crop and to grow the same crop in the same area year after year. This is **monoculture** (*mono* means *one*). Afforestation by pines alone is an example of monoculture. Figure 12.1 of hectares of land under a cereal crop, is another example. In East England many hedges have been removed in order to merge fields and allow even greater areas of monoculture. Since the hedges were distinctive habitats, the number of species living in these regions has been reduced. Some of the lost species preyed on **pests**, organisms that harm crops. There has therefore been an increase in pests.

Growing the same crop on the same land year after year means a constant demand on the same natural resources, such as certain inorganic ions in the soil. Before monoculture was introduced, farmers

Fig. 12.1 Monoculture

practised mixed farming and rotated their crops so that different demands were made on the land's natural resources in different years. The rotation included a 'grass break', which meant sowing grass and clover mixtures to provide grazing for sheep and cattle, whose dung returned organic matter to the soil. In addition the clover, a leguminous plant, contained nitrogen-fixing bacteria in root nodules. When the turf was ploughed in before the next crop was sown, the plants and dung rotted, improving the physical structure of the soil and increasing its water-holding and nutrient-holding ability as well as its nitrogen compounds. During the winter the cattle were housed indoors and fed on hay, made from the grass, and other products of the farm. Their dung was heaped to rot before being spread on the fields to return natural resources, notably nitrates, to the soil.

Monoculture requires both an elaborate system of chemical weed and pest control and the extensive use of artificial fertilisers to provide the nutrients a crop needs in order to grow. Both these practices are expensive and some farmers are now wondering if monoculture is worthwhile.

Non-renewable resources

Minerals we dig out of the earth, such as iron, tin and gold, cannot be renewed. Once the mines are exhausted we shall have no more supplies of them. The same is true of our **fossil fuels**: coal, oil and gas. Coal is the chemically changed remains of forests that grew in swamps and did not fully decompose; oil and gas are the chemically changed remains of plants, and animals that fed on them, which fell to the bottom of the sea. What will we do when all these fossil fuels are used up? Traditional sources of power, such as animals, windmills, water-mills and the sun, do not provide enough energy for modern needs. Nuclear power is an important new source of energy, but it is expensive and its dangerous waste products present problems of disposal.

Pollution

Pollution is virtually anything in the wrong place, but the term's usual meaning is the harmful effects of human activities on their environment. **Chemical pollution** is too many chemicals (created by humans) in the wrong place; **sewage pollution** is too much sewage in the wrong place; **thermal pollution** is too much heat (created by humans) in the wrong place; **noise pollution** is too much noise (created by humans) in the wrong place. The newspapers report serious examples of pollution such as oil spillage from ships, leakage from farm silos and accidents at chemical works.

There is concern about damage to vegetation throughout Europe from sulphur dioxide, a gas released from burning fuel. Higher chimneys, installed to disperse waste gases away from towns, have caused pollution in parts of the countryside not previously affected.

Acid rain is formed when sulphur dioxide (mainly from industrial smoke) and oxides of nitrogen (mainly from lorry and car exhausts) dissolve in rainwater and fall to earth. At Pitlochrie in Scotland acid rain has been recorded with a pH of 2.4, about six times more acid than vinegar (pH 7 is neutral and the lower the pH the higher the acidity); a scientist has estimated that the famous London fog in 1952, which killed thousands, had a pH of 1.8; pH levels of 4.6 are common in Britain. Acid rain not only kills trees it falls on but affects plants and animals in lakes and streams it drains into.

Fig. 12.2 Effects of acid rain

Sewage need not be a pollutant: urine, faeces and other biological products it contains are completely **biodegradable**, i.e. they can be decomposed by saprophytes. But complete decomposition by saprophytes takes time and needs oxygen for aerobic respiration. When a river receives large amounts of untreated concentrated sewage, there are soon so many saprophytic fungi and bacteria that they use up all the oxygen in the water. After that they can no longer complete the breakdown of sewage, and animals, including fish, are starved of oxygen and die. Untreated sewage may also contain human parasitic organisms such as bacteria causing diarrhoea and typhoid.

Noise pollution can damage hearing. Loud noise in factories and from gunfire has long been known to be harmful, but the problem used to be a small one. Noise pollution has increased with the widespread use of amplifiers in the transmission of music.

Whole books are written on pollution: sulphur dioxide, sewage and noise are just three examples. *Friends of the Earth Guide to Pollution* by Brian Price (Temple Smith, 1983) is a short book that explains the problems in an interesting way.

Pest and weed control

We now have an enormous range of chemicals used to kill such unwanted organisms as parasites that cause human and animal diseases, animal pests that eat food crops, and weeds that compete with food crops for land resources. **Pesticides**, chemicals that kill pests, **fungicides**, chemicals that kill fungi such as rusts and mildews, and **herbicides**, chemicals that kill plants, have increased food-crop yields and are easy to use, but they are not selective enough. They kill not only pests, unwanted fungi and weeds but also organisms that benefit us: for example, the chemicals that kill aphids (blackfly and greenfly) may also kill pollinating insects and bees that provide honey. Pesticides, fungicides and herbicides all reduce the number and variety of natural populations. We cannot be certain that their use on our food crops does not harm us.

It is true there are 'selective products' that kill only certain pests, fungi or weeds provided that they are given in exactly the right dose at exactly the right time under exactly the right weather conditions for them. In practice, since weather conditions are not always exactly right at the right time, even selective pesticides, fungicides and herbicides sometimes damage the crop and kill useful organisms.

DDT is an example of a pesticide that has been used too much. The trouble with DDT is that it is almost indestructible. Because it dissolves in fat, it becomes concentrated in the fat-storage tissues of animals in the food chain. Top carnivores which eat the concentrated DDT in their prey accumulate even higher concentrations in their own bodies and may be killed by them: birds of prey in particular have been killed by DDT. Moreover the widespread use of DDT, while killing off many pests, has allowed those resistant to DDT to thrive. In short, DDT has done more and more harm and less and less good. Its use in Britain is now banned.

Chemical control of pests and weeds may be replaced by **biological control**, which is safer and more selective. Biological control of pests and weeds makes use of other organisms, often natural predators. For example, it is known that ladybirds eat aphids. In future we may put ladybirds into crops to eat the aphids on them instead of spraying the crops with pesticides which kill not only aphids but bees and other insects. Although biological control is still at the research stage, successful examples of it have been known for some time. Fish have been put into lakes to eat mosquito larvae and thus prevent the spread of malaria by adult mosquitoes. Myxomatosis, a rabbit disease, was deliberately, though not officially, spread in Britain in order to reduce the vast number of rabbits eating food crops. A subtle method of biological control is the introduction of sterilised males into natural populations of insects: females mate with them in the normal way and, since their eggs are infertile, the population is reduced.

Recycling

When substances are not renewable, the ideal solution is to **recycle** them, i.e. to use them over and over again. This cannot be done with fossil fuels, but it can be done with minerals and with metals extracted from them. The reason there is not much recycling of minerals is that the expense of collecting them and cleaning them can be greater than the expense of buying them new. When minerals get scarce, and consequently expensive, more efforts will be made to recycle them.

One substance already scarce enough and expensive enough to make recycling worthwhile is water. Although in Britain we are surrounded by water and have a high rainfall, it is too expensive to remove the salt from sea water or to collect all the water we need in the form of pure rain. For this reason we recycle water that we collect off roads and gutters, that we pour down our sinks and that we use to flush our toilets. Reading is a town higher up the River Thames than London: it is said jokingly that what Reading drinks today London drinks tomorrow.

Sewage arrives at a sewage works mixed with water. The mixture is first filtered to remove large solids. It is then left to settle in large tanks in which the remaining solid matter sinks to the bottom and forms a sludge which is slowly decomposed by anaerobic organisms (those that can respire without oxygen). Without oxygen, organic compounds in the sludge cannot be fully broken down to carbon dioxide and water: instead they form methane gas and other organic products which, together with the anaerobic organisms, can be dried and sold as organic soil fertiliser, i.e. are recycled. The methane still contains usable energy and can also be sold or can be used at the sewage works as fuel to drive the machinery.

In a traditional sewage works the liquid in the tanks goes into moving bars from which it is sprinkled over broken stones which have large surface areas and plenty of air spaces between them. Figure 12.3 shows sprinkler bars and stone beds at a sewage works. Millions of saprophytic micro-organisms form a slimy film over the stones. As the sewage liquid trickles through the bed of stones, saprophytes and small animals digest and absorb the organic matter, including bacteria and other parasitic organisms harmful to humans. It is this aerobic-decomposition process which, occurring in a lake or river that receives untreated sewage, may use up all the oxygen and kill the animals. When sewage passes through a sewage works, only the purified liquid which collects below the stone beds is discharged directly into rivers and lakes. It is from the same rivers and lakes that our tap water is collected and stored in reservoirs before it is piped to our homes. Many towns have their sewage works on their outskirts near a river into which they discharge the purified water. You can recognise a traditional sewage works by the stone beds and the slowly moving sprinkler bars. Some towns have modern sewage works in which the digestion and absorption of organic matter by saprophytes and small animals takes place in aerated tanks instead of stone beds.

Although carbon dioxide is given off as a gas from the stone beds, numerous salts, including phosphates and nitrates, remain dissolved in the purified water released into rivers and lakes. Excess nitrates from artificial fertilisers are also washed into rivers and lakes. Figure 12.4 shows the rise in the level of nitrates in the Thames over fifty years. The nitrate level is approaching the World Health Organisation's recommended maximum (of 11.3 mg per dm^3) and will soon have to be brought down by an expensive extraction process.

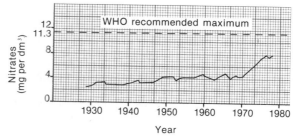

Fig. 12.4 Nitrate levels in the Thames

Fig. 12.3 Stone sprinkler beds at sewage works

Questions

Q 12.1 (a) From the data in Figure 12.5 calculate the approximate percentage increases and decreases in the various uses of world land in the seventy years between 1882 and 1952.

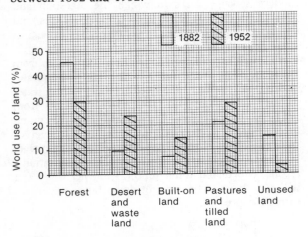

Fig. 12.5 Changes in world use of land

(b) Suggest reasons for these changes.

Q 12.2 Lichens are plants that grow over rocks, houses and the barks of trees. The table shows the percentage cover of such surfaces by lichens at different distances from the centre of a city.

Distance from city centre (km)	Lichen cover (%)
5	3
8	22
11	45
16	75
19	75

(a) Draw a graph plotting distance from the city centre on the horizontal axis and percentage cover on the vertical axis.
(b) (i) Explain the trend in percentage lichen cover between 5 and 16 km from the city centre. (ii) Suggest why the trend is not continued between 16 and 19 km from the city centre.
(c) Suggest how lichen growth might be used to monitor pollution levels.

Q 12.3 Explain how a commercial plant grower can benefit from (a) a greenhouse, (b) strip lighting, (c) a commercial brand of potting compost, (d) pesticide sprays.

PRACTICAL WORK

Note for teachers
There is no practical work in this unit. Pollution is likely to stimulate discussion and there are plenty of questions in this unit and in the related Unit 11.

UNIT 13

The cell

Animals and plants are made up of small units of living matter called **cells**. A small animal or plant may consist of only one cell whereas a large animal or plant consists of millions. A new-born baby has about six million million cells.

Cell structure

Cells are made of colourless jelly-like matter called **protoplasm** which is usually divided into two distinct regions: a small mass of **cytoplasm** containing a ball-like **nucleus**. The cytoplasm is surrounded and kept intact by the **cell membrane**, the nucleus by the **nuclear membrane**. These are shown in Figure 13.1.

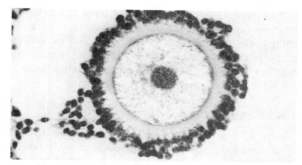

Fig. 13.2 Egg cell of a cat

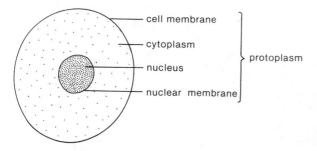

Fig. 13.1 Animal cell

The human egg, a very large cell more than 0.1 mm across, looks like Figure 13.1. Most cells are different because they are **specialised** to do different jobs in the body. The human egg does not need to be specialised because it is surrounded by cells doing jobs for it: feeding it and moving it along. Figure 13.2 shows a similar cell, an egg cell of a cat, surrounded by the cells that feed it.

Cell size

The human egg cell is not the largest kind of cell. The yolk of a bird's egg is a very large cell because it contains enough food for a bird to grow from: the yolk of an ostrich's egg is almost 10 cm across. (As the term is used in biology, the egg cell is the yolk alone.) By contrast the human sperm (if one ignores its tail) is among the smallest animal cells: less than 0.01 mm across. Figure 13.3 shows that most cells of large animals and plants are somewhere between the size of the human sperm and egg, while most single-celled animals and plants are larger than the human egg. Bacteria are the smallest organisms of all.

Cell functions

All living things have to be able to carry out many activities in order to keep themselves alive, and most of these activities need energy. A cell must therefore have a source of food which it can use to release energy. We do not fully understand the difference between a living cell and a dead cell, but it has to do

0.001 mm	0.01 mm	0.1 mm		1.0 mm		1 cm	10 cm
bacteria	human sperm	most cells of large animals and plants	human egg	most single-celled animals and plants			yolk of ostrich egg

Fig. 13.3 Cell diameters

with the use of energy to keep the cell parts organised. A cell that has no source of energy soon dies.

Most of the living processes are carried out in the cytoplasm, though cells without nuclei cannot reproduce. Human red blood cells, which have no nuclei, can live for about 100 days; other cells can live for a short time after their nuclei have been removed artificially.

While the cytoplasm carries out most of the living processes and often contains food stores as a source of energy, the nucleus controls reproduction as well as many of the processes that the cytoplasm is carrying out. For example, the nucleus contains coded instructions for making enzymes in the cytoplasm.

Plant and animal cells

Figure 13.1 is an animal cell. A plant cell is a more rigid structure since it is surrounded by its own skeleton in the form of a **wall**. Plants (which include trees) can grow so much bigger than animals because each cell has its own supporting wall. Figure 13.4

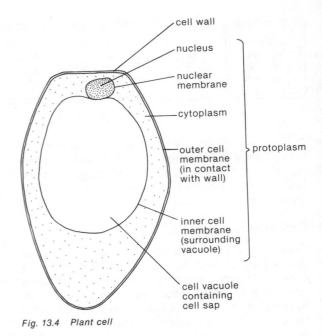

Fig. 13.4 Plant cell

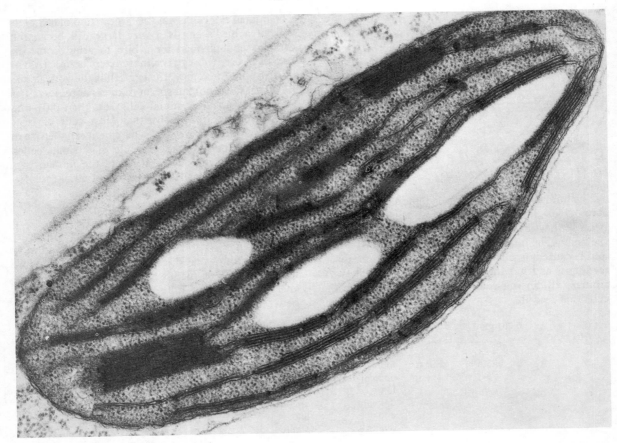

Fig. 13.5 Chloroplast: electron micrograph

shows this cell wall and a large fluid-filled cavity inside the cytoplasm called a **vacuole**. The fluid in the vacuole is a watery solution called **cell sap**. Most plant cells have a large vacuole.

The cell sap in the vacuole also helps to support the cell. If you fill a plastic bag with water and make it airtight, it becomes quite firm: though you can change its shape, you cannot squash it flat. If you squeeze this plastic bag into a cardboard box, it will no longer be able to change its shape. The water in the plastic bag in the cardboard box is like the vacuole's cell sap inside the protoplasm inside the cell wall.

In the green parts of cytoplasm, plants can make their own food by absorbing light for photosynthesis. These green parts, called **chloroplasts**, usually take the form of small discs and are scattered among the rest of the cytoplasm. Leaves are green because they contain chloroplasts.

Chloroplasts

Figure 13.5 is an electron micrograph of a chloroplast inside the cytoplasm of a plant cell. At the side of the chloroplast you can see the cell wall; around the chloroplast you can see a thin layer of cytoplasm. The white patches in the chloroplast are starch grains; the dark patches contain the green pigment. Electron micrographs are photographs taken through an electron microscope. The electron microscope can magnify much more than a light microscope does, but it does not enable us to look at objects directly: we have to take photographs of them or project them on a television screen.

The parts of a plant cell which do not exist in an animal cell are:
  the wall, which supports the cell;
  the vacuole, containing cell sap, which also supports the cell;
  the chloroplasts, which absorb light for photosynthesis.

Figure 13.6 is a green plant cell, i.e. one containing chloroplasts.

Plant cells are usually bigger than animal cells, not because they have more cytoplasm, but because they have large vacuoles containing cell sap and because they are surrounded by walls.

The cell membrane and nuclear membrane are common to plant and animal cells. In a plant cell a third membrane around the vacuole separates the cytoplasm from the sap. None of these membranes in plant or animal cells can be seen: their width is a hundred times smaller than the smallest bacterium. Nonetheless it is usually clear where the cytoplasm ends.

Diagrams

Figure 13.7 shows the surface view of an entire plant cell. The vacuole is at the centre and cannot be seen. Figure 13.6 is what you see if you cut a slice longways through it: it is a **longitudinal section**, which is abbreviated to L.S. Figure 13.8 is what you see if you cut a slice across the entire cell: it is a **transverse section**, which is abbreviated to T.S.

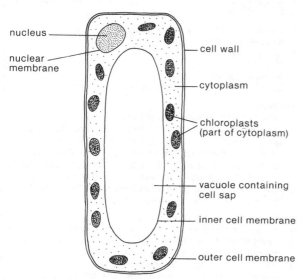

Fig. 13.6 Green plant cell: L.S.

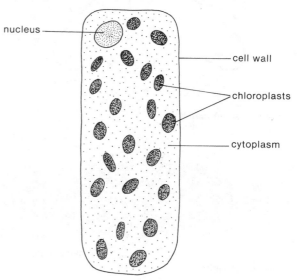

Fig. 13.7 Entire plant cell: surface view

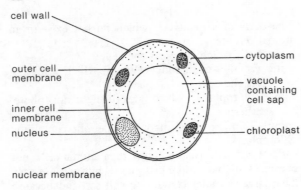

Fig. 13.8 Plant cell: T.S.

Chemical composition

The most common chemical elements in a cell are carbon, hydrogen, oxygen, nitrogen, phosphorus and sulphur. The five important compounds in protoplasm are water (the most abundant), carbohydrates, lipids, proteins and nucleic acids.

Proteins form the basis of protoplasm (enzymes are proteins);

lipids form special structures, especially membranes, and are an important food reserve;

carbohydrates form food reserves (as starch in plants, as glycogen in animals) and the walls of plant cells (as cellulose);

DNA (deoxyribonucleic acid) lies in the nucleus and forms the coded controlling instructions for making enzymes.

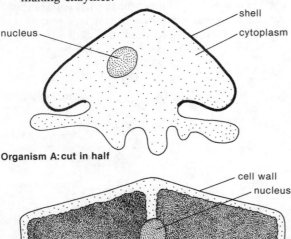

Fig. 13.9

Questions

Q 13.1 Draw a simple animal cell and a simple plant cell next to one another. Label their parts. Make a list of (a) the parts they have in common and (b) the parts of a plant cell which are not found in an animal cell.

Q 13.2 Figure 13.9 shows organisms A and B, which are single-celled and can be seen only with a microscope. They live in freshwater ponds.

Give your reasons for deciding whether each of the organisms is an animal or a plant and describe how each differs from the animal or plant cells you learnt about in this unit.

PRACTICAL WORK

Note for teachers
The two experiments in this unit require a compound microscope with an objective lens allowing at least × 10 magnification. These experiments can be performed adequately with one microscope for every two students. If there are not enough microscopes, the teacher can demonstrate the investigation, allowing the students to look down the microscope, or use photographs, or do both.

Experiment 13.1 To investigate the structure of human cheek cells

Materials required by each student
2 dropper pipettes
1 microscope slide
1 microscope
absorbent paper
a few drops of methylene-blue solution (a stain)

Method
1. Having first washed your hands, firmly rub a finger against the inside of your cheek. Put the sticky liquid you collect on to the clean slide and leave it to dry.
2. Put two drops of methylene-blue solution on the dried liquid and leave it for about one minute. Use a pipette full of water to wash away the surplus stain.
3. Carefully dry the slide and study the dried liquid from your cheek under the microscope.

Interpretation of results
1. Look for patches of blue on your slide. These are cells that you scraped from inside your cheek. They have been stained by the methylene-blue solution.
2. Search the slide until you find a cell which has been separated from a group and whose structure you can clearly see.

3 Inside this cell you will probably see a roundish structure which is a much darker blue than the rest of the cell. What is this? Do all cells on the slide have one?
4 Are all the cells on the slide exactly the same shape? Can you *see* any structure on the outside of the cells which gives them their shape?
5 Can you see any other structures in your selected cell? If so, is their appearance clear?
6 Draw your selected cell. Label your drawing with the names of the parts of the cell which you have been able to see.

Experiment 13.2 To investigate the structure of onion epidermis cells

Materials required by each student
1 dropper pipette
1 microscope slide and coverslip
1 pair of fine forceps
1 scalpel
1 microscope
absorbent paper
a few drops of iodine in potassium-iodide solution
1 slice of fleshy leaf from a fresh onion bulb

Method
1 Cut a square of about 5 mm × 5 mm in the inner (concave) surface of the fleshy onion leaf. Use the forceps to peel away its epidermis (skin).
2 Put two drops of iodine solution on a slide. Put the piece of epidermis into the two drops of iodine solution. Hold the coverslip with the forceps and gently lower it over the epidermis: if you put it down on one side first, and perhaps use the scalpel to stop it from slipping, you will get fewer air bubbles trapped in the iodine solution.
3 Blot away any surplus stain from your slide and study the onion epidermis cells under the microscope.

Interpretation of results
1 Are the onion cells the same shape? If so, can you *see* any structure on the outside of the cells which gives them this shape?
2 Do the cells have a roundish structure in them which has stained a darker yellow than the rest of the cell? Can you name this structure and say what its function is?
3 Do the contents of the cells look the same as those of the cheek cells you examined? If not, describe any ways in which they look different.
4 Make a drawing of one cell of the onion epidermis. Label your drawing with the names and functions of the parts of the cell which you have been able to see.

UNIT 14

Cell specialisation

Most plant and animal cells are specialised to carry out different functions. Animal cells are more specialised in appearance than plant cells and can look very different from the animal cell in Unit 13.

Plant cells

Figure 14.1 is a photograph of a section through the leaf of a plant taken through a microscope. It shows a number of cells that look somewhat different from one another. Remember that the major task of the leaf is to make food in the form of sugar for the whole plant. It does this by photosynthesis: by trapping the energy in light; by using that energy to break down water to combine its hydrogen with carbon dioxide to make sugar. Now look at the photograph of the leaf and see how the cells are specialised to do this.

The leaf is enclosed in a continuous layer in which the cells are attached to one another so that there are virtually no spaces between them. This layer is called the **epidermis**. In fact there *are* tiny spaces or pores between some of the cells of the epidermis on the lower side, though none is shown clearly in the photograph. The size of each pore, or **stoma**, is

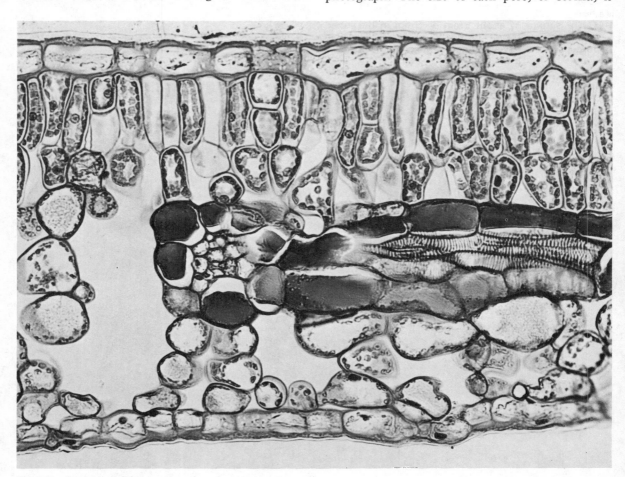

Fig. 14.1   Part of leaf: T.S.

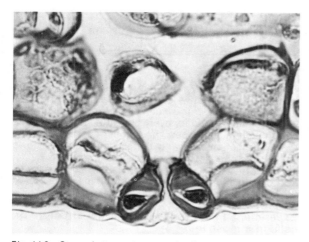

Fig. 14.2 Stoma between two guard cells

controlled by two small **guard cells**, one on each side of it. Figure 14.2 shows a single stoma between two guard cells.

The epidermis protects the food-making cells inside the leaf and prevents them drying out. Its cells make a waxy covering which forms a waterproof surface called the **cuticle**. Leaves with thick waxy cuticles may look shiny. On the upper side of the leaf, which is in direct sunlight and gets warmer, the cuticle is always thicker. But carbon dioxide can reach the food-making cells through the hundreds of tiny **stomata** (plural of *stoma*) between the guard cells under the leaf. You should have seen cells of onion epidermis in surface view in the practical work in Unit 13. The food-making cells in the leaf, the **mesophyll** cells, are also specialised. On the upper side of the leaf, where the sun's rays will be more powerful, the chloroplasts are much more numerous and the cells are elongated and close together, catching as much light as possible as it strikes the leaf. On the lower side the chloroplasts are fewer and the cells have much larger air spaces between them: gases, particularly carbon dioxide, are able to move freely between the cells; carbon dioxide can also reach the compact upper layer. All cells of the mesophyll which contain chloroplasts can carry out photosynthesis, but food-making will be more efficient on the upper side, where light is brighter and where there are more chloroplasts.

The cells of the epidermis do not have chloroplasts: their contents are colourless cytoplasm, nuclei and vacuoles containing cell sap; in the photograph (Figure 14.1) the small dark dots in the epidermis are possibly a food reserve.

The section in Figure 14.1 passes through a small **vein** which is called a **vascular bundle** and contains the cells that bring water to the leaf. These cells are the ones with dark parallel lines, which are ridges of wall-thickening helping to keep the cells open. Other cells in the vascular bundle take the sugars that have been made in the leaf to other parts of the plant which need them.

Animal cells

Very few animal cells are as simple in appearance as a human egg cell. In the human body most cells have very different structures suitable for the special jobs they do.

Look at the photograph of a **white blood cell** and **red blood cells** in Figure 14.3. This type of white blood cell has cytoplasm able to spread around and capture invading bacteria; its nucleus is often branched instead of roundish. A red blood cell is specialised to contain the red pigment **haemoglobin**, which carries oxygen around the body: haemoglobin is spread throughout the cytoplasm; the cell is without a nucleus.

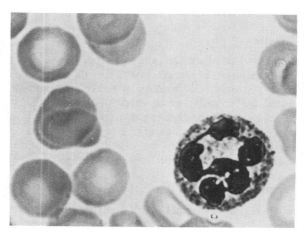

Fig. 14.3 Blood cells

Look at the diagram of a **sperm cell** in Figure 14.4. A sperm cell is very specialised because it has to be able to swim several centimetres to reach an egg. Its cytoplasm is drawn out into a long tail which can lash to drive the sperm through fluid towards the egg.

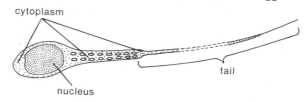

Fig. 14.4 Sperm cell

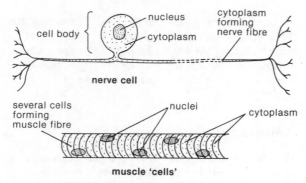

Fig. 14.5 *Nerve cell and muscle fibre*

Nerve cells and **muscle cells** are shown in Figure 14.5. A nerve cell has its cytoplasm drawn out into long thin fibres that may even reach from the spine to the toes, a length of almost a metre. Electrical nerve impulses can pass along these fibres from the toes to the spine. Muscle cells may not even be separate cells. Some are merged with one another so that they form muscle fibres in which many cells act together to shorten the muscle and do the work of moving the skeleton.

Tissues

When groups of similar cells work together, they are called **tissues**. Epidermis is a tissue; mesophyll is a tissue. Muscle cells working together form muscle tissue. Nerve cells, which meet at the tips of their fibres and work together to pass nerve impulses, form nerve tissue. Blood cells working together form blood tissue. Cheek cells are an example of **epithelial** tissue, found on body surfaces.

Organs

When different tissues work together in compact masses they are called **organs**. The leaf is an organ to make food for the plant. The heart is an organ which contains muscle tissue to drive the blood around the body, nerve tissue to control the heartbeat and connective tissue to bind the other tissues together.

Questions

Q 14.1 Describe three leaf cells performing different functions and explain how they are suitable for the functions they perform.

Q 14.2 Describe three human cells performing different functions and explain how they are suitable for the functions they perform.

PRACTICAL WORK

Note for teachers
The experiment in this unit requires a compound microscope with an objective lens allowing at least × 10 magnification. The experiment can be performed adequately with one microscope for every two students. If there are not enough microscopes, the teacher can demonstrate the investigation, allowing the students to look down the microscope, or use photographs, or do both.

Students should also see slides of leaf transverse sections (in addition to the photograph in Figure 14.1) if they are available.

Experiment 14.1 To investigate the structure of cells in a moss leaf

Materials required by each student
1 dropper pipette
1 microscope slide and coverslip
1 microscope
absorbent paper
a few moss leaves

Method
1. Put two drops of water on the slide. Put one whole moss leaf into the two drops of water. Gently lower a coverslip over the moss leaf as in Experiment 13.2.
2. Blot away any surplus water and study the moss-leaf cells under the microscope.

Interpretation of results
1. The cells of the moss leaf will not be as clear as those from your cheek or from the onion epidermis. If you adjust the focus of the microscope, you will find that as some cells come into focus others go out of focus. Can you explain this? Remember that you have mounted the *whole* leaf.
2. Focus the microscope so that you can see one of the green cells clearly. Does this cell have the same structures that you saw in the onion cells? If any structures appear to be absent, can you explain why? Remember that you have not stained these cells.
3. The cell you study will have a number of small roundish bodies within it which were not present in the onion cells. Name these structures.
4. Draw the moss-leaf cell you have studied. Label your drawing with the names and functions of the parts of the cell which you have been able to see.
5. Make a table like the one illustrated to compare the structures of the three types of cell you have studied. List all the structures you have found in the different cells. Indicate whether they are present or absent in each type of cell.

Cell structure	Type of cell		
	Human cheek	Onion epidermis	Moss leaf

UNIT 15 — Diffusion and active transport

Diffusion

When you cut open an onion, you soon smell it. People standing further away smell it a bit later than you and smell it less strongly. People still further away will not smell it at all.

When you cut open the onion, the gas that forms the onion smell escapes into the air. The smallest particles (molecules) of this gas pass into the air and move off in all directions. As they escape from the onion, they get further apart, which is why people further away smell the onion less strongly. The onion molecules reach people even further away, but by this time there are so few of them that the sense cells of the nose do not detect them.

This is **diffusion**. The molecules spread out from regions where there are lots of them, i.e. where they are **concentrated**, to regions where there are few of them, and this spreading out goes on until they are everywhere equal. Even then they go on moving about but, since all of them are moving about at more or less the same speed in all directions, they remain more or less equally spread out. You can see that diffusion occurs in the ammonia experiment at the end of this unit.

This spreading out, or diffusion, goes on easily when one gas is spreading through another. It also goes on easily when a substance is spreading through a liquid provided that the spreading substance dissolves, but it will not go on at all through solids. If you cut up the onion in an airtight room, the smell will not escape because the gas cannot pass through the solid walls and doors. As soon as you open a window or a door, the onion smell escapes.

Diffusion in plants and animals

Gases

The diffusion of gases is important in plants because they make their food from the gas carbon dioxide. There is very little carbon dioxide in the air – only about 3 parts in every 10 000 parts of air, or 0.03 % – yet enough finds its way by diffusion into a plant, particularly into the leaves, through the stomata. When the cells of the mesophyll are carrying out photosynthesis, they use carbon dioxide from the air spaces in the leaf. Hence the amount of carbon dioxide, i.e. its concentration, in the air spaces near the mesophyll cells gets less. Carbon dioxide immediately diffuses through the continuous air spaces in the leaf towards the mesophyll cells, reducing its concentration everywhere inside the leaf. When its concentration inside the leaf is reduced near the stomata, carbon dioxide diffuses into the leaf through the stomata from the air outside, where it is now in a higher concentration.

This is how most diffusion in plants and animals occurs: substances get used up or taken away, their concentration falls by comparison with the concentration elsewhere and immediately diffusion occurs to equalise the concentration again. When the concentration is low in one place and high in another, we say a **concentration gradient** exists.

If the leaf needs oxygen for respiration, and if photosynthesis is not providing oxygen, it gets it in the same way: by diffusion through the stomata from the air.

Animals do not need carbon dioxide, but they do need oxygen. In fact they often need so much that they cannot wait for it all to come by diffusion. Large active animals, including humans, have a ventilation system (breathing) to bring air containing oxygen into their bodies. But they rely also on diffusion to move oxygen inside their bodies.

One problem of diffusion is that it is limited by the amount of surface in contact with the diffusing substance. Where diffusion is vital, surfaces are often enlarged. The insides of the lungs are an example of this principle. Mesophyll cells have a large surface area in contact with the internal air-space system of a leaf. You will read about many examples in other units.

Liquids

Most diffusion in plants and animals is not of gases but of substances dissolved in water and of water itself. Remember that water forms the major part of protoplasm: substances that can dissolve in water, and water itself, diffuse throughout the protoplasm of all cells. Dissolved in water, the sugar made by a chloroplast in photosynthesis can diffuse throughout the protoplasm of the cells.

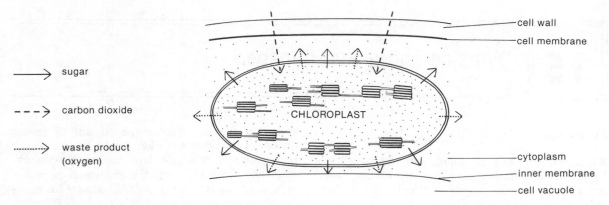

Fig. 15.1 Diffusion into and out of a chloroplast

Anything dissolved in water can diffuse throughout the protoplasm. This applies equally to the substances the cell needs, such as oxygen and food, and to the waste substances the cell needs to get rid of. A substance will move from a high-concentration region to a low-concentration region (along the diffusion gradient) until the concentrations everywhere are equal.

Figure 15.1 shows diffusion into and out of a chloroplast as a result of photosynthesis: carbon dioxide diffuses in; sugar and waste products diffuse out. Concentration gradients are set up in both directions.

At the end of this unit is an experiment showing diffusion through gelatin, which, like protoplasm, is a jelly-like substance.

Entry and exit

So far we have described movement inside the cell and through air spaces. To get in and out of a cell, substances must be dissolved and must pass through the cell membrane. The substances we have mentioned so far – sugar, carbon dioxide, oxygen and water itself – can usually move freely through a cell membrane along a diffusion gradient. The thin cell membrane is **permeable** to these substances.

If, however, the cell membrane allowed all dissolved substances to enter and leave simply according to their diffusion gradients, all cells would have the same dissolved contents, which would be the same as any surrounding liquid's. Roots would have the same dissolved substances as the water in the soil. When it rained, plants would lose their dissolved contents as the water in the soil got more dilute. Fish would have the same contents as the water around them, which means that fish in fresh water would have fresh water inside them and fish in sea water would have sea water inside them. Obviously this does not happen: the membrane is able to keep some substances out and some substances in. The membrane is said to be **selectively permeable**: it lets some substances pass through it at certain times but not others. The substances it lets pass through tend to be ones with small molecules, though factors other than size are also involved.

A plant cell has a wall as well as a membrane round it. But the wall is a much more permeable structure than the membrane and presents no barrier to diffusion. A cell wall is not the solid airtight barrier that it looks even under the microscope, or the plant cell would die. There are spaces in the wall which may not be seen under a microscope but which are easily big enough to let molecules through.

Remember that:
 gases and dissolved substances diffuse;
 diffusion of gases takes place through gases;
 diffusion of dissolved substances takes place through liquids;
 solids do not diffuse;
 there is no diffusion through solids (even when set, gelatin, used in Experiment 15.2, is a liquid, not a solid);
 diffusion follows a gradient from regions of high concentration to regions of low concentration.

Permeability of the cell membrane

As soon as a cell is dead, its membrane no longer stops substances getting in and out and the whole organisation of the cell collapses. You can often see this when coloured vegetables or fruits are boiled in water: when the vegetables or fruits are still alive, the water is colourless; when they are boiled and dead, the colour leaks out of the cells; after boiling, cabbage water turns green, redcurrant water red, blackcurrant water deep purple.

During life anything that the cell needs is prevented from diffusing out, even when the concentration is far lower outside the cell. Energy is needed for this. Similarly the membrane can take substances into the cell against the concentration gradient, and again energy is needed. With the use of energy the laws of physical diffusion can be overcome. Moving a substance against the concentration gradient is called **active transport** and always needs energy.

Look at Table 15.1, which shows some of the contents of equal parts of the fluid of human blood (plasma) and of the fluid in an adjoining body cell.

Table 15.1

Contents	Concentration (arbitrary units)	
	Blood fluid	Cell fluid
Chlorine ions	36	3
Sodium ions	50	13
Potassium ions	2	40
Dissolved protein	5.5	24

Chlorine and sodium ions are more concentrated in blood fluid than in cell fluid. They would pass through the membrane into the cell by diffusion if only the cell would let them. But the cell keeps them out by the use of energy. It also uses energy to bring potassium ions into the cell by active transport against the concentration gradient. We know little about how the cell uses energy to do all this.

What about the dissolved protein? Its molecules are too big to get through the membrane: the cell does not need to use energy to get them in or to stop them getting out, because the membrane is **impermeable** to protein molecules even when they are dissolved.

How do protein molecules get into and out of living cells? You will read about this in Unit 22.

Questions

Q 15.1 Describe the diffusion path that is set up throughout the leaf between the chloroplast and the air outside when a chloroplast uses carbon dioxide in photosynthesis.

Q 15.2 Look at Table 15.2, which shows some of the contents of a large seaweed cell and of the sea water that surrounds it.
(a) Which ions are more concentrated in the sea than in the cell?
(b) Which ions are more concentrated in the cell than in the sea?

Table 15.2

Ions	Parts per thousand	
	Sea water	Seaweed cell
Chlorine	19.6	21.2
Sodium	10.9	2.1
Potassium	0.46	20.14

(c) Energy is used by the cell to keep some ions out. Which are they?
(d) Energy is used by the cell to keep some ions in. Which are they?
(e) Which ions would enter the cell by diffusion if they were needed?
(f) Which ions would enter the cell only by active transport if they were needed?

PRACTICAL WORK

Experiment 15.1 To investigate the rate of diffusion of ammonia

Note for teachers
Students must wear safety spectacles and use a pipette filler with the pipette when handling ammonia solution.

Materials required by each student
1 100 cm^3 measuring cylinder with a glass cover
1 1 cm^3 graduated pipette fitted with a pipette filler
1 30 cm glass rod
1 stopclock
1 pair of safety spectacles
a few strips of indicator paper (e.g. red litmus or Universal)
absorbent cotton wool
adhesive tape
2 cm^3 9 mole dm^{-3} ammonia solution (0.88 ammonia solution diluted with an equal volume of water)

Method
1 Make a table like the one illustrated to show the volumes occupied by the ammonia at various time intervals.

Time (min)	Position of indicator paper which has just changed colour showing the presence of ammonia	Volume of measuring cylinder occupied by the ammonia gas (cm^3)
0	100 cm^3 mark	0
2	.	.
.	.	.
	0 cm^3 mark	100

2 Moisten the strips of indicator paper with water. Using the glass rod, stick them lengthwise to the inside of the measuring cylinder so that they form a continuous chain of paper from the bottom up to the 100 cm^3 mark.
3 With a strip of adhesive tape attach a small ball of cotton wool to one side of the glass cover.

4. Wearing safety spectacles to protect your eyes, use the pipette fitted with a pipette filler to put 1 cm^3 of ammonia solution on the cotton wool.
5. Place the glass cover over the top of the measuring cylinder with the cotton wool downwards.
6. As soon as the very top of the indicator paper (at the 100 cm^3 mark on the cylinder) changes colour, start the stopclock. Every two minutes use the marks on the outside of the cylinder to record in your table where the indicator paper is just beginning to change colour.

Interpretation of results
1. Use the data you have recorded in the middle column of your table to calculate the volumes occupied by the ammonia gas and enter them in the third column.
2. On a sheet of graph paper draw a jagged-line graph to show time on the x-axis and volume occupied by ammonia gas on the y-axis.
3. Explain what happened to the molecules of ammonia in the *solution* on the cotton wool to cause the colour change in the indicator paper.
4. Did the ammonia gas spread along the cylinder at the same rate throughout the experiment?
5. Suggest a reason why you were not instructed to extend the indicator paper to the very top of the cylinder but left a space in which no measurement was made.
6. Suggest why the ammonia would have spread more quickly (a) if a stronger solution of ammonia had been used and (b) if the room had been warmer.

Experiment 15.2 To investigate the diffusion of methylene blue in gelatin

Note for teachers
This experiment takes about a week, during which methylene blue diffuses from an agar cube into a container full of solidified gelatin. Before the experiment the teacher must prepare the agar cubes and gelatin.

Agar blocks impregnated with methylene blue: Dissolve 15 g of agar powder in 100 cm^3 of tap water in a flask by heating them in a boiling-water bath. Mix in 2 cm^3 of a 1% aqueous methylene-blue solution. Pour this mixture into a suitable container to a depth of 0.5 cm. Once it has set, cut it into cubes of side 0.5 cm. You will need one cube per student.

Gelatin: This is best prepared just before the practical session in order to reduce the risk of contamination by micro-organisms. You will need 80 cm^3 of water and 8 g of gelatin crystals per student. Put the crystals in the water in a beaker. Dissolve the crystals by stirring them continuously as you heat the water to boiling. Allow the gelatin to cool to 60°C before adding a few drops of Nipagin solution to discourage microbial growth. Pour 40 cm^3 into equal numbers of 100 cm^3 beakers and boiling tubes (so that you have enough for every student to have one of each). Allow the gelatin to set.

Materials required by each student
1 250 cm^3 beaker of hot water
1 100 cm^3 beaker containing 40 cm^3 solidified gelatin
1 boiling tube containing 40 cm^3 solidified gelatin
1 pair of forceps
1 ruler graduated in mm
1 spirit marker
1 piece of cling film to cover the 100 cm^3 beaker
1 agar cube impregnated with methylene blue

Method
1. Put the boiling tube containing the gelatin into the beaker of hot water and leave the gelatin to melt.
2. Using the forceps, put the agar cube on to the centre of the set gelatin in the glass beaker.
3. Pour the melted gelatin from the boiling tube into the beaker containing the set gelatin and the agar cube and leave it to set. You will then have a blue-stained agar cube in the middle of a beaker full of gelatin.
4. When all the gelatin has set, cover the top of the beaker with cling film.
5. Draw a diagram of the contents of the beaker.
6. Leave the beaker where it will not be disturbed. After a week draw a second diagram to show both the position and intensity of the blue colour. Use the ruler to measure the distances over which the methylene blue has diffused.

Interpretation of results
1. The methylene blue should have diffused from the agar cube into the gelatin in the beaker. What property must the gelatin have for this diffusion of methylene blue to occur?
2. Has the methylene blue diffused equally in all directions? What can you tell about the effect of gravity on diffusion?
3. Explain the different intensities of blue colour in the gelatin.
4. Methylene blue diffused much more slowly in gelatin than did ammonia in air (Experiment 15.1). Suggest reasons to explain this.
5. Suggest why gelatin rather than water was used to show the diffusion of methylene blue.

UNIT 16 Diffusion of water: osmosis

Water with something dissolved in it is a **solution**. Unit 15 was about the diffusion of substances dissolved in water. This unit is about the diffusion of water itself.

If you dissolved 5.0 g of sugar in 1 dm³ of water, you would hardly taste the sugar. You could call this a very dilute solution of sugar. Or you could call it a solution in which water is very concentrated. If you dissolved 50 g of sugar in 1 dm³ of water, you would get a definite taste of sugar. By comparison with the first solution you could call this one of four things:
- a less dilute solution of sugar;
- a solution in which water is less concentrated;
- a more concentrated solution of sugar;
- a solution in which water is more dilute.

You already know that a substance dissolved in water diffuses to places where it is less concentrated (or more dilute). Similarly water itself diffuses to places where it is less concentrated (or more dilute).

	impermeable membrane
Solution A	Solution B
high concentration of sugar,	low concentration of sugar,
low concentration of water	high concentration of water

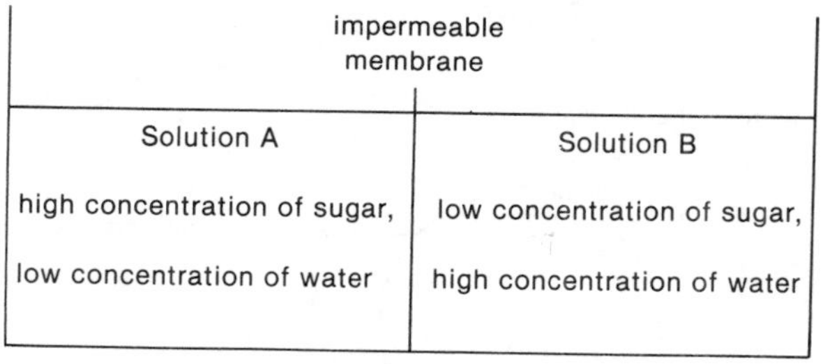

Fig. 16.1 Sugar solutions separated by an impermeable membrane

Figure 16.1 shows two sugar solutions separated by an *impermeable* membrane: neither water nor sugar can pass through it. Solution A has a high concentration of sugar and a low concentration of water. Solution B has a low concentration of sugar and a high concentration of water. This means that in the same volume solution A has more sugar molecules than solution B, while solution B has more water molecules than solution A. This is shown in Figure 16.2.

If we now cut large holes in the membrane between the two solutions, more sugar molecules will diffuse from solution A to solution B than from solution B to solution A and more water molecules from solution B to solution A than from solution A to solution B.

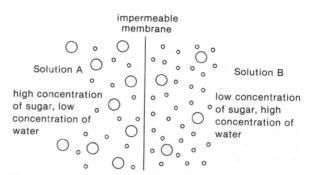

Fig. 16.2 Sugar and water 'molecules'

Diffusion will continue until sugar and water molecules are evenly spread throughout the two solutions, i.e. until the two solutions are the same, as shown in Figure 16.3.

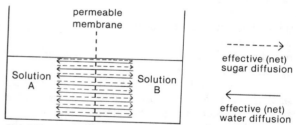

Fig. 16.3 Diffusion through a permeable membrane

When we cut large holes in the membrane, it becomes *completely permeable*. Suppose that, instead of cutting large holes in the membrane, we are able to pierce it with lots of tiny holes that water molecules can pass through but not sugar molecules. Because it lets some molecules through but not others, we can now call the membrane *selectively permeable*. Figures 16.4 and 16.5 show diffusion of water across a selectively permeable membrane.

When the two sugar solutions are separated by a selectively permeable membrane, more water moves from solution B to solution A than from solution A to solution B. More water moves from where *water* is in high concentration to where it is in low concentration than in the opposite direction. Another way of saying this is that more water moves from a dilute solution (of sugar) to a concentrated solution (of sugar) than moves in the opposite direction. When the concen-

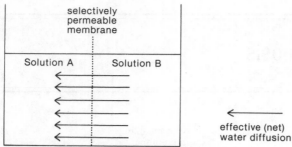

Fig. 16.4 Diffusion through a selectively permeable membrane

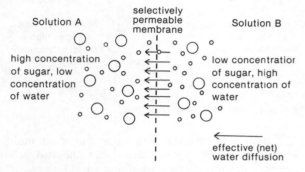

Fig. 16.5 Diffusion of water 'molecules' through a selectively permeable membrane

trations on the two sides of the membrane are equal, water will go on moving but the amount moving in one direction will equal the amount moving in the opposite direction.

As more water leaves solution B, its volume gets smaller. As more water enters solution A, its volume gets bigger. Since the sugar molecules cannot move across, the only way the two solutions can reach the same concentration is by changing their volumes. Look at Figure 16.6 and you will see that diffusion is more powerful than gravity. It has to be, to move substances upwards inside plants.

We can say that water has moved from solution B, in which water was concentrated, to solution A, in which water was dilute. This is short for saying that

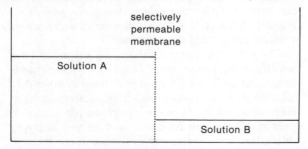

Fig. 16.6 Concentration of solution A = concentration of solution B

more water has moved from solution B to solution A than from solution A to solution B. The movement of water by diffusion through a selectively permeable membrane from a dilute to a concentrated solution is given a special name: **osmosis**.

Osmosis in an animal cell

In Unit 15 you learnt that the cell membrane of an animal or plant cell is selectively permeable. Suppose that we put an animal cell into a solution that is more dilute than the cell's fluid contents (i.e. one in which the water is more concentrated than in the cell's fluid contents) and that its cell membrane is permeable only to water. Water will diffuse into the cell from the more dilute solution outside. The cell will increase in volume and the cell membrane will stretch. If the solution is very dilute, or if it is pure water, the cell membrane will soon be stretched to breaking point: it will burst and the cell will die. This is shown in Figure 16.7.

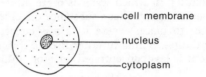

(i) animal cell

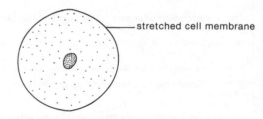

(ii) animal cell in a dilute solution

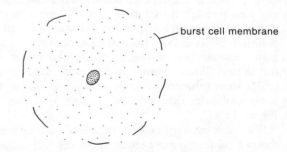

(iii) animal cell in a very dilute solution bursts

Fig. 16.7 Osmosis in an animal cell

Fig. 16.8 Animal cell in a concentrated solution

Now suppose we put an animal cell into a solution that is more concentrated than the cell's fluid contents (one in which the water is more dilute than in the cell's fluid contents). Osmosis will occur in the opposite direction. More water will diffuse from the cell to the more concentrated solution outside than will diffuse into the cell. The cell will decrease in volume and the cell membrane will shrink. When the membrane can shrink no more, it will wrinkle. This is shown in Figure 16.8. Figure 16.9 shows what happens to red blood cells that have been put into a solution more concentrated than their fluid contents.

Fig. 16.9 Red blood cells: scanning electron micrograph

Such problems are encountered by animal cells in real life. Animals that live in fresh water – in ponds, lakes, rivers and canals – are surrounded by a solution more dilute than their cells' contents. Water is constantly entering their cells by osmosis. Nor is the problem of excess water confined to freshwater animals. Virtually all animals drink water. The water drunk by land animals is enough to create a problem. Only simple sea-water animals are free of the problem of excess water. This is because sea water is roughly the same concentration (though it does not have the same chemical contents) as the fluid in their body cells. All other animals have a method of regulating their water content; otherwise they would die.

Single-celled freshwater animals are able to pump the water that keeps coming in by osmosis into small spaces called **vacuoles**. When a vacuole gets to a certain size, it is pushed to the cell membrane where it

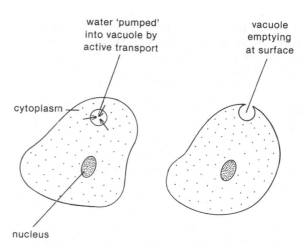

Fig. 16.10 Contractile vacuole filling and emptying

contracts to empty water inside it back into the outside water, as shown in Figure 16.10. Filling the vacuole with water is active transport (because the water must move against the concentration gradient) and needs energy. Pushing the vacuole to the surface needs energy and opening the membrane to empty the water out needs energy. Only a living cell can do all this. This kind of vacuole is a **contractile vacuole**.

Larger animals have different ways of coping with excess water: some get rid of it through a blood system and kidneys (as is described in Unit 30).

Osmosis in a plant cell

Look at Figure 16.11. Again solution A has a higher sugar concentration than solution B and solution B has a higher water concentration than solution A. Again there is a selectively permeable membrane between them. But this time there is a tight cover on the surface of solution A which prevents it increasing in volume. It is impossible for more water to diffuse from solution B to solution A. Solution A continues to have the same higher sugar concentration and solution B to have the same higher water concentration.

	selectively permeable membrane	
	COVER	
cover prevents increase in volume of Solution A therefore no diffusion	Solution A high concentration of sugar, low concentration of water	Solution B low concentration of sugar, high concentration of water

Fig. 16.11 Water unable to enter solution A

This is how plants, as well as bacteria and fungi, cope with the entry of water by osmosis whenever they are in contact with fresh water or with a dilute solution. The cell wall stops too much water getting into the cell, as in Figure 16.11 the cover, the sides and bottom of the container stop water getting into solution A from solution B. The cell wall also stops the cell bursting like the one shown in Figure 16.7(iii). Unlike the surroundings of solution A in Figure 16.11, the cell wall does expand a little. But beyond a certain point, when the cell has filled with water and slightly expanded, the wall stops any more water getting in. Such a cell, full of water and fully expanded, is described as **turgid**. A turgid cell is firm like a fully blown-up balloon (while a partly blown-up balloon is a flimsy weak structure). Osmosis is able to keep the cells of a land plant turgid: turgid cells help to support the plant.

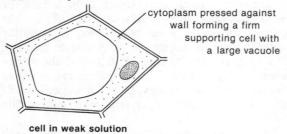

Fig. 16.12 Osmosis into a plant cell

A plant cell has more fluid contents than an animal cell because it contains, in addition to fluid in the protoplasm, a lot of fluid in the cell vacuole.

If a plant cell is put into a solution more concentrated than its fluid contents, osmosis again occurs. This time water leaves the cell and diffuses into the concentrated solution outside. (Remember that the cell wall is no barrier to the diffusion of water or of any dissolved substances.) The cell contents shrink in size, as did those in the animal cell in Figure 16.8, and the membrane is able to shrink with the cell contents. But the cell wall cannot shrink. The cell contents therefore shrink away from the wall, as shown in Figure 16.13. Such a shrunken cell is described as **plasmolysed**. It continues to lose water by osmosis until the cell fluid has the same concentration as the solution outside. The space left between the cell membrane and the wall is filled as the concentrated solution outside passes in through the wall. Thus the concentrated solution stays in contact with the membrane.

Does this ever happen to plant cells in real life? Yes, it does. Plants that live in fresh water (which is very dilute) have turgid cells in which the cellulose wall prevents further expansion and bursting. When plants that live on land have their roots in contact with very dilute soil water, their cells too are turgid. By contrast, plants surrounded by sea water are neither plasmolysed nor turgid because they have a cell fluid concentration similar to that of the sea water around them (though you saw in Table 15.2 that the chemical nature of the contents is not the same). If plant cells lose too much water, or cannot get enough water, their cells plasmolyse and the plants **wilt** and droop. You can sometimes see this in hot dry weather, or when someone has forgotten to water a pot plant.

Figure 16.14 shows that cells of onion epidermis become turgid in a dilute solution and plasmolysed in a concentrated solution.

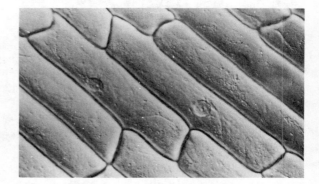

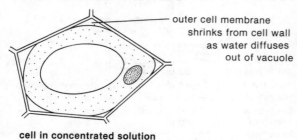

Fig. 16.13 Plasmolysed plant cell in a concentrated solution

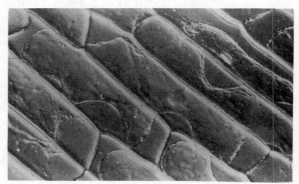

Fig. 16.14 Onion epidermis: turgid and plasmolysed

Questions

Q 16.1 How would you expect organisms A and B in **Q 13.2** to cope with osmosis in their freshwater habitat?

Q 16.2 Figure 16.15 shows three compartments, Q, R and S, each filled with pure water. Compartment Q is separated from compartment R by an impermeable membrane. Compartment R is separated from compartment S by a permeable membrane which allows both water and dissolved substances to pass through it. Substance X is placed in compartment Q and substance Y in compartment S. Both X and Y are able to dissolve in water. The fluid in each compartment is free to increase or decrease in volume.

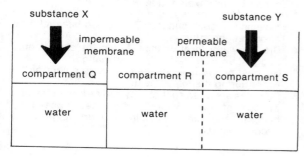

Fig. 16.15

(a) Describe what will happen to the water and to the two substances X and Y in compartments Q, R and S when they are put into the water.

(b) After some time the impermeable membrane between Q and R is made selectively permeable to allow water but neither dissolved substance X nor Y to pass through. What will now happen to the solutions in compartments Q, R and S?

PRACTICAL WORK

Experiment 16.1 To investigate the diffusion of water (osmosis) through a selectively permeable membrane

Note for teachers
Students must use a balance to weigh lengths of dialysis tubing filled with different solutions: any type of dialysis tubing, such as Visking tubing, is suitable. Many students find it difficult to knot the end of soaked strips of dialysis tubing: the teacher can reduce the time needed for this practical by knotting one end of each piece of tubing before the class begins.

The 1% starch suspension should be made within 24 hours of its use by adding a little distilled water from a measuring cylinder containing 100 cm^3 to 1 g of starch powder. Boil this to obtain a clear suspension. Add the rest of the water while stirring.

Materials required by each student
2 250 cm^3 beakers
1 5 cm^3 pipette fitted with a pipette filler
1 pair of scissors
1 bowl
1 stopclock
1 ruler graduated in mm
1 piece of absorbent paper
30 cm dialysis tubing
10 cm^3 1% starch suspension

Materials required by the class
balances to weigh up to 200 g (1 per 4 students)

Method
1. Copy the table at the foot of the page.
2. Cut two pieces of dialysis tubing 15 cm long and put them in water in the bowl for about one minute.
3. While the lengths of tubing are soaking, label one beaker A and the other B and half fill them both with water. Be sure to put the same volume of water in each.
4. Pipette 5 cm^3 of starch suspension into a soaked length of tubing with a knot in one end. Squeeze the tube above the starch suspension to remove any air from it and tie a knot in the open end.
5. Weigh this tube containing starch suspension, record its mass in your table and put it in the beaker labelled A.
6. Repeat instruction 4 with the second length of tubing, pipetting 5 cm^3 of water instead of starch suspension.
7. Weigh this tube containing water and record its mass in your table. Put it in the beaker labelled B. Leave the tubes in the beakers for about 30 minutes.

Contents of tube	Mass at start (g)	Mass at end (g)	Change in mass (g)	% change in mass $\dfrac{\text{change in mass}}{\text{mass at start}} \times 100$
Starch suspension				
Water				

8 Remove the tube of starch suspension from the beaker labelled A and carefully blot away any water on the outside of it. Weigh the tube and record its mass in your table.
9 Remove the tube of water from the beaker labelled B and carefully blot away any water on the outside of it. Weigh the tube and record its mass in your table.

Interpretation of results
1 Complete your table by calculating the change in mass of each tube and its percentage change in mass.
2 Did the tube containing water show any change in mass? Suggest an explanation of your answer.
3 Did the tube containing starch suspension show any change in mass? Suggest an explanation of your answer, bearing in mind that starch molecules are too big to get through the small pores in the tubing.
4 Why was the tube containing water used in this experiment?

Experiment 16.2 To investigate the effect of diffusion of water (osmosis) in dried sultanas

Note for teachers
A period of at least one hour must be allowed between setting up this experiment and taking the results. It may be convenient to leave the sultanas overnight and to finish the experiment during the next class.

Materials required by each student
1 100 cm^3 measuring cylinder
1 250 cm^3 beaker
1 stopclock
1 spirit marker
1 piece of absorbent paper
1 piece of cling film to cover the beaker
6 dried sultanas

Materials required by the class
balances to weigh 0.1 g (1 per 4 students)
distilled water

Method
1 Weigh the six dried sultanas together and record their mass.
2 Put the sultanas in the beaker and add 100 cm^3 of distilled water from the measuring cylinder. Mark the outside of the beaker to show the water level and then cover the top of the beaker with cling film.
3 After at least one hour note the position of the water level in the beaker. Remove the sultanas from the beaker and gently blot any surplus water from their surface. Weigh the sultanas together and record their mass.

Interpretation of results
1 What was the change in mass of the sultanas during the experiment? Explain why this change in mass occurred.
2 What change in the appearance of the sultanas supports your explanation in 1?
3 Explain why you would not expect the sultanas to have shown such changes if they had been left in a strong sugar solution.
4 Did the water level change during the experiment? Suggest an explanation for any change.
5 Suggest one reason for the change in colour of the water during the experiment.
6 Sultanas are dried so that they can be stored for long periods of time without being decomposed by bacteria and fungi. Explain how drying prevents decomposition.

UNIT 17 Enzymes

If you put some hydrogen-peroxide solution in a beaker, nothing seems to happen. Even after half an hour nothing will seem to have happened.

If you add some manganese(IV) oxide, almost immediately bubbles will be given off.

You may think there is nothing extraordinary about this. What has happened, it may seem obvious, is that the manganese(IV) oxide has reacted chemically with the hydrogen peroxide.

In fact this is not what has happened. If you collect all the gas given off you will find that it is oxygen. If you examine what remains in the beaker you will find that it is water and manganese(IV) oxide. What is more, you will find the same amount of manganese(IV) oxide as you put in. Manganese(IV) oxide has not been used up in the chemical reaction.

Yet it is obvious that the manganese(IV) oxide has influenced the chemical reaction whereby hydrogen peroxide changed into water and oxygen. It may seem to you that the manganese(IV) oxide has caused the chemical reaction, but that is not so. The truth is that, even before you added the manganese(IV) oxide, the hydrogen peroxide was changing into water and oxygen, but so slowly that you could not notice it. The effect of the manganese(IV) oxide has been to speed up this reaction.

A substance that influences the speed of a chemical reaction without being used up in it is known as a **catalyst**.

Now suppose that, instead of manganese(IV) oxide, you add small pieces of liver to hydrogen peroxide in a beaker. Again almost immediately bubbles will be given off. There is no manganese(IV) oxide in liver, but there is a biological catalyst which also speeds the change of hydrogen peroxide (a cell poison) into water and oxygen.

If you had heated the hydrogen peroxide, it would also have changed quickly into water and oxygen: both the manganese(IV) oxide and the liver enabled the change to take place quickly without extra heat. Since chemical reactions in living cells must always take place at moderate temperatures, at moderate pressures and under near-neutral conditions, living cells have their own catalysts. Inside any living cell there are at least a thousand different catalysts enabling chemical reactions to proceed at speed.

More than seventy different catalysts have been identified controlling the reactions that take place when energy is released from sugar molecules. Biological catalysts are made inside the cytoplasm of the living cell on instructions from the DNA in the nucleus: they are called **enzymes**.

Enzymes, unlike (non-biological) chemical catalysts, are very **specific**. Each one controls one chemical reaction or one stage of a chemical reaction. Many chemical reactions in a living cell take place through a number of stages and a different enzyme is needed to control each stage. This is why so many different enzymes are needed in any living cell.

Over a period of a minute one molecule of one of the fastest-acting enzymes can control up to a million molecules of reacting substances. At the other extreme one molecule of one of the slowest-acting enzymes, which controls the conversion of nitrogen to ammonium in nitrogen-fixing bacteria, takes one and a half seconds to convert one molecule.

Like chemical catalysts, enzymes:
  can alter the speed of a reaction (usually they make it faster);
  can enable a reaction to take place at a different temperature (usually they allow it to take place at a lower temperature);
  do not alter the kind of reaction;
  do not form part of the end-products of a reaction;
  are not used up by or chemically changed in the reaction;
  are needed only in small amounts.

Unlike chemical catalysts, enzymes:
  are protein molecules;
  act best under moderate conditions of temperature and pressure, such as those found in living cells;
  are highly sensitive to temperature and pH (acidity and alkalinity) and act well over only small ranges of temperature and pH;
  are very specific, controlling only one reaction or one stage of a reaction.

Making enzymes

Enzymes are made in the cytoplasm. Most enzymes are made in small quantities, act inside the cell and never leave the cell. Only a few enzymes are made in vast quantities. These are secreted by cells to act outside the cell (**extracellularly**). In humans such enzymes are secreted on to food as it passes through the gut. These **digestive** enzymes change insoluble food into soluble food, with much smaller molecules, which can pass through the gut wall and into the blood stream. The saprophytes that cause decay and decomposition also secrete large quantities of digestive enzymes on to their food.

Enzymes are proteins. Enzymes therefore contain the elements carbon, hydrogen, oxygen and nitrogen in the form of amino-acid chains. They may also contain other elements. Messenger molecules pass out from the DNA in the nucleus with instructions on how different amino acids are to be linked together to form the proteins of enzymes. The complexity of an enzyme molecule is enormous: even the smallest enzyme molecules are about 12 000 times heavier than an atom of hydrogen while the largest are over a million times heavier. Figure 17.1 shows a model of an enzyme molecule.

The chains of amino acids which form enzymes are rolled up in a highly organised way to form what are called **globular protein** molecules. Within each molecule some amino acids that lie near one another become chemically linked even though they may be distant from one another along the chain: the linkages are called **disulphide bridges** because they include sulphur atoms. Imagine a loose ball of wool in which you have clipped together some of the threads lying next to one another.

The parts of an enzyme molecule involved in catalysing a particular reaction are near the outside of the globular shape. Working out the structure of enzyme molecules is the kind of research which takes years. The Nobel Prize for Chemistry has been awarded for such work.

Naming enzymes

The enzymes that were first studied, because they were in large supply and easy to obtain, were those of the gut of mammals and those of micro-organisms that make alcohol. All sorts of different names were given to them (e.g. rennin, ptyalin, pepsin, zymin) and, as more and more were discovered, the naming of enzymes got into a muddle. In 1964 a Commission on Enzymes brought order to their naming: the ending -ase is now added to the names of all enzymes and groups of enzymes.

One of the six large groups of enzymes is the **hydrolases**, which include the digestive enzymes. The hydrolases control the splitting of large molecules into smaller ones by adding water (hydrolysis). **Amylase** (the name comes from the Latin for *starch*, *amylum*) is a hydrolase that splits starch; you will work with it in Experiment 17.1 at the end of this unit. Other hydrolases are **lipases**, which act on lipids, and **proteases**, which act on proteins. **Glucose phosphorylase**, which transfers the phosphate in glucose phosphate to another molecule and builds up glucose into starch, is another enzyme you will work with in Experiment 17.2 at the end of this unit. This is not a hydrolase but belongs to another large group (the tranferases).

Have you noticed that enzymes are named by adding -ase to the substances they act on, while groups of enzymes are named by adding -ase to the processes they carry out?

How do enzymes work?

Not much is known about how enzymes work. Remember that they cannot determine what sorts of chemical reactions take place: they can only alter the speed of reactions (often making them a million times faster) which would take place anyway. One hypothesis (untested theory) is that an enzyme acts like a glove into which the substances of the reaction fit like a hand. Bringing the reacting substances, the **substrates**, into the controlled environment of the large enzyme molecule, the glove, encourages the

Fig. 17.1 *An enzyme molecule: model*

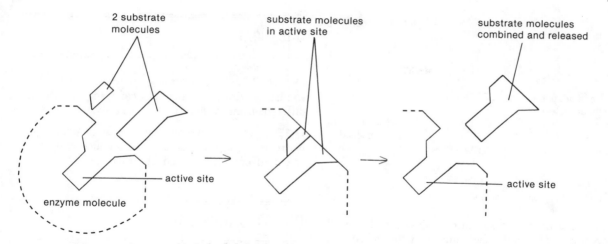

Fig. 17.2 Enzyme action: substrates combined

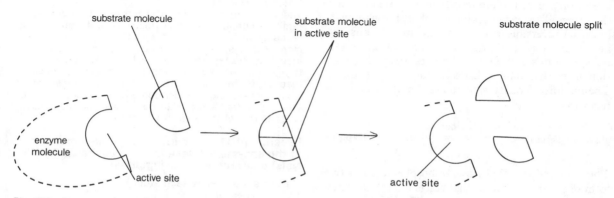

Fig. 17.3 Enzyme action: substrate split

reaction and makes it easier. The products of the reaction are released from the glove so that more substrates can take their place, react and be released. This is shown in Figure 17.2. The region of the enzyme where this catalytic activity takes place is called the **active site**. Only substrate molecules that fit the active site exactly can react there. This explains the specific property of enzymes, i.e. it explains why each enzyme controls only one reaction or one stage of a reaction.

Figure 17.2 shows a reaction in which two substrates have been combined. In other enzyme-catalysed reactions one substrate may split to form two substances. Figure 17.3 shows a reaction of this type at the active site.

This hypothesis of enzyme action explains another fact: certain substances can stop the action of enzymes temporarily or permanently. They are called **inhibitors**. An inhibitor can occupy the active site temporarily if its molecule has the right 'shape': such inhibitors can often be removed from the active site of the enzyme by the correct substrate and are called **competitive** inhibitors because they compete for the same space as the substrates. Or an inhibitor, without occupying the active site, may position itself where it prevents the substrate from occupying it: such inhibitors are usually permanent and are called **non-competitive** inhibitors because they do not compete for the same space as the substrates. Inhibition is shown in Figures 17.4 and 17.5.

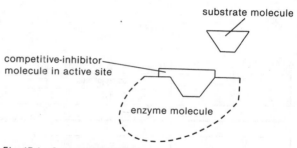

Fig. 17.4 Competitive inhibition

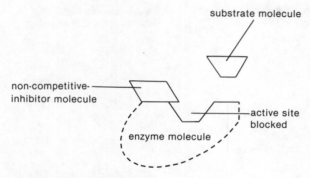

Fig. 17.5 Non-competitive inhibition

The reactions that enzymes catalyse are almost all reversible. Moreover the same enzyme catalyses a reaction in both directions. Under certain conditions a reaction proceeds more readily in one direction than the other. The enzyme glucose phosphorylase catalyses the reversible reaction shown in the equation, i.e. it catalyses both the conversion of glucose-1-phosphate to starch and phosphates and the conversion of starch and phosphates to glucose-1-phosphate. The modified 'equals' sign shows that the reaction is reversible.

$$\text{glucose-1-phosphate} \xrightleftharpoons{\text{glucose phosphorylase}} \text{starch + phosphates}$$

The conversion of glucose-1-phosphate, an active form of sugar (i.e. a form that reacts easily), to starch and phosphates proceeds more readily if the concentration of glucose-1-phosphate is high and the concentrations of starch and phosphates are low. If the concentration of glucose-1-phosphate is low and the concentrations of starch and phosphates are high, the reverse reaction from starch and phosphates to glucose-1-phosphate proceeds more readily. Because the enzyme also catalyses this reverse reaction, its name is sometimes given as glucose (starch) phosphorylase. You will work with glucose phosphorylase in Experiment 17.2.

Questions

Q 17.1 List the properties of enzymes which (a) are similar to those of chemical catalysts and (b) are different from those of chemical catalysts.

Q 17.2 You are told an enzyme is a member of the hydrolase group. What can you deduce about this enzyme from such limited information?

PRACTICAL WORK

Note for teachers
The following experiments assume that students are familiar with the chemical tests for both starch and reducing sugar, details of which are given in Experiment 21.1 (page 98). The starch suspension should be made within 24 hours of its use. If kept longer, even in a refrigerator, it may hydrolyse and give a positive reaction for reducing sugar. Add a little distilled water from a measuring cylinder containing 100 cm³ to 1 g of starch powder. Boil this to obtain a clear suspension. Add the rest of the water while stirring.

Experiment 17.1 To investigate the importance of the enzyme amylase in the breakdown (hydrolysis) of starch to reducing sugar

Materials required by each student
1 250 cm³ beaker
2 5 cm³ pipettes fitted with pipette fillers
2 dropper pipettes
1 test-tube rack with 2 test-tubes
1 stopclock
1 white dimple tile
1 Bunsen burner, tripod and gauze
1 spirit marker
5 cm³ 1 % amylase solution
5 cm³ Benedict's solution
5 cm³ iodine in potassium-iodide solution
10 cm³ freshly prepared 1 % starch suspension

Method
1. Copy the table.
2. Label one tube A and the other B and place them in the test-tube rack.
3. Pipette 5 cm³ of starch suspension into each tube.
4. Pipette 5 cm³ of water into tube A only. Mix the contents of the tube.

Time (min)	Colour of iodine solution		Colour of Benedict's solution	
	Tube A	Tube B	Tube A	Tube B
5			no test	no test
10			no test	no test
15				

5   Pipette 5 cm³ of amylase solution into tube B only. Mix the contents of the tube and start the stopclock.
6   Put one drop of iodine solution in each of six wells of the tile so that you have two rows of three iodine wells.
7   Five minutes after starting the stopclock use a dropper pipette to take one drop of the mixture from tube A and add it to the left-hand drop of iodine solution in the upper row on the tile. Record the colour of the drop of iodine solution in your table.
8   Immediately use the second dropper pipette to take one drop of the mixture from tube B and add it to the left-hand drop of iodine solution in the lower row on the tile. Record the colour of the drop of iodine solution in your table.
9   Ten and fifteen minutes after starting the stopclock repeat instructions 7 and 8 using the middle and right-hand drops in each row and taking care not to get any iodine solution into the two tubes.
10 Pipette 2 cm³ of Benedict's solution into each tube. Put both tubes in a beaker of water and boil the water for one minute. Record in your table the colour produced in each tube.

Interpretation of results
1   What do the results of your iodine tests indicate about the presence of starch in tubes A and B throughout the experiment?
2   What do the results of the Benedict's tests indicate about the presence of reducing sugar in tubes A and B at the end of the experiment?
3   Using your answers to 1 and 2, explain the action of amylase.
4   Why was tube A (without amylase solution) used?
5   If you had accidentally mixed some of the iodine solution into the contents of tube B, the reaction you observed would have stopped. Suggest why.

Experiment 17.2   To investigate the importance of the enzyme glucose (starch) phosphorylase in the synthesis of starch from reducing sugar

Note for teachers
At the time of writing glucose phosphorylase is not commercially available as a purified extract. It is therefore necessary to extract it from living tissues. While the extraction increases the time needed for this experiment, it can be used to emphasise the importance of intracellular enzymes in physiological processes.

Materials required by each student
1 250 cm³ beaker
1 filter funnel
1 10 cm³ pipette fitted with a pipette filler
3 dropper pipettes
1 test-tube rack with 2 test-tubes
1 scalpel
1 stopclock
1 white dimple tile
1 Bunsen burner, tripod and gauze
1 pestle and mortar
10 cm³ distilled water
10 cm³ 5 % glucose-1-phosphate solution
5 cm³ iodine in potassium-iodide solution
clean fine sand
1 potato cube approximately 20 mm × 20 mm × 20 mm

Materials required by the class
fluted filter papers

Method
1   Use the scalpel to cut the potato cube into small pieces and place them in the mortar. Add a small amount of clean sand and grind the potato pieces with the pestle.
2   Pipette 10 cm³ of distilled water into the mortar and continue grinding for half a minute.
3   Label a clean tube EXTRACT. Filter the extract from the mortar through the fluted paper into this tube.
4   Take one drop of extract from the tube and mix it with one drop of iodine solution in one of the wells of the tile. If a blue-black colour develops, use a new fluted filter paper to filter the extract for a second time. Use one drop of iodine solution to test one drop of this extract for the presence of starch.
5   While you are waiting for the extract to filter, copy the table.

Time (min)	Colour of iodine solution	
	Top row	Bottom row
5		
10		
15		
20		

6   When the potato extract no longer turns iodine solution blue-black, pour half of it into the second tube. Put this tube in a beaker of water and boil the water and the extract for two minutes.
7   Use a clean dropper pipette to put one drop of the 5% glucose-1-phosphate solution into each of eight wells of the tile so that you have two rows of four wells.
8   Use a clean dropper pipette to put one drop of unboiled extract into each of the four wells of glucose phosphate on the upper row. Use a clean dropper pipette to put one drop of the boiled extract into each of the drops of glucose phosphate in the lower row on the tile. Start the stopclock.
9   After five minutes add a drop of iodine solution to the first drop of the mixture in each row and record the colours in your table. Repeat these tests at five-minute intervals for a further fifteeen minutes, recording the colours in your table.
10 At the end of the experiment use one of the spare wells in the dimple tile to test a sample of the glucose-phosphate solution with iodine.

Interpretation of results
1   Describe and explain the colour changes that occurred when iodine solution was added to (a) glucose phosphate and unboiled extract and (b) glucose phosphate and boiled extract.
2   Explain the importance of filtering the potato extract so that it failed to turn iodine solution blue-black.
3   The extract you prepared contained a number of enzymes, only one of which was used in the experiment.
    (a) What part would glucose phosphorylase play in the development of a potato tuber?
    (b) Suggest probable functions of two of the enzymes in the extract which were not investigated.

UNIT 18 Enzyme activity

Over a period of one minute an enzyme can catalyse up to a million molecules of substrate. This is the peak of efficiency and needs **optimum** conditions. The optimum is the best when the best is neither the minimum nor the maximum. The best score for a football team is the highest possible and is *not* an optimum. But the best pitch to play on is one neither too soft nor too hard. We may therefore call the best pitch the optimum pitch.

Temperature

In general enzymes work best at a moderate temperature. For example, enzymes in the human body work well at temperatures of about 37°C. Even a few degrees either way will decrease the efficiency of the enzymes.

Too low a temperature

Temperature affects the speed at which molecules move about in gases and liquids. They move faster as the temperature is increased and slower as it is decreased. The speed of a reaction may depend partly on how quickly the reacting molecules move into and out of the active site of the enzyme. If cytoplasm is frozen, i.e. solid, there is no movement of the molecules at all. As the temperature rises above freezing point, the reacting molecules begin to move and then move faster and faster.

Too high a temperature

Temperature affects the structure of proteins. You can see this for yourself when an egg is cooking. The white of the egg, the albumen, is a globular protein which, as it heats up, changes from a colourless jelly-like substance to a semi-solid and finally to the solid white of a hardboiled egg. In its solid white form albumen is **denatured**, which means that its globular protein molecules have lost their highly organised structure. All proteins become denatured if heated beyond a certain temperature, and of course all enzymes are proteins. As the temperature rises, an enzyme becomes denatured, the shape of its active site is destroyed and it is deactivated.

Figure 18.1 shows the typical relation between temperature and enzyme activity. Below a certain temperature (such as 0°C) there is no reaction; above this certain temperature the rate of the reaction gradually increases until the enzyme protein is denatured, when it falls off rapidly. The temperature at which the rate of reaction is fastest is the optimum.

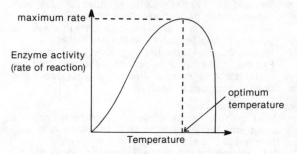

Fig. 18.1 Effect of temperature on enzyme activity

pH

Similarly, enzymes have narrow pH ranges over which they work best. The graph in Figure 18.2 shows an enzyme whose optimum pH is 7.5 (slightly alkaline). Slightly above and slightly below pH 7.5 the enzyme still works well, but by pH 6.5 (slightly acid) and pH 8.5 (fairly alkaline) it hardly works at all. The fall in activity, whether in acid or alkaline conditions, is again due to damage to the structure of the globular protein molecule and to the active site of the enzyme, i.e. to the denaturing of the protein.

Different enzymes work best at different temperatures. While the enzymes in the bodies of mammals and birds, which are warm-blooded, work well between 35 and 40°C, the enzymes in a cold-blooded

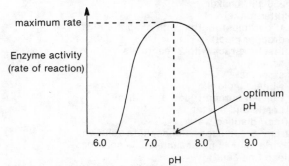

Fig. 18.2 Effect of pH on enzyme activity

animal such as a fish work well at lower temperatures.

Different enzymes also work best at different pH values. Most cytoplasm has a near neutral pH (pH 7.0) but there are enzymes secreted into the human gut which work best at an acid or alkaline pH: protein-digesting enzymes of the stomach at pH 2.0 and of the small intestine at pH 8.5.

Substrate and enzyme concentrations

Other factors affecting the speed of enzyme action are the concentrations of the substrate and the enzyme. Up to a certain point, the greater the concentration of substrate the more molecules there will be near the active site and the shorter will be the time before the next substrate molecule moves into the active site. But there is a limit to the rate of the reaction because the active site cannot cope with more than about a million substrate molecules a minute. This is shown by the graph in Figure 18.3. Beyond a certain substrate concentration the rate of the reaction cannot be increased: the line levels off along a maximum rate on the graph. Substrate concentration is not an optimum condition because increases never reduce activity.

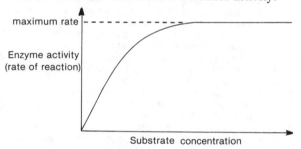

Fig. 18.3 Effect of substrate concentration on enzyme activity

Although enzymes are needed in only small quantities, enzyme concentration also affects the rate of the reaction: the more enzyme molecules, the more active sites and the more substrate molecules that can react in them.

Inhibitors

If molecules other than the substrate fit into the active sites, they can block them. If some of the active sites are blocked by competitive inhibitors, the rate of reaction is sure to slow down. If enough non-competitive inhibitors prevent substrates from occupying active sites, the enzyme will not work at all.

Food preservation

Several methods of preventing foods from decaying depend on the properties of enzymes.

Dehydration

Saprophytic organisms, bacteria and fungi that cause decay, do so by secreting enzymes from their cells on to surrounding food substances. Water must be present for digestive enzymes to work and hydrolyse the large insoluble molecules to smaller soluble molecules. Foods are liable to decay merely as the result of the working of their own enzymes, but again water must be present.

Certain foods, such as cereal grains and flours, are therefore preserved merely by being kept perfectly dry. Other foods preserved by being kept dry must first have their own water removed by some special treatment (usually heating). Fruits such as currants and apricots, fish and milk are all preserved in a dehydrated form. When it is time to eat them, water is added to them. After that their own enzymes can begin to work again and they may also be attacked by bacteria and fungi, which means they must be eaten soon or they will decompose.

Salting

Surrounding goods by salt is a way of dehydrating them. Water is in higher concentration in the foods than in the surrounding salt. Hence it passes from the food to the salt by osmosis. Dehydration prevents activity by the food's own enzymes and by bacteria and fungi. Salting was formerly used a great deal, particularly to preserve beans and pork, and is still used wherever refrigerators are scarce.

Syruping

Syruping, soaking foods in a concentrated solution of sugar, uses the same principle as salting. Because the surrounding solution is concentrated (the water in the solution is in low concentration), water passes out of the tissues by osmosis. Fruits are sometimes preserved in syrup: crystallised fruits are fruits preserved in such a high concentration of sugar that the sugar crystallises as the water of the syrup evaporates.

Low and high temperatures

Deep-freezing and refrigeration prevent enzymes from working at all or allow them to work only slowly. High temperatures denature proteins and hence deactivate enzymes: blanching, which means dipping food in very hot, not necessarily boiling, water for only a few minutes, is a means of deactivating enzymes even though it does not cook the food. Even deep-frozen food is often blanched to prevent slow enzyme action and discolouration. Enzymes are also deactivated before food is canned.

Pickling

Pickling food by soaking it in vinegar, a mild acid, prevents action by enzymes that need a neutral or alkaline medium. The principle of pickling is thus to provide an unsuitable pH for enzymes. Pickling is often combined with salting. Salt is added to the vinegar to make a concentrated solution (one in which water is in low concentration): as in pure salting, water then passes out of the food by osmosis.

Questions

Q 18.1 If we want to compare the speeds at which amylase acts on starch at 15°C and at 30°C, what must we keep the same in the two experiments?

Q 18.2 Give the names of three foods preserved in different ways. Explain how decay is prevented in each.

PRACTICAL WORK

Experiment 18.1 To investigate the effect of different temperatures on the activity of amylase

Materials required by each student
5 250 cm³ beakers
2 5 cm³ pipettes fitted with pipette fillers
5 dropper pipettes
1 test-tube rack with 5 test-tubes
1 stopclock
1 thermometer
1 white dimple tile
1 Bunsen burner, tripod and gauze
1 spirit marker
30 cm³ 1 % amylase solution
10 cm³ iodine in potassium-iodide solution
30 cm³ freshly prepared 1 % starch suspension
4 ice cubes

Method
This experiment involves a complicated sampling method. It is important to read the whole method before you start work so that you can plan what you are going to do.
1. Label five beakers A to E.
2. Using ice, cold water and warm water, half fill each beaker with water at the temperatures indicated:

Beaker	Temperature (°C)
A	10
B	20
C	30
D	40
E	50

3. Pipette 5 cm³ of the 1 % starch suspension into each of the five tubes. Place one tube into each beaker of water and leave them for five minutes.
4. Copy the table.

Time (min)	Colour when added to iodine solution				
	Tube in A	Tube in B	Tube in C	Tube in D	Tube in E
1					
2					
3					
4					
⋮					
15					

5. Put one drop of iodine solution in each well of the tile.
6. When the tubes have been in their water baths for at least five minutes, quickly pipette 5 cm³ of amylase solution into each one. This should take you only about one minute. Start the stopclock as soon as you have added the amylase to the last tube.
7. Use different dropper pipettes to remove one drop of each mixture from each tube at one-minute intervals. Immediately test each drop for the presence of starch by adding it to a drop of iodine on the tile and record the colour in your table.
8. Stop all your tests after fifteen minutes even if the contents of some of the tubes continue to turn the iodine solution blue-black.
9. Add one drop of the starch suspension and one drop of the amylase solution to unused drops of iodine solution on the tile and record the colours.

Interpretation of results
1. Summarise your results in a table like the one illustrated.

Tube	Temperature (°C)	Time taken for blue-black colour to fail to appear with iodine (min)
A		
B		
C		
D		
E		

2. Explain why the blue-black colour failed to appear when drops of mixture from some of the tubes were added to iodine solution.
3. At which temperature did the blue-black colour first fail to appear?
4. Why were you instructed to leave each tube in its water bath for five minutes before adding the amylase solution?

5 The rate (speed) of the reaction involving amylase can be found by dividing 1 by the time taken for the mixture to fail to turn iodine blue-black. Using the times in your table, calculate to two decimal places the rate of reaction at each temperature. Take as 0 the reaction rate of tubes whose contents continued to turn iodine blue-black throughout.
6 Plot a jagged-line graph to show temperature on the x-axis and rate of reaction on the y-axis. Suggest reasons to explain any changes in the gradient of your graph.

Experiment 18.2   To investigate the effect of pH on the activity of amylase

Note for teachers
Students must wear safety spectacles and use a pipette filler with the pipette when handling acids and alkalis.

Materials required by each student
1 5 cm^3 pipette fitted with a pipette filler
3 dropper pipettes
1 test-tube rack with 3 test-tubes
1 stopclock
1 white dimple tile
1 spirit marker
1 pair of safety spectacles
3 strips of pH paper and 1 colour identification chart
15 cm^3 1% amylase solution
10 cm^3 dilute hydrochloric acid
10 cm^3 iodine in potassium-iodide solution
10 cm^3 dilute sodium-carbonate solution
15 cm^3 freshly prepared 1% starch suspension

Method
This experiment uses a sampling technique similar to that used in Experiment 18.1. It is important to read the whole method before you start work so that you can plan what you are going to do.
1 Label three test-tubes A, B and C and pipette 5 cm^3 of the starch suspension into each.
2 Wearing safety spectacles and using the pipette filler on the pipette, add 5 cm^3 of dilute hydrochloric acid to tube A. (See Figure 18.4.)
3 Taking the same precautions, add 5 cm^3 of dilute sodium-carbonate solution to tube B.
4 Add 5 cm^3 of water to tube C.
5 Use pH paper to find and record the pH of the contents of each tube.
6 Copy the table.

Time (min)	Colour when added to iodine solution		
	Tube A	Tube B	Tube C
1			
2			
3			
•			
•			
•			
15			

7 Put one drop of iodine solution into each well of the tile.
8 Quickly add 5 cm^3 of amylase solution to each tube and start the stopclock.

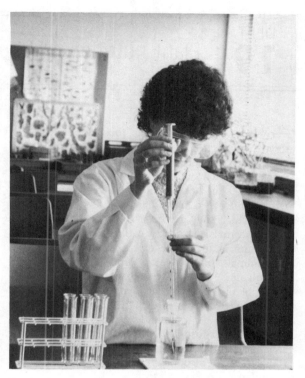

Fig. 18.4 Pipetting hydrochloric acid: safety spectacles protect the eyes; a pipette filler on the pipette protects the mouth

9 Use different dropper pipettes to remove one drop of each mixture from each tube at one-minute intervals. Immediately test each drop for the presence of starch by adding it to a drop of iodine on the tile and record the colour in your table.
10 Stop all your tests after fifteen minutes even if the contents of some of the tubes continue to turn the iodine solution blue-black.

Interpretation of results
1 Summarise your results in a table like the one illustrated.

Tube	pH of contents	Time taken for blue-black colour to fail to appear with iodine (min)
A		
B		
C		

2 Explain why the blue-black colour failed to appear in some tubes.
3 At which pH did the blue-black colour first fail to appear?
4 What effect does pH have on the activity of this enzyme?
5 Assuming that other enzymes are affected in a similar way to amylase, suggest why pickled onions are not decomposed by bacteria.

UNIT 19 Plant nutrition 1: photosynthesis

Every living organism needs food for energy, for growth and for repair of its body. Green plants are different from other organisms because they are able to make their own food. Green plants are **autotrophic**: self-feeding, i.e. not feeding on other organisms or their products. To make their own food, plants need energy, which they get from sunlight by **photosynthesis**. Animals get food by eating plants or by eating other animals that have eaten plants or by eating other animals that have eaten other animals that have eaten plants.... Saprophytes (bacteria and fungi) also feed on plants or on animals that have eaten plants or on animals that have eaten other animals that have eaten plants.... Animals and saprophytes are **heterotrophic**: other-feeding, i.e. feeding on other organisms or their products. Whereas autotrophs are producers, heterotrophs are consumers. Food chains are explained in Units 5 to 8.

In photosynthesis the gas carbon dioxide and the liquid water combine, using energy from light, into a simple sugar (glucose). A waste gas, oxygen, is produced in this process. The green pigment chlorophyll is needed to trap the energy in sunlight. We can sum up photosynthesis as:

$$\text{carbon dioxide} + \text{water} + \text{light energy} \xrightarrow{\text{chlorophyll}} \text{sugar (glucose)} + \text{oxygen}$$

The balanced chemical equation is:

$$6CO_2 + 6H_2O + \text{light energy} \xrightarrow{\text{chlorophyll}} C_6H_{12}O_6 + 6O_2$$

Photosynthesis takes place where healthy plant cells containing chlorophyll have access to carbon dioxide, water and light. In Unit 14 there is a photograph of a section of a leaf where photosynthesis takes place. But any green plant cell, given the necessary raw materials (carbon dioxide, water and light), can photosynthesise: for example, the algae that look like a green powder on the damp surfaces of walls and tree trunks can photosynthesise. Photosynthesis occurs in artificial light as well as in sunlight.

Leaves

A leaf is usually wide and flat, held by its skeleton of veins, with its broad surface towards the sun. A broad flat leaf presents the maximum area to the sun for photosynthesis. Figure 19.1 shows a French-bean leaf in surface view: it consists of three separate parts or **leaflets**. The skeleton of veins spreads throughout the leaf and supports the weaker tissues of the leaf blade or **lamina**.

Fig. 19.1 French-bean leaf: surface view

Unit 14 shows how cells of a leaf are specialised to carry out photosynthesis:

mesophyll cells have chloroplasts containing chlorophyll to absorb sunlight; on the upper side of the leaf the **palisade mesophyll** cells have a regular shape and are closely packed, which concentrates chloroplasts near the upper surface where light is strongest; on the lower side of the leaf **spongy mesophyll** cells are loosely packed and rounded in shape, which allows carbon dioxide and other gases to diffuse easily between them;

veins of the leaf, apart from supporting the lamina, bring water from the soil for photosynthesis; veins also remove sugar after photosynthesis;

guard cells in the epidermis allow diffusion of carbon dioxide into the leaf through the **stomata** (the pores that open between guard cells) and allow diffusion of oxygen from photosynthesis out of the leaf.

Much is now known about the trapping of light energy by chlorophyll and about the use of that energy to split water and to form ATP. Hydrogen ions from water, together with energy from ATP, are transferred to compounds built up from carbon dioxide. By a number of stages, each involving a different enzyme, sugar is formed. Thus photosynthesis, by trapping energy in chemical compounds, provides a source of energy, i.e. food, for plants (and indirectly for animals and saprophytes). Photosynthesis is also vital in removing from the air the carbon dioxide continually formed and in replacing in the air the oxygen continually used up in respiration (and in the burning of fuel).

The sugar produced in photosynthesis could dissolve in water in chloroplasts, in the rest of the cytoplasm and in the cell vacuole to form a concentrated solution. Unchecked, such a process would encourage osmosis: water would enter the photosynthesising cells from surrounding cells which were not photosynthesising. Inside the chloroplast an immediate conversion of sugar to starch (involving the enzyme glucose phosphorylase) prevents sugar accumulating in the cells. The electron micrograph of a chloroplast, Fig. 13.5 in Unit 13, shows several large starch grains formed from sugar made by photosynthesis.

But not every leaf converts sugar immediately to starch: the iris leaves used for Experiment 7.1 retain sugar after photosynthesis, though they also convert some sugar to starch. Chloroplasts would soon become full of starch grains except that, in the dark,

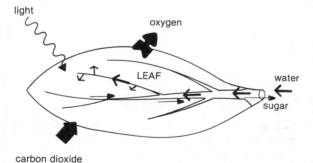

Fig. 19.2 Photosynthesis in a leaf

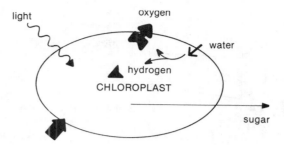

Fig. 19.3 Photosynthesis in a chloroplast

starch is reconverted to sugar and removed from the leaf to permanent storage regions in the stem or root of the plant. Sugar moves out of the leaves via **phloem** tissue in the veins.

Waste oxygen, made throughout photosynthesis, diffuses out of the leaf via the stomata.

Figure 19.2 shows, diagramatically, photosynthesis in a leaf. Figure 19.3 shows, diagramatically, photosynthesis in a chloroplast.

Photosynthesis in water

There are plenty of plants living in water, though those in the open sea and large lakes are usually so small that they can be seen only under a microscope. All water plants photosynthesise: producers are as essential in water as on land for trapping energy and forming the basis of food chains, for removing carbon dioxide and for replenishing oxygen. The process of photosynthesis is the same in water as on land but the conditions are different.

Because light is quickly dimmed as it passes through water, only the top layers have sufficient light for photosynthesis. Hence plants survive only in the top layers of water.

Carbon dioxide dissolves readily in water and there is more available in water than in air. Carbon dioxide can be released from ions of hydrogencarbonate, which are usually plentiful in water. While air contains about 300 ppm (parts per million) of carbon dioxide and about 200 000 ppm of oxygen, water contains about 400 ppm of carbon dioxide but only about 10 ppm of oxygen. Obviously plants living in water have all the water they need for photosynthesis. The only thing they may be short of for photosynthesis is light.

The oxygen produced in photosynthesis soon saturates the water, because little of it dissolves, and oxygen bubbles are given off. You can see these bubbles coming off the water plant *Elodea* in

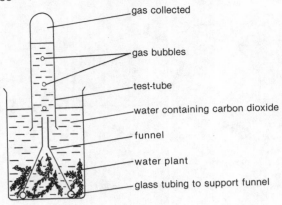

Fig. 19.4 Collecting gas from a water plant

Experiment 7.2. Figure 19.4 shows a method of collecting the gas bubbles in a test-tube. The gas bubbles are not in fact pure oxygen but have a higher proportion of oxygen than normal air.

The rate of photosynthesis

The rate at which photosynthesis takes place depends on the amounts present of all the necessary factors – light, carbon dioxide, water and chlorophyll – as well as on factors affecting enzyme activity and on the amounts of sugars and starch already in the plant.

To measure the rate of photosynthesis we need a standard of comparison. One such standard is the rate at which oxygen is given off by a plant: the faster the rate of photosynthesis, the more oxygen produced. Unfortunately, measuring the rate at which oxygen is given off, like all the other ways of measuring the rate of photosynthesis, is difficult.

But the gas bubbling up from a water plant can easily be collected and measured. Though the gas collected is not all oxygen, it contains a fairly constant proportion of oxygen. The gas bubbling up from a water plant therefore provides a relative measure of its rate of photosynthesis at different times and in different conditions. It does not make possible comparison of the rates of photosynthesis of different plants. But it enables us to discover which conditions favour photosynthesis and which do not.

A method of measuring the gas given off by plants photosynthesising under water is described in Experiment 7.2. A rough and ready measure of the rate of photosynthesis can also be obtained simply by counting the bubbles given off by an isolated part of a water plant. Figure 19.5 shows a bubbler by which bubbles are kept to about the same size.

Questions

Q 19.1 List everything a plant needs to carry out photosynthesis. Why is each item you list needed?

Q 19.2 Explain why plants do not photosynthesise at night or at winter temperatures below about 2°C.

Q 19.3 Look at Figure 19.5.
(a) Why do you think hydrogencarbonate is added to the water containing the plant?
(b) Why do you think the water is saturated with oxygen?

Q 19.4 (a) Graph A shows the effect of light intensity on the rate of photosynthesis.

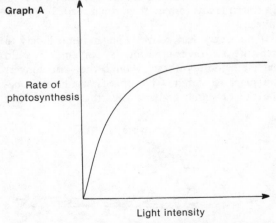

Suggest explanations of the fact that the rate of photosynthesis does not increase after a certain point even though the light intensity continues to increase.
(b) Graph B shows the effect of carbon-dioxide concentration on the rate of photosynthesis.

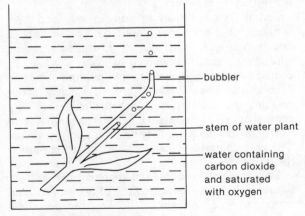

Fig. 19.5 A bubbler

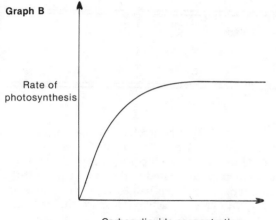

Graph B

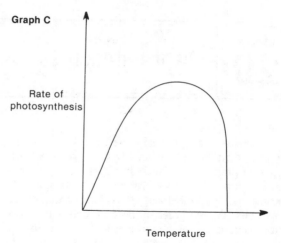

Graph C

Suggest explanations of the fact that the rate of photosynthesis does not increase after a certain point even though the carbon-dioxide concentration continues to increase.

(c) Graph C shows the effect of temperature on the rate of photosynthesis.

Suggest explanations of the fact that above a certain temperature the rate of photosynthesis drops rapidly until it stops entirely.

Q 19.5 Figure 19.6 shows the production of plant biomass (in grams per square metre per day) in different regions of the world.

(a) Suggest reasons for low rates of photosynthesis in each of the regions A, B, E and F.

(b) Suggest reasons for the high rate of photosynthesis in region D.

PRACTICAL WORK

Note for teachers
Experiment 7.2, in which the volume of oxygen produced by a submerged aquatic plant is measured, may be repeated here to determine the effect on the rate of photosynthesis of changes in (a) light intensity, (b) carbon-dioxide concentration, (c) temperature.

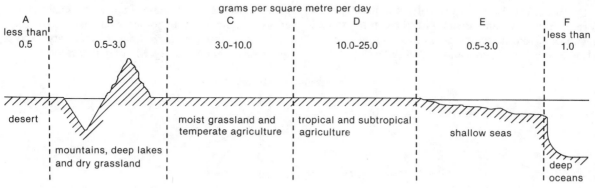

Fig. 19.6

UNIT 20 Plant nutrition 2

Plants consist of a great deal more than the sugar and starch formed initially from carbon, hydrogen and oxygen in photosynthesis. Straightforward rearrangements of sugar molecules produce not only starch (a food reserve) but **cellulose**, which forms the walls of plant cells. Many other important compounds in plants contain carbon, hydrogen and oxygen in yet further arrangements, usually in combination with other elements.

Nitrogen

Nitrogen is present in all proteins. Enzymes, because they are proteins, cannot be made without a supply of nitrogen. DNA, the vital chemical in every nucleus, also contains nitrogen.

Although about 80% of air is nitrogen gas, plants cannot use it in this form. The only way plants can get nitrogen is as ions of nitrate (NO_3^-) and ammonium (NH_4^+) dissolved in water. Water containing dissolved ions enters land plants by elongated cells, called **root hairs**, protruding from their roots. Figure 20.1 shows a root hair surrounded by particles of soil.

Soil water is a dilute solution of soil minerals in the form of ions. By comparison with the cytoplasm and vacuoles of root hairs, soil water is a more dilute solution (water is more concentrated in soil water than in the root hairs). Water therefore enters root hairs by osmosis. Ions enter root hairs by diffusion if they are more concentrated in the soil water. If ions are already more concentrated in the root hairs (or if the concentrations are equal) and if still more ions are needed, they enter by active transport: the root hairs use energy to get the ions in against the concentration gradient.

Nitrogen in the form of nitrate or ammonium ions passes through the plant, either by diffusion or by active transport, from cell to cell. Ions are also carried along in a passive way in the stream of water that passes up the veins in **xylem** tissue from roots to leaves. Once they have entered the plant, nitrate and ammonium ions soon combine with sugar to form amino acids; thereafter they pass through the veins in **phloem** tissue as amino acids.

Wherever proteins and DNA are being made, sugars, nitrate and ammonium ions and amino acids combine in complex chemical reactions, all involving the use of enzymes and energy. Proteins and DNA are made at all the growing points of a plant: in a young plant these would be mainly at the stem and root tips and in the unfolding leaves of young buds. Lack of nitrate or ammonium ions in soil results in poor growth because protoplasm cannot be made at the growing points. Many other elements, apart from nitrogen, carbon, hydrogen and oxygen, are needed by growing plants. They all enter plants as ions via

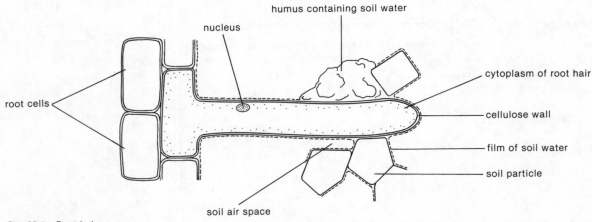

Fig. 20.1 Root hair

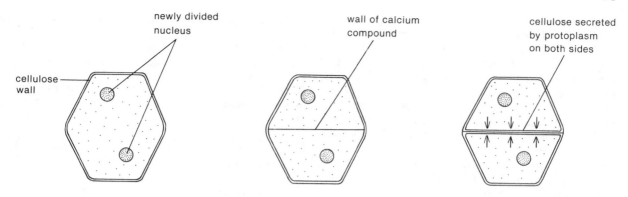

Fig. 20.2 Wall of calcium compound between cells

the root hairs as nitrogen does. These are the elements that form the basis of the fertilisers added to soils where crops are grown.

Calcium

Calcium is an element needed by plants for growth. It is usually available in soils in the form of limestone or chalk: if the soil lacks calcium, powdered limestone can be added as fertiliser. Calcium enters root hairs as ions, by diffusion if the concentration is greater in the soil water than in the cytoplasm and vacuole, by active transport if it is not.

Calcium is needed as a framework on which protoplasm secretes the cellulose wall between newly divided cells. This is shown in Figure 20.2. The thin wall of calcium compound can sometimes still be seen between the cellulose walls of individual cells in tissues where the cells remain attached to one another. Lack of calcium in soil results in poor growth because the cells at the growing points cannot make cell walls.

Iron

Iron is vital yet, unlike nitrogen and calcium, is needed only in small amounts. Iron is found in the molecules of certain enzymes and, although not part of the chlorophyll molecule, is necessary for the formation of chlorophyll. Plants growing in soils that cannot supply iron have yellowish leaves because chlorophyll does not form properly.

Nitrogen, calcium and iron are only three examples of the inorganic elements needed by plants. Nitrogen, calcium and at least half a dozen others are needed in large amounts; iron and probably a dozen others, described as **trace elements**, are needed in small amounts.

Farmers and gardeners who harvest crops must ensure that the elements they thereby remove from the soil are replaced. Carbon dioxide and water are freely available and are continually replaced. Other elements are replaced in the form of fertilisers.

Fertilisers

There are different forms of fertilisers: organic and inorganic. Organic fertilisers are any form of plant or animal residues: dung from farm animals is widely used, especially where animals are kept indoors and the dung is easily collected; decaying seaweed and leaves are also used; on a smaller scale, gardeners make compost from decaying plant debris, kitchen waste and weeds.

Inorganic fertilisers are usually made in factories and the elements in them carefully controlled to provide a complete 'food' in suitable proportions for different crops. Limestone is an exception: it is marketed in the chemical form in which it is quarried, though it has to be crushed. Inorganic fertilisers are usually in a granular or powdered form and are much easier to handle than heavy smelly organic fertilisers. But they are quickly washed out of the soil in rain water and they can harm plants and animals if they are applied in too high a dosage. If inorganic fertilisers are too concentrated in the soil water, osmosis occurs *outwards* from plant and animal tissues: water leaves cells to pass from dilute solutions (a high concentration of water) in the cytoplasm and vacuoles to concentrated solutions (a low concentration of water) in the soil. Plant and animal cells become dehydrated and die.

Plants can absorb mineral ions through their leaves and stems. This has led to the development of 'foliar feeds', solutions of mineral ions sprayed on to leaves and stems. This is an unnatural and wasteful way of 'feeding' plants since most of the ions are washed off the leaves and are not absorbed. It is more suitable for gardeners than for farmers.

In a natural ecosystem there is no need for fertilisers. When plants die, the elements they used in growth are returned to the soil. Even if the plants are eaten by animals, their elements are eventually returned to the soil in the form of faeces, of urine or of decaying corpses. Some elements may be leached away from the soil in natural rainwater and carried to rivers, lakes and the sea. But the rocks from which soil is made can make up for this loss by providing more natural ions. Nitrogen is also replenished by nitrogen-fixing bacteria. The fixation of nitrogen is described in Unit 9.

Figure 20.3 shows the calcium cycle in one year in a hectare of a young oak and beech forest. The annual increase of calcium in this forest biomass is 74 kg per hectare (10 000 square metres or a square of side 100 metres). Can you see how this is worked out? If 201 kg are absorbed and only 127 kg are returned to the soil, the difference is the amount of calcium retained in the plant tissues. Of the 127 kg of calcium returned to the soil, most will have been in the leaves that fell in autumn, some will have been in fruits and seeds dispersed from the trees, and some will have been eaten by animals and will have been returned in faeces, urine and dead bodies.

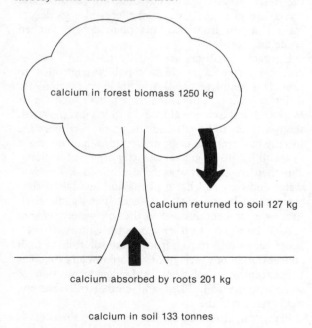

Fig. 20.3   Annual calcium cycle

Questions

Q 20.1 The table shows the nitrogen cycle and the nitrogen content of a mixed oak and beech forest in 1970.

Oak and beech forest (1970)	Mass per hectare
Nitrogen in soil	4500 kg
Nitrogen in biomass	540 kg
Nitrogen absorbed	90 kg
Nitrogen returned to soil	60 kg
Nitrogen increase in biomass	?

(a) What is the figure missing from the table?
(b) Draw a diagram similar to that of Figure 20.3 showing the values for nitrogen instead of calcium.
(c) In what form and how is nitrogen absorbed?
(d) Suggest as many ways as you can by which nitrogen is returned to the soil.
(e) In what compounds will nitrogen be found in the increased biomass?
(f) What proportion of the nitrogen absorbed was returned to the soil in 1970?
(g) How do you explain the fact that trees contain more calcium than nitrogen?

PRACTICAL WORK

Experiment 20.1   To investigate the effects of nitrate and iron deficiencies on the growth of plant seedlings

Note for teachers
This experiment lasts for several weeks and two weeks are required for its preparation.
Although wheat is recommended, any small-seeded plant may be used. The seeds should be sown in clean sand about fourteen days before the experiment is due to start. The sand is kept moistened by the full culture medium. About twice as many seeds should be sown as will be needed. The culture media should be made up prior to the experiment using the formulae below:

(a) Full culture medium
distilled water	1 dm^3
$CaSO_4 \cdot 2H_2O$	0.25 g
$Ca(H_2PO_4)_2 \cdot H_2O$	0.25 g
$MgSO_4 \cdot 7H_2O$	0.25 g
NaCl	0.08 g
KNO_3	0.70 g
$FeCl_3 \cdot 6H_2O$	0.005 g

(b) Culture medium lacking iron
The same as (a) except that the $FeCl_3 \cdot 6H_2O$ is omitted

(c) Culture medium lacking nitrate
The same as (a) except that the KNO_3 is replaced by 0.52 g KCl

Materials required by each student
1 test-tube rack with 3 test-tubes (preferably sterile)
1 spirit marker
aluminium foil to cover the test tubes
cotton wool
distilled water
20 cm³ full culture medium
20 cm³ iron-deficient medium
20 cm³ nitrate-deficient medium
3 healthy wheat seedlings

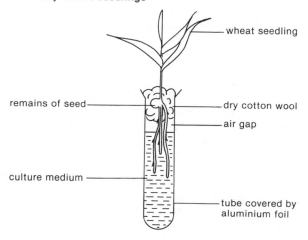

Fig. 20.4 Seedling used in the culture-medium experiment

Method
1 Cover each of the test-tubes with aluminium foil so that no light can pass through the sides.
2 Label the tubes F, ID and ND on the aluminium foil and fill them (to within 2 cm of their rims) as follows:

Tube	Contents
F	full culture medium
ID	iron-deficient medium
ND	nitrate-deficient medium

3 Gently wrap each seedling in a ball of cotton wool so that its leaves and roots stick out as in Figure 20.4.
4 Place one of these seedlings in each tube taking care not to let the cotton wool become moist. Replace it with dry cotton wool if it does.
5 Leave the tubes in a rack in a well-lit place such as your laboratory window.
6 Every now and then remove each seedling from its tube and wash its roots in distilled water to remove any deposits of minerals. At the same time replace any culture medium lost by evaporation. Take care to use the same medium as is already in the tube. Replace the seedlings as before.
7 After a few weeks compare the growth of the three seedlings and record your results in a table like the one illustrated.

Feature	Seedling in F	Seedling in ID	Seedling in ND
Stem length			
Leaf number			
Leaf colour			
Length of longest root			
Number of roots			

Interpretation of results
1 Compare the growth of the seedling in the iron-deficient medium with the growth of the seedling in the full culture medium. Suggest one use of iron in the seedlings.
2 Compare the growth of the seedling in the nitrogen-deficient medium with the growth of the seedling in the full culture medium. Explain the function of nitrogen in the seedlings.
3 Explain why:
(a) the tubes were covered with aluminium foil to exclude light;
(b) the seedlings were left in a well-lit place.
4 Suggest two places at which water was lost by evaporation with the result that the volume of the culture medium decreased.

UNIT 21 Human diet

Some rats were encouraged to overeat by being given unlimited supplies of foods such as chocolates and cakes over a three-week period. These rats were called cafeteria rats. Over the same three-week period another group of rats, forming a control, were fed on unlimited supplies of their natural food. Table 21.1 shows the average energy content of food eaten, the average gain in body mass, the average gain in body lipid, the average gain in body energy and the average energy loss by the rats in each of the two groups after twenty-one days.

Table 21.1 Energy balance in rats

	Averages after 21 days	
	Cafeteria rats	Control rats
Energy content of food eaten (kJ)	11 670	6 480
Gain in body mass (g)	131	103
Gain in body lipid (g)	66	40
Gain in body energy (kJ)	2 230	1 790
Energy loss (kJ)	9 440	4 690

What can we learn from these data? On a cafeteria diet rats eat almost twice as much as on a diet of their natural food. (Work out the difference in food eaten.) But they do not put on twice as much weight. They put on more weight, but only about a third as much as on a natural diet. (Work out the difference in body-mass gain.) Where does their extra weight go? Almost all of it goes into body lipid. (Work out the difference in body-lipid gain and compare it with the difference in body-mass gain.) What has happened to the extra energy content of the food eaten by the cafeteria rats which did not go into increased energy in their body tissues? It has been lost in a number of different ways. Some has been lost as undigested food in faeces; some may have been lost by extra activity in the cafeteria rats (because they are heavier, they have used more energy to move about); most has been converted in respiration to heat and lost from the body in that form.

What can we learn from this about human diet? The data cannot prove anything: only similar experiments on humans could do that. But we can develop hypotheses (untested theories) about similar effects in humans. Overeating is more likely on sweet, unnatural foods; the excess weight is likely to be deposited as lipid; healthy people are usually able to lose most but not all of the excess energy in their food, mainly in the form of heat.

What are natural foods for humans? A diet that keeps close to the foods eaten by our human ancestors is likely to be ideal: these are the foods that humans evolved to eat. Like the present-day gorilla, our human ancestor of a few million years ago, *Australopithecus africanus*, probably had a diet of fruits, seeds and vegetation occasionally supplemented by meat and fish. *Homo erectus*, probably the immediate ancestor of *H. sapiens*, was more of a hunter. Living half a million years ago, our *H. erectus* ancestors used fire. As hunter-gatherers, they ate both meat and plants – much more meat than their ancestors.

But, because it was difficult to kill animals with simple stone weapons, they will have eaten much less meat than most people do in Britain today (unless, like Eskimos, they happened to live where it was easy to catch animals). Nor will they have had other sources of large quantities of lipids (unless they happened to live near naturally occurring plants with lipid-storing seeds). Certainly they will not have eaten large quantities of sugar (unless they happened to live near naturally occurring sugar cane or sugar beet or similar plants).

Fig. 21.1 *Homo erectus*

Plants were not cultivated till 8000 BC; ordered farming and irrigation developed in about 5000 BC; oxen and ploughs were not used till 3000 BC. A few thousand years is a very short time in evolution. Our natural diet must be closely related to what was eaten by *H. erectus* and the early humans (*H. sapiens*) who came next. Since human teeth are not well adapted to eating only meat (like those of a carnivore such as a lion) or to eating only vegetation (like those of a herbivore such as a cow), we can feel confident it is natural for us to eat both. We can also feel confident that our natural diet includes both less meat and less sugar than most people in Britain eat.

Figure 21.1 is a photograph of a reconstruction of our human ancestors.

Human dietary needs

We can classify most of the foods needed by humans into three large groups: **carbohydrates**, **lipids** and **proteins**. We also need small amounts of **inorganic ions**, such as calcium and iron, and of special complex substances called **vitamins**. In addition we must have water.

Carbohydrates

Carbohydrates contain only the elements carbon, hydrogen and oxygen. Four you have already come across are sugar, starch, cellulose and **glycogen**. Glycogen is the form in which animals store carbohydrate and is equivalent to starch, the form in which plants store carbohydrate. With a natural diet humans ate plenty of starch and cellulose (both found only in plants), small amounts of sugar (from plant tissues) and some glycogen (found particularly in animal liver and to some extent in animal muscle). Nowadays we eat a lot of starch in the form of flour (derived from corn and wheat seeds) and sugar (derived from sugar cane and sugar beet).

Carbohydrate is usually a source of energy. Excess carbohydrate is converted into glycogen and stored in the liver and muscles or converted into lipid and stored under the skin or around body organs such as the kidneys. Cellulose is a carbohydrate that is *not* a great source of energy for humans: they cannot digest it themselves; it is only because in the end part of the gut (the colon) there are bacteria which digest cellulose that humans absorb any digested cellulose products at all. Much of it passes through the gut chemically unchanged. Such indigestible food, called **roughage** or **fibre**, is valuable because it gives the muscles of the gut something to act on and prevents constipation.

Lipids

Lipids are in two forms as a source of food: **oils**, which are liquid and generally come from plant tissues, particularly seeds; **fats**, which are solid and generally come from animal tissues such as meat. (Oils may be converted to fat in the form of margarine.) Like carbohydrates, lipids contain only the elements carbon, hydrogen and oxygen and are mainly a source of energy; but some lipids are essential in forming special parts of the body such as cell membranes. Lipids are valuable storage foods because, for the same mass, they contain more energy than carbohydrates. In Britain we eat far more lipids than our ancestors did.

Proteins

Proteins are obtained by eating plants and animals. Protoplasm, the substance of living organisms, is largely protein and water. In addition to carbon, hydrogen, and oxygen, protein always contains nitrogen and often other important elements as well. Protein is essential for growth of the human body, since we too are made up of protoplasm. Because protein cannot be stored by the body, it should be eaten regularly, particularly by growing children, pregnant women and people trying to develop muscle (which is largely protein). Meat and fish, since they consist mainly of muscle, and animal products, such as milk and cheese, are excellent sources of protein, as are certain seeds (which contain protein for the developing plant embryo). In many parts of the world, diets are short of protein. Protein deficiency causes a disease called **kwashiorkor**.

Vitamins

Though vitamins were unheard of until the beginning of this century, it was known earlier that people who appeared to be eating enough food could nonetheless suffer from **deficiency diseases**. In other words, it was known that certain kinds of food were needed as well as a certain amount of food. Even ways of curing or preventing deficiency diseases were known before their cause was known. For example, people on long sea voyages, who ate little or no fresh food, suffered from **scurvy**, a disease which makes tissues, especially gums, bleed easily. Sea captains arranged for long-lasting fresh vegetables, such as onions, to be taken on voyages and these prevented scurvy.

The substance that prevents scurvy is needed only in small amounts (100 g of onion contains only 10 mg of the substance). When the substance was isolated

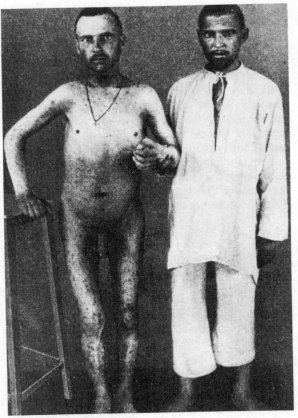

Fig. 21.2 Scurvy victim

early this century, other similar substances, also needed only in small amounts, had been discovered and had been called vitamins. The vitamin preventing scurvy is called vitamin **C** or **ascorbic acid**. Although there is very little vitamin C in onion, two medium-sized onions a day are enough to prevent scurvy. Better sources of vitamin C are fresh citrus fruits (lemons, oranges, grapefruits and limes), strawberries, blackcurrants and green vegetables. A lot of vitamin C is lost in storing and in cooking foods. Figure 21.2 shows a victim of scurvy.

Minerals

Like plants, animals need certain elements for specific purposes. Animals get these elements from their food both in simple inorganic form (as ions) and in complex organic substances. Two elements, calcium and iron, are particularly needed in the human diet: usually they are eaten as part of complex foods.

Calcium is the most abundant mineral in the human body: 2.5 % of the human-body mass is calcium. Most of this calcium (99 %) hardens bones and teeth. The remaining 1 % is needed for the contraction of muscles, the passing of nerve impulses, the activity of enzymes and the clotting of blood. Too little calcium in the diet results in bone deformity in growing children and weakened bones in adults. Milk and cheese (which is made from milk) are good sources of calcium: you would expect this since milk is the food of all young growing mammals. (Milk and cheese also have plenty of protein.) In Britain flour and bread also contain calcium because flour is fortified by the artificial addition of minerals. Hard water is another source of calcium.

Iron is needed only in small amounts but is vitally important. Half the iron in the body is in the red blood pigment, **haemoglobin**; some is present in the red muscle pigment (myoglobin); some is stored in the liver. Iron is not usually lost from the body except by bleeding: at menstruation in women; or because of injury or internal ulcers. Although red blood cells are continually broken down, the iron is usually stored and used again. Lack of iron causes anaemia, a serious condition because it reduces the haemoglobin in the blood and hence the transport of oxygen for respiration. Meat, especially liver and kidney, and cocoa are good sources of iron. Iron is not present in milk: when a young mammal is born, its liver contains a store of iron which it has received from its mother during its development inside her body; it does not get any more iron until it starts to eat solid food.

Water

Water forms about 80 % of protoplasm and is continually lost from the body in sweat, urine and expired air. It must therefore be continually replaced by drinking and by eating foods that contain water. Since food is digested in solution and absorbed from the gut in solution, water is also needed to mix with food in the gut.

Additives

All kinds of substances are now added to our foods to preserve them, to maintain their colour and to give them a pleasant smell and taste. Animals have hormones added to their feeds to speed their growth. All these additives are tested and thought to be harmless. As a general rule, however, the more natural the foods we eat, the less harmful they should be.

Excess

Excess food of any kind can be unhealthy. Carbohydrates and lipids eaten in excess will put on weight and will lead to deposits of fatty material around organs and to narrowing of the cavities of blood vessels (**atheroma**).

Sugar causes tooth decay because it encourages bacteria on the surfaces of teeth, in the spaces between them and at their junctions with the gums. Bacteria produce acids which damage the surface of teeth by removing calcium and phosphates (**demineralisation**) and eventually lead to tooth cavities called **dental caries**. Fluoride is included in toothpaste and drinking water because it encourages **remineralisation** where acids have eaten into the tooth surface. Bacteria and their products form a soft white or colourless deposit on teeth, especially at the junctions with the gums, called **plaque**. Plaque causes inflammation of the gums leading eventually to loosening of the teeth. Correct brushing of teeth and use of dental floss prevent or reduce the accumulation of plaque. You can also protect your teeth by eating sugar less often (rather than by eating less sugar) and by cleaning your teeth or rinsing your mouth out with water immediately after eating sugar.

Different diets

Table 21.2 shows: the different daily dietary needs of moderately active people at different ages; the different dietary needs of males and females from the age of ten onwards; the special dietary needs of pregnant women and of lactating women (those producing milk to breast-feed their children). Only the foods mentioned in this unit are shown: there are many other vital minerals and vitamins.

Table 21.2 has been reproduced from a World Health Organisation book on human nutrition and is meant to apply world-wide. Most people in Britain eat more than they should and most people in the Third World eat less than they should. The data are not for you to learn, but you should be able to suggest sensible reasons for the differences between the recommended diets of people of different ages and sexes and of women who are and are not childbearing. These figures show that in early adolescence girls are heavier than boys but that from the age of twelve onwards boys put on weight much faster than girls. On average boys not only reach a greater height than girls but also develop more muscle, all of which requires more protein. The figures for energy food suggest that males lose energy faster or are more active than females (at least from ten onwards). More calcium is needed for the development of bones and teeth in the first year of childhood than in later years. Boys need more iron for an increase in muscle pigment and blood volume from thirteen to fifteen. Females need more iron from thirteen to fifty to make up for the iron lost each month at menstruation. Look at the figures for pregnant and lactating women: they are all related to food provided by the mother for the growth first of the embryo and then of the baby. *Extra* iron is not needed in pregnancy if the mother is in good health and has had plenty of iron, because the

Table 21.2 Recommended daily intake

	Body mass (kg)	Energy food (carbohydrate) and lipid) (MJ)	Protein (g)	Vitamin C (ascorbic acid) (mg)	Calcium (g)	Iron (mg)
Children						
under 1	7	3.4	14	20	0.5–0.6	5–10
1–3	13	5.7	16	20	0.4–0.5	5–10
4–6	20	7.6	20	20	0.4–0.5	5–10
7–9	28	9.2	25	20	0.4–0.5	5–10
Males						
10–12	37	10.9	30	20	0.6–0.7	5–10
13–15	51	12.1	37	30	0.6–0.7	9–18
16–19	63	12.8	38	30	0.5–0.6	5–9
adult	65	12.6	37	30	0.4–0.5	5–9
Females						
10–12	38	9.8	29	20	0.6–0.7	5–10
13–15	50	10.4	31	30	0.6–0.7	12–24
16–19	54	9.7	30	30	0.5–0.6	14–28
adult	55	9.2	29	30	0.4–0.5	14–28
Pregnant female (last 5 months)		10.7	38	30	1.0–1.2	14–28
Lactating female (first 6 months)		11.5	46	30	1.0–1.2	14–28

Source: *Handbook on human nutritional requirements* (WHO, 1974)

loss of iron at menstruation does not occur in pregnancy.

There are other differences, not shown in the table, in the diets of very old people (who use far less energy); of people recovering from accidents (who need more protein to repair tissues, more calcium to repair bones and more iron to make up for loss of blood); and of people with particular health problems (for example, diabetics who should not eat sugar).

Questions

Q 21.1 Make a table like the one illustrated, but larger, and complete it to compare the three major groups of food.

	Carbohydrate	Lipid	Protein
Elements contained			
Examples			
Food sources			
Function in humans			
Where found in humans			

Q 21.2 Make a table like the one illustrated, but larger, and complete it to compare the importance of vitamin C (ascorbic acid), calcium and iron in the diet.

	Vitamin C (ascorbic acid)	Calcium	Iron
Food sources			
Function in humans			
Deficiency symptom			

Q 21.3 Explain how eating sugary foods damages teeth and gums.

Q 21.4 Identify from Table 21.2 the group or groups who need the most of each substance and suggest an explanation of each group's need.

Q 21.5 (a) Use Table 21.2 to calculate the daily requirement of protein, in units of g of protein per kg of body mass, of females aged 0.5, 2, 5, 8, 11, 14, 17.5, 21 and 30. Treat the figures given for children as figures for females aged 0.5, 2, 5 and 8. Use the figure you have calculated for female adults for females aged both 21 and 30.
(b) Draw a jagged-line graph with age (years) on the x-axis and daily requirement of protein (g per kg of body mass) on the y-axis.
(c) Explain the changes in the protein requirement of females shown by your graph.

PRACTICAL WORK

Experiment 21.1 To carry out qualitative tests for biologically important chemicals

Note for teachers
This experiment familiarises students with tests for starch, reducing sugar, protein, lipid and vitamin C (as a reducing agent). Though the test for protein is not required by the syllabus, it is included because of the importance of protein in the diet. Having established the tests, students often enjoy finding the composition of 'unknown' test solutions and food items.

The starch suspension must be tested to ensure that it does not give a positive result for reducing sugar before class use. See 'Note for teachers', Experiment 17.1.

Materials required by each student
1 250 cm^3 beaker
1 5 cm^3 graduated pipette fitted with a pipette filler
1 dropper pipette
1 test-tube rack with 6 test-tubes
1 stopclock
1 Bunsen burner, tripod and gauze
1 pair of safety spectacles
2 cm^3 1 % albumen solution
2 cm^3 0.1 % ascorbic-acid solution (vitamin C)
2 cm^3 Benedict's solution
1 cm^3 0.5 % copper(II)-sulphate solution
5 cm^3 10 % glucose solution
2 cm^3 iodine in potassium-iodide solution
1 cm^3 0.1 % phenol-indo-2,6-dichlorophenol solution (known as PIPDC or DCPIP)
5 cm^3 propan-2-ol
2 cm^3 10 % sodium-hydroxide solution
5 cm^3 freshly prepared 1 % starch suspension
1 cm^3 vegetable oil

Method
1 Copy the table.

Name of substance	Nature of test	Result of test
Starch		
Glucose (a reducing sugar)		
Albumen (a protein)		
Oil (a lipid)		
Vitamin C		

2 **Starch test.** Add a few drops of iodine solution to 5 cm^3 of starch suspension in a tube. Record the colour in your table.
3 **Reducing-sugar test.** Add 2 cm^3 of Benedict's solution to 5 cm^3 of glucose solution in a tube. Record the colour in your table. Place the tube in a beaker of water and boil the water over a Bunsen burner for two minutes. Record the colour in your table. The reducing-sugar test gives a positive result with all common sugars except sucrose. Sucrose, or cane sugar, which is the white sugar used domestically, is a non-reducing sugar and gives a negative result with Benedict's solution.

4 **Protein test**. Wearing the safety spectacles, add 2 cm³ of the sodium-hydroxide solution to 2 cm³ of albumen solution in a tube. Next add a few drops of the copper(II)-sulphate solution and record the colour in your table.

5 **Lipid test**. Dissolve a little of the vegetable oil in 5 cm³ of propan-2-ol. Since this alcohol is inflammable, you must not use it near a lighted Bunsen burner. Add the mixture to 5 cm³ of water in a second tube and record in your table the appearance of the mixture that forms.

6 **Vitamin-C test**. Pipette 1 cm³ of the DCPIP solution into a tube. Add a few drops of vitamin-C solution and record the colour.

Interpretation of results

1 Does the experiment you have performed prove that the tests are specific to the substances you tested? What other tests could you perform to find out if they are specific?

2 Plan how you could determine the composition of one of your favourite items of food. Your plan should include the way in which you would prepare the food for testing, the tests you would use and the way in which you would record your results.

3 Suggest how the test you have used for vitamin C could be used to compare the amount of vitamin C in different fruit juices.

UNIT 22

Digestion and absorption

Humans

To be of use to the body, food must be in a form that can enter cells. To pass through the cell membrane by diffusion or active transport, food must have small soluble molecules. Molecules of complex carbohydrates, lipids and proteins must all be digested, i.e. broken down by enzyme action into small soluble molecules which, once dissolved, can be absorbed by the cells. Since there is water in the gut, food only has to be made soluble to be automatically dissolved. The two processes that concern us in humans are therefore digestion and absorption.

Enzymes

The enzymes produced in the largest quantities are those involved in the digestion of food. In humans the enzymes are secreted by special cells often grouped together in tissues called **glands**. Different glands secrete different enzymes into different parts of the gut with different pHs: each enzyme is secreted where the pH enables it to work best.

The three major groups of food (carbohydrates, lipids and proteins) are digested by three major groups of enzymes (carbohydrases, lipases and proteases) into sugars, fatty acids and glycerol, and amino acids. Amylase, the enzyme you study, is one of the carbohydrases. Vitamins, inorganic ions and water enter cells without being broken down.

The human gut has five major parts:
  mouth cavity
  gullet or oesophagus – lying behind the windpipe, lungs and heart in the chest cavity
  stomach ⎫
  small intestine ⎬ lying in the abdominal cavity
  large intestine ⎭

These parts are shown in the human body in Figure 22.1. When you read about each part, make sure you can find it in the diagram.

Mouth cavity

Here food is crushed and chewed by the teeth, moved around by the tongue and swallowed. Chewing is important because it increases the surface area of food and allows enzymes to make greater contact with it. Saliva, secreted by salivary glands, moistens food in the mouth and makes it easier to swallow. Most people's saliva contains amylase, the enzyme that catalyses the conversion of starch to sugar. It does not matter greatly if there is no amylase in your saliva because amylase is produced lower down in the gut by the pancreas. When saliva moistens food, the taste cells in the mouth are stimulated. The taste of food, the smell of food and the memory of past food stimulate other secretions of the gut, particularly those of the stomach. Saliva has a near neutral pH of about 7, a suitable pH for the action of amylase.

Gullet (oesophagus)

Food spends little time passing through the gullet, which is merely a tube from the mouth to the stomach through the back of the chest cavity (**thorax**). The gullet interferes little with breathing and heartbeat, which take place in the chest cavity. The circular muscles of the gullet wall contract behind food and relax in front of it: they thus push food through the gullet as you would push a marble through a tube by squeezing the tube behind it.

Stomach

The stomach is a greatly enlarged part of the gut closed at both ends by valves. A valve is a device which allows substances to pass in one direction only. Hydrochloric acid, secreted by glands in the wall of the stomach, reduces the pH to 2, a very acid pH indeed. If this acid escapes back into the gullet, it causes a painful form of indigestion called heartburn. It has nothing to do with the heart, but the pain is near the heart. The wall of the stomach itself is protected from this acid by mucus, a slimy substance also secreted by glands in the wall of the stomach. The acid kills bacteria in food and produces the ideal pH for the action of a protease that is secreted by the wall of the stomach and begins the digestion of protein. Food is churned around in the stomach, which secretes water as well as a great deal of mucus. When food leaves the stomach, about two or three hours later, the mixture, called **chyme**, is fluid, partly digested and very acid. Amylase from the

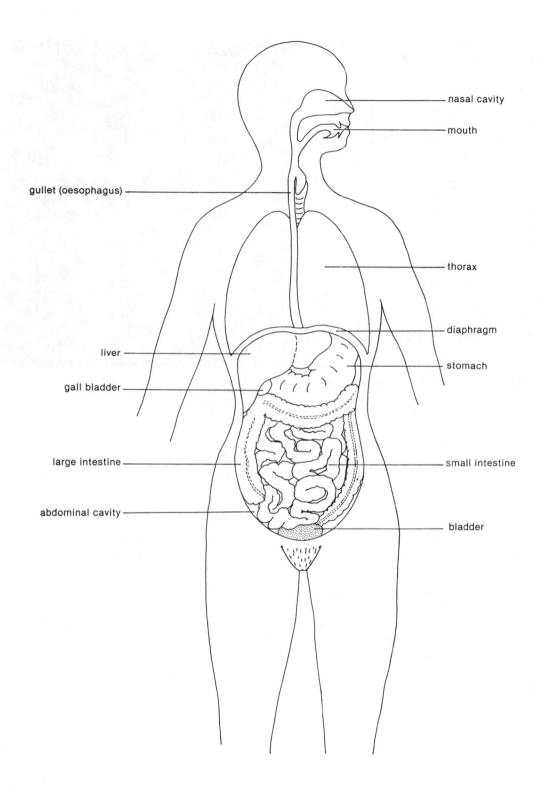

Fig. 22.1 *Human gut*

mouth is deactivated and stops working in the acid conditions of the stomach.

Small intestine

This coiled tube, about six metres long in an adult, occupies most of the abdominal cavity. Digestion is completed in the first part of the small intestine, after which the mixture is no longer called chyme; absorption takes place throughout the small intestine. When the very acid chyme arrives from the stomach, it is first neutralised and then made alkaline (to a pH of about 8.5) by bile and by secretions from glands in the wall of the first part of the small intestine. The chyme contains lipids, partly digested protein and partly digested carbohydrates.

More enzymes are secreted on to the chyme in the first half metre of the small intestine. They come not only from the wall of the small intestine but also from a special gland, the **pancreas**. Enzymes from both sources contain carbohydrases (including amylase), lipases and proteases different from the protease in the stomach. Figure 22.2 shows the pancreas, which lies behind the stomach and the colon. **Bile**, a bitter green fluid made in the **liver** and stored in the **gall**

Fig. 22.3 Villi: scanning electron micrograph

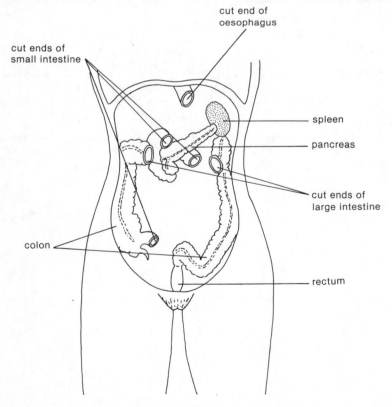

Fig. 22.2 Human abdominal cavity: liver, stomach and some other parts removed

bladder, is also added to the chyme in the first part of the small intestine. Bile breaks lipid into small droplets (i.e. **emulsifies** lipid) and thus gives the lipase enzymes more surface contact.

The liquid contents are wafted along the small intestine by thousands of 1 mm-long fingerlike processes, called **villi**, which line its wall. Villi help to mix the fluid contents and provide a large surface area through which the digested food – simple sugars, fatty acids and glycerol, and amino acids – can be absorbed. By the time the contents reach the valve separating the small intestine from the large intestine, the digested food has been absorbed. The contents remaining in the gut at this stage are indigestible food and fluid, particularly the cellulose (fibre) of plant tissues, certain inorganic ions and water.

Large intestine

In an adult, between 300 and 500 g of undigested food and fluid pass into the first part of the large intestine, the **colon**, daily. The average daily mass of faeces expelled from the end part of the large intestine, the **rectum**, is about 150 g, of which about 100 g is water. Between 150 and 350 g of water (and products of bacterial cellulose digestion) are therefore absorbed daily from the contents of the colon. Inorganic ions are also absorbed. The rectum is normally empty; when the contents of the colon pass into it, the need to defaecate is stimulated. As faeces pass out of the rectum, the anus muscles relax. Figure 22.2 shows the rectum, which lies behind the coiled small intestine at the back of the abdomen.

Fungi and bacteria

Internal digestion and absorption of food are typical of animals. Animals take in complex food (**ingestion**), carry out digestion and absorption and get rid of indigestible food (**egestion**). Plants, described in Units 19 and 20, have a completely different way of feeding: by photosynthesis and the absorption of inorganic ions to make the complex molecules of their bodies. Their kind of feeding is **autotrophic** (self-feeding).

Another method of feeding is **saprophytic**. Saprophytes, organisms which decompose the remains of all dead organisms, are described in Unit 10. Saprophytes cannot make their own carbohydrates by photosynthesis as plants can; nor can they eat complex foods as animals do; they feed on complex foods but in a manner all their own.

Bread, jam or fruit, if left in a damp warm place, soon develops thin white hyphae of saprophytic fungi or **moulds**. Since they have nothing like a mouth, saprophytic organisms cannot take in solid food. Instead they digest their food *outside* their bodies by secreting enzymes on to it. This process is called **external digestion**. When digestion has converted the food into small enough molecules and made it soluble, it must be dissolved. Some food, such as certain fruits, has enough water in it to do the dissolving. Other food can be dissolved only if water is available from rain or condensation or some other source: saprophytes cannot feed on dry food. When the food is dissolved, it diffuses (or enters by active transport depending on the concentration gradient) into the saprophyte. Exactly the same groups of enzymes – carbohydrases, lipases and proteases – are involved in saprophytes' digestion as in animals' digestion.

But saprophytes do not necessarily need *all* their food in a complex form. Having acquired sugars by external digestion and inorganic ions from water and from their food, some saprophytes can make proteins from them in the same way as plants. (They cannot of course make the sugars by photosynthesis as plants do.)

The processes involved in saprophytic feeding are described in Unit 10. Inside a saprophyte the food is used, as all food is used, either as a source of energy or as a raw material for making more protoplasm. If a saprophyte's food is not used immediately, it is stored as droplets of lipid or as glycogen.

Because animals and saprophytes cannot make their own food, their feeding is described as **heterotrophic** (other-feeding).

Questions

Q 22.1 What is digestion?

Q 22.2 Summarise the digestive processes in the mouth, stomach, small intestine and large intestine in a table like the one illustrated, but larger.

	Mouth	Stomach	Small intestine	Large intestine
Function				
Enzymes present				
pH				
Substances digested				

Q 22.3 In what ways is a hypha of a fungus similar to and different from a palisade-mesophyll cell of a leaf?

PRACTICAL WORK

Experiment 22.1 To investigate the hypothesis that saliva contains an amylase enzyme

Note for teachers
The amylase activity in human saliva varies from person to person and at different times of the day. Some people have no amylase in their saliva.

Materials required by each student
1 250 cm^3 beaker
1 100 cm^3 beaker
1 5 cm^3 graduated pipette fitted with a pipette filler
3 dropper pipettes
1 test-tube rack with 4 test-tubes
1 stopclock
1 white dimple tile
1 Bunsen burner, tripod and gauze
1 spirit marker
1 clean elastic band
6 cm^3 Benedict's solution
4 cm^3 iodine in potassium-iodide solution
15 cm^3 freshly prepared 1% starch suspension tested to ensure that it gives no positive reaction for reducing sugar

Method
1. Collect some of your own saliva by spitting into the 100 cm^3 beaker. It is easier to produce saliva if you first chew the clean elastic band.
2. When you have collected about 5 cm^3 of saliva, pour about half of it into a tube, put the tube in a beaker of water and boil the water for two minutes.
3. Label three tubes A, B and C and pipette 5 cm^3 of starch suspension into each.
4. Pipette 1 cm^3 of unboiled saliva into tube A, 1 cm^3 of boiled saliva into tube B and 1 cm^3 of water into tube C. Start the stopclock and leave the tubes in the test-tube rack.
5. Copy the table at the foot of the page.
6. Ten minutes after starting the stopclock, use separate dropper pipettes to transfer one drop of mixture from each tube to a drop of iodine solution in the well of a dimple tile. Record in your table the colour produced by each mixture.
7. Add 2 cm^3 of Benedict's solution to each tube. Put the tubes in a beaker of water and boil the water. Record in your table the colour in each tube.

Interpretation of results
1. In which tubes was starch still present after ten minutes?
2. In which tubes was reducing sugar present after ten minutes?
3. Explain the presence of reducing sugar wherever it was found.
4. Explain the value of tubes B and C in this experiment.

Experiment 22.2 To investigate the hypothesis that fungi secrete digestive enzymes on to their substratum

Note for teachers
Pure cultures of fungi such as *Aspergillus oryzae* or *Mucor hiemalis*, which can be obtained from biological suppliers, are suitable for this experiment.

Starch agar used in this experiment can be obtained from biological suppliers or it can be made by shaking 1 g of agar and 0.3 g of starch in 100 cm^3 of distilled water and heating the mixture until it boils. While the mixture is still liquid, pour it into a number of sterile petri dishes to an even depth of about 5 mm and cover them: 100 cm^3 of agar will be enough for fifteen to twenty plates. The agar will set in about ten minutes and should preferably be left upside down and open in a drying cabinet to allow the excess moisture to evaporate. Since the experiment has to run for more than two days, the agar should be autoclaved at a gauge pressure of 100 kPa (15 lb per in^2) for fifteen minutes before being poured into the petri dishes.

Materials required by each student
1 petri dish of starch agar
1 inoculating loop
1 spirit marker
1 Bunsen burner
10 cm^3 iodine in potassium-iodide solution

Material required by the class
agar culture of a hyphal fungus

Method
1. Write your initials on the bottom of the petri dish containing starch agar.
2. Hold the inoculating loop in the Bunsen flame until the wire at the end glows. Remove the loop from the flame and allow it to cool without letting it touch anything.
3. Use the loop to transfer a small piece of the fungal growth from its culture to the middle of your own starch-agar dish. While transferring the fungus, hold the dish lid at an angle, as in Figure 22.4, to reduce the chance that airborne organisms will land on the starch agar.
4. Again hold the inoculating loop in the Bunsen flame until its wire glows red. Allow it to cool and then leave it on the bench.
5. After about a week make a drawing of the starch agar to show the pattern of growth of the fungus over its surface.

Tube	Contents	Result after 10 minutes	
		Iodine solution	Benedict's solution
A	Starch and unboiled saliva		
B	Starch and boiled saliva		
C	Starch and water		

Fig. 22.4 Transfer of fungal culture to sterile starch agar

6 Remove the petri-dish lid and pour iodine solution over the whole surface of the agar. After two minutes or so wash the iodine solution away and examine the agar surface against a light background.
7 Make a drawing to show the position of the blue-black colour in the agar.

Interpretation of results
1 Why did some of the agar turn blue-black?
2 Why did some of the agar fail to turn blue-black?
3 What was the relation between the position of the blue-black colour and the growth of the fungus?
4 What does this tell you about the fungus? Suggest one way in which your conclusion could be tested.

UNIT 23

Digestion and transport in plants

Digestion in plants

Animals and saprophytes are not the only organisms to digest their food. Although plants make their own food entirely from simple compounds (carbon dioxide and water) and inorganic ions, they may immediately store it in the form of large insoluble molecules which will not interfere with the cells' activities. Sugar (glucose), which is formed by photosynthesis, is unsuitable for storage: it increases the concentration of the cells' contents and encourages water to enter by osmosis. Sugar is soon converted to insoluble and inactive (inert) starch in the chloroplast. You can see the starch in the chloroplast in Figure 13.5.

Nor is starch the only insoluble food stored in a plant. Wherever a plant forms food storage organs, sugar may be converted into starch, fatty acids and glycerol into lipids, and amino acids into proteins. Starch, lipids and proteins all consist of large inactive molecules and are insoluble. All this insoluble food must be digested before it can be transported and used by the plant. Seeds are always food-storage organs because they must be able to grow into independent seedlings on the food that they store within themselves. Other food-storage organs, such as potatoes, onions and carrots, are parts of plants that die down in unfavourable weather and need a store of food for growth of new leaves and roots before they will be able to make their own food again.

Like animals and saprophytes, plants use enzymes to digest their insoluble food. They have amylase to digest starch, lipases to digest lipids, proteases to digest protein. Digestion occurs in cells where there is water to dissolve the food as soon as it is digested, i.e. as soon as it is made soluble.

Let us consider starch. Amylase is present to convert it to soluble sugar when the plant needs to use it. The sugar is immediately dissolved and moved to the regions of the plant that need it.

$$\text{starch} + \text{water} \xrightarrow{\text{amylase}} \text{sugar}$$

Transport of sugar in plants

Where is sugar needed in plants? It is needed by all living cells but particularly by those at the growing points. Wherever there is growth, sugar is needed:
- to provide energy to make the many complex compounds that form new cells;
- to make the cellulose walls of the new cells;
- to combine with inorganic ions to make amino acids which form proteins;
- to combine with inorganic ions to form other complex compounds in protoplasm;
- to provide energy to move substances by active transport.

The sugar that moves through the plant is not a simple sugar (glucose) but a more complex one (sucrose). One hypothesis to explain this is that glucose is used too easily by the cells through which the sugar has to pass: it might never get to its destination.

Whether sucrose moves from one plant cell to another by diffusion or by active transport depends on the concentrations of sucrose in the two neighbouring cells. Figure 23.1 shows diagrammatically how sucrose diffuses from a cell where it is in higher concentration to one where it is in lower concentration along the concentration gradient, but moves by active transport against the concentration gradient. Movement of sucrose takes place dissolved in the water of the cytoplasm and cell sap.

Movement by diffusion or active transport is slow. In plants there are certain specialised cells that enable sucrose to move more quickly – but we do not know exactly how they do it. There is a tissue within the vascular bundles or veins called **phloem**, which contains tube-like cells. In these tube-like cells there are cross-walls like sieves (in fact these tubes

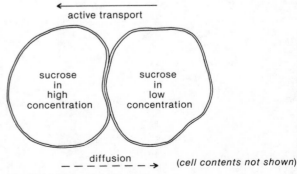

Fig. 23.1 Movement of sucrose

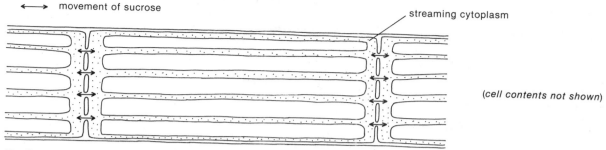

Fig. 23.2 Phloem cell: L.S. sieve tube

are called **sieve tubes**) with large holes in them. A section through one of these tubes is shown in Figure 23.2. Each cell of the tube contains cytoplasm (but not a nucleus) which is continuous (through the holes of the end walls) with the cytoplasm of the next cells. The activities of the cell are probably controlled by other living cells around the tube. Sucrose moves quickly in either direction through these tubes: active transport is involved and so is movement of the cytoplasm (called streaming), but they do not explain the speed that is achieved.

Phloem is present in all the veins. Veins are continuous throughout a plant from leaves through stems to roots. Amino acids also travel through sieve tubes.

Transport of water in plants

Water enters plants through their roots. Some cells on the surface near the root tips are specialised to form single-celled outgrowths called **root hairs**. Water enters mainly through these root hairs, which are yet another example of an increase in surface area for absorption. A root-hair cell is shown in Figure 20.1. The outer wall of the cell is extended: the root hair is so thin that it can pass between particles of soil and can contact the film of water that surrounds soil particles and the water that is present in decaying remains of plants and animals in the soil (called **humus**).

It is often just possible to see thousands of these root hairs near the tip of each root. They absorb water by osmosis because the soil water is a weaker solution (i.e. has a greater concentration of water) than the contents of the root-hair cell.

Once a concentration gradient is set up throughout the cells of a plant, water moves from one cell to the next by osmosis. This happens both in the roots and in the leaves, though the causes of the gradient are different.

Gradient in roots

Figure 23.3 shows how the gradient is formed in the roots. Water enters a root-hair cell because there is a higher concentration of water molecules in the soil outside. Because it receives a supply of water directly from the soil, the root-hair cell has a higher concentration of water molecules than the next cell in the root; water therefore diffuses from the root-hair cell to the next cell by osmosis, and so on to the next-but-one cell and the cell after that, establishing a concentration gradient in which water molecules are most concentrated in the root-hair cell and least concentrated in the cell furthest away from it.

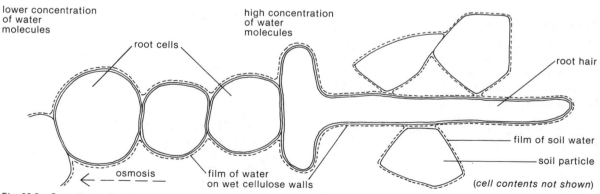

Fig. 23.3 Osmotic gradient in root

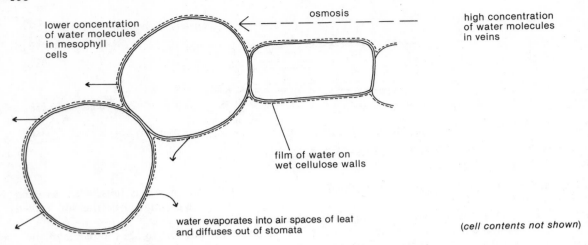

Fig. 23.4 Osmotic gradient in leaf

Water also moves in an unbroken film around and through the wet cellulose walls of the cells. If water is removed from the wet cell surface of one cell, it is immediately replaced either from within the cell cytoplasm or from the wall of a neighbouring cell. This process, which occurs because molecules of water stay together (by cohesion), helps to move water across the root via the root hairs which are in contact with water in the soil.

Gradient in leaves

Figure 23.4 shows how the gradient is formed in the leaves. Water vapour diffuses out of the open stomata; water evaporates from the mesophyll cells to replace it; because these mesophyll cells now have a lower water concentration than their neighbours, water passes into them by osmosis; and thus water passes on from cell to cell. Water also moves in an unbroken film through and around the wet cellulose walls of the mesophyll cells. The loss of water by evaporation from the leaves is called **transpiration**.

Transport through xylem

Movement of water by osmosis is slow. Just as there are specialised cells that allow sucrose to move quickly, so there are specialised cells that allow water to do so. These cells (called **vessels**) are tubes without any form of cross-wall. (The original cross-walls and the living protoplasm completely disintegrate.) These tubes have thick walls or they would collapse. The tissue formed by these cells is called **xylem**. Together with phloem, xylem is contained in the vascular bundles or veins.

Since the xylem tubes contain no selectively permeable membranes (because there is no cytoplasm or cell membrane), water cannot move through them by osmosis. How then does water move through the xylem? Mainly it is 'pulled' up the plant. This sounds impossible, but you have probably 'pulled' a drink up through a straw. Because water molecules stay together (by cohesion), they can be pulled up airtight columns.

What is the force that pulls water up a plant? It is the evaporation of water from the leaves, i.e. transpiration. If you tie a plastic bag around a few leaves of a plant, water will soon collect inside it: this is water lost from the leaves by transpiration. Once water vapour *diffuses* out through the stomata from the air spaces in the leaf, a concentration gradient is formed. More water vapour *evaporates* from the cells into the air spaces. This sets up the gradient in the cells of the leaf which is explained earlier in this unit in the section on 'Gradient in leaves'. On a damp misty day, when the air outside a plant also contains concentrated water vapour, there is no gradient and diffusion of water vapour does not take place. Transpiration then stops and will not start again until a gradient develops.

While water is pulled up the xylem by transpiration from the leaves, two other factors may help its upward movement:

water moves up any fine tube by capillary action, the attraction between water and the wall of a tube;

water is secreted into the xylem tubes by living cells surrounding the xylem in the root (this is called **root pressure**).

Figure 23.5 shows the extent of transpiration. Of water that enters a maize plant, 1.8 % is held in the growing cells, a mere 0.2 % is used in photosynthesis, and the remainder, 98 %, is transpired.

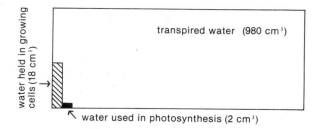

Fig. 23.5 Fate of 1 dm³ of absorbed water in a maize plant

Transport of inorganic ions and amino acids

Ions enter the plant, dissolved in the water of the soil solution, by the root hairs. The entry of nitrate and ammonium ions is described in Unit 20: other ions enter in the same way. Whether they enter by diffusion or active transport depends on their concentrations in the root-hair cell and in the soil solution. Dissolved in water, they pass to the xylem in the root, again by diffusion or active transport depending on the concentration gradient. While most dissolved inorganic ions travel through the plant in the xylem, some travel in the phloem.

Transport of amino acids, made in the plant, is usually in the phloem.

Transport to growing points

Movement of sugars, amino acids and inorganic ions from the xylem and phloem to growing points is by diffusion or active transport depending on the concentration gradient. Movement of water is by osmosis. Xylem and phloem take time to form: they do not usually reach all the way to the growing points.

Questions

Q 23.1 Figure 23.6 shows a French-bean seedling before the leaves have opened. Where is starch being digested in the seedling? What will happen to the sugar that is formed?

Q 23.2 Make a table like the one illustrated, but larger, and complete it to show the differences between xylem and phloem.

	Xylem	Phloem
Description of transporting cells		
Substances carried		
Direction in which substances are carried		
Mechanism by which substances are carried		

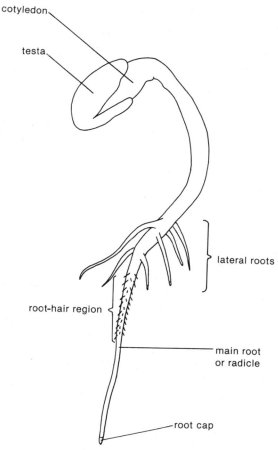

Fig. 23.6

Q 23.3 In what ways are root hairs and villi similar to one another and in what ways are they different from one another?

Q 23.4 The table shows the relative absorption of potassium ions by carrot roots at different oxygen concentrations.

Oxygen concentration	Relative absorption of potassium ions (arbitrary units)
2.7	22
12.2	96
20.8	100
28.5	110
43.4	117

(a) Draw a jagged-line graph with oxygen concentration on the x-axis and relative absorption of potassium ions on the y-axis.
(b) What conclusions can you draw from the graph about the absorption of potassium ions in these roots?

PRACTICAL WORK

Experiment 23.1 To investigate the ability of a potato tuber and a germinating French-bean seed to break down starch

Note for teachers
Starch agar can be bought or can be made as explained in Experiment 22.2. If the experiment is to continue for longer than two days, the agar should be autoclaved at a gauge pressure of 100 kPa for fifteen minutes.

Materials required by each student
1 250 cm^3 beaker
1 pair of forceps
1 scalpel
2 petri dishes of starch agar
1 Bunsen burner, tripod and gauze
1 spirit marker
10 cm^3 iodine in potassium-iodide solution
2 unsoaked French-bean seeds
2 French-bean seeds which have been soaked for 24 hours
1 3 cm cube of raw potato tuber

Method
1. Label the bottoms of the two petri dishes of starch agar with your initials.
2. Cut the cube of potato into two and boil one piece in the beaker of water for five minutes.
3. While the potato is boiling, cut the soaked bean seed longitudinally through the middle and place the two halves about 4 cm apart cut-face downwards at one end of the agar plates.
4. Cut the unsoaked bean seed longitudinally through the middle and place the two halves cut-face downwards at the other end of the same agar plate about 1 cm apart so that you will be able to recognise the soaked and unsoaked seeds later.
5. Remove the boiled potato from the beaker and allow it to cool. Place it on the surface of the agar at one side of the second petri dish. Place the piece of raw potato on the surface of the agar at the other side of the same petri dish.
6. Leave both petri dishes for one to seven days.
7. Remove the seeds and potato pieces and flood both agar plates with iodine solution. After two minutes wash the iodine away and examine both plates against a light background.
8. Make a drawing of both plates to show the positions of the blue-black colour and of the different seeds and potato cubes.

Interpretation of results
1. Why did the iodine solution turn parts of the starch agar blue-black?
2. Which parts of the starch agar did the iodine solution not turn blue-black?
3. Is it likely that the seeds and the potato pieces have simply absorbed the starch from the agar? From your knowledge of Unit 15 explain your answer.
4. Suggest a hypothesis to relate the colourless areas of the agar and the positions of the boiled and unboiled potato.
5. Suggest a hypothesis to relate the colourless areas of the agar and the positions of the soaked and unsoaked French-bean seeds.

Experiment 23.2 To investigate the movement of water through a plant leaf stalk or stem

Note for teachers
Leaf stalks of celery or stems of Busy Lizzy (*Impatiens walleriana*) should be used. The advantage of using celery is that many leaf stalks can be obtained from a single plant purchased from a greengrocer. The advantage of using Busy Lizzy is that the movement of water in its stems tends to be faster, because its leaves provide a larger surface area for water loss.

Materials required by each student
1 250 cm^3 beaker
1 scalpel
1 hand-lens
1 ruler graduated in mm
100 cm^3 1 % methylene blue or eosin solution
1 fresh leaf stalk of celery or stem of Busy Lizzy

Method
1. Cut off about 5 mm from the bottom of the leaf stalk of celery or the stem of Busy Lizzy and put the cut end in the beaker.
2. Pour the coloured water into the beaker so that it covers the bottom of the plant part and leave the beaker in the light for about 30 minutes.
3. Remove the plant part from the coloured water. Measure the length of the part which shows the colour of the dye inside it.
4. Cut across the part coloured from the water. Examine the cut end with a hand-lens and locate the exact position of the colour.

Interpretation of results
1. To what height had the colour travelled?
2. What assumption must be made when drawing a conclusion from your result about the movement of water?
3. Calculate the rate of movement of the coloured water by using the formula

$$\frac{\text{distance travelled by the coloured water (mm)}}{\text{time (min)}}$$

4. Compare the result of your calculation with other results in your class. Was the rate of movement always the same? If not, suggest an explanation of the different rates.
5. Draw a diagram of a cross-section of the leaf stalk of celery or the stem of Busy Lizzy and use it to show the precise positions of the coloured water. What plant *tissue* is found in the positions you indicate?

UNIT 24 Transport in humans

Scientists of a few centuries ago realised that the veins of plants transported substances. That is how they got their name. Scientists thought they were like the veins and blood vessels in humans which, they correctly assumed, also transported substances. That both are transport systems is about the only similarity between the veins of plants and of humans.

Major differences between the transport systems of plants and humans are shown in Figure 24.1. In plants the veins, consisting of tubes (xylem) and sieve tubes (phloem), begin in one place and end in another. In humans the veins are only part of a closed system of tubes containing a transport medium, **blood**, which flows endlessly round the body pumped by an organ, the **heart**. **Veins** are tubes that return blood to the heart from the body. Tubes that carry blood to the body from the heart are **arteries**.

Blood is a red mixture that transports substances to and from the living cells of the human body. (Over short distances substances usually move by diffusion or active transport; in certain special cases substances are moved by lymph.) Blood contains 55 % fluid, called **plasma**, and 45 % cells. Most of the cells, red in colour, carry oxygen and carbon dioxide. All the other transported substances are carried in the fluid plasma. Among the most important substances to be transported in the plasma is food. Every living cell needs a supply of food as a source of energy to maintain the organised processes that keep it alive.

Food is therefore transported from the small intestine, where it is absorbed, to all the living cells.

If you have three meals a day, there will be times when a lot of food is absorbed in the small intestine and times when nothing is absorbed. Cells could not survive on such irregular food supplies. The supply of food, especially of sugar, is regulated so that approximately the same amounts arrive at the living cells all the time. The regulator is the **liver**.

Look at Figure 22.1 and you will see that the liver lies in the abdomen, partly over and to the right of the stomach (to your right that is – to the left of the stomach in the diagram). All the blood from the parts of the gut which lie in the abdomen – stomach, small intestine and large intestine – passes to the liver (by the hepatic portal vein).

In the liver the sugar (glucose) is regulated so that only $0.6\,g$ per dm^3 of blood passes out to the rest of the body. The liver also regulates the lipids and amino acids so that only small amounts pass out to the rest of the body. The liver acts under the direction of hormones, especially insulin. The movement of lipids from the small intestine is complicated: having passed through the villus, fatty acids and glycerol form up again as very fine droplets of lipid and do not immediately go into the blood.

Blood from the liver, containing regulated supplies of food, passes out into the main circulation just before it enters the heart.

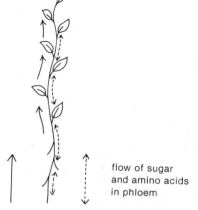
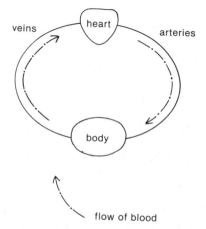

Fig. 24.1 Transport systems: plants and humans

Blood circulation

The force driving blood around the body is developed by a pump, the heart. Its walls are made of muscle; in a person at rest it pumps about 75 times a minute, about 40 million times a year. Its two sides, left and right, are completely separate: on one side it pumps blood to the lungs; on the other side it pumps blood to every other part of the body. It is therefore a double pump resulting in a double circulation. This is shown in Figure 24.2.

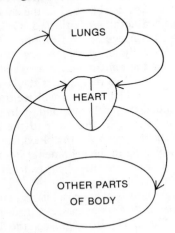

Fig. 24.2 Double circulation

The fact that the circulation to the lungs is separate from the circulation to the rest of the body allows the pump to work at two different pressures. The lungs are delicate structures, with a wall only one cell thick in parts, which would be damaged by pressures high enough to force blood right out to the toes and fingers. The pressure of blood passing from the heart to the lungs is only about 3 kilopascals, while that of blood passing to the other parts of the body leaves the heart at about 16 kilopascals. By the time blood gets back to the heart the pressure on both sides is almost zero.

Another advantage of the double circulation is that all the blood that comes back from the body goes to the lungs, where it is given oxygen and is freed of toxic carbon dioxide before it goes out to the body again. In a complete circulation blood goes twice through the heart before it ends up where it started.

Since blood leaving the heart is always at a higher pressure than blood entering the heart, the arteries that carry blood away from the heart are thicker-walled and tougher than the veins that bring blood back to the heart.

What happens between the arteries and the veins in the lungs and the body? This is where the functions of blood as a transport system are carried out. Arteries, which take blood from the heart, branch again and again until they are so narrow that blood cells have to pass along them in single file. These narrow blood vessels, called **capillaries**, have walls only one cell thick: they allow substances in the blood to diffuse out to the tissues and allow other substances to diffuse into the blood from the tissues wherever there is a concentration gradient. Figure 24.3 is similar to Figure 24.2 but shows the capillaries in the circulation.

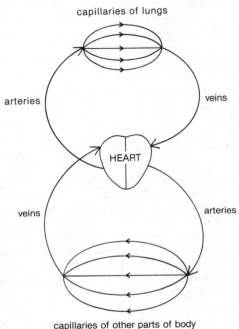

Fig. 24.3 Double circulation showing capillaries

Figure 24.4 shows more detail of the blood supply to the gut. Arteries bring blood to the gut, where it receives sugars, amino acids, water and other substances such as inorganic ions and vitamins. Blood then goes, not straight into the circulation, but to the liver. Also shown is the liver's supply of arterial blood. From the liver blood goes back to the heart by the veins.

The two **kidneys**, where urea, salts and water are removed from the blood, lie at the back of the abdomen, each supplied by an artery which breaks up into capillaries. The capillaries rejoin to form veins. Figure 24.5 includes the blood supply to the kidneys. You need to know the names of a few of the blood vessels entering and leaving the heart:

the **pulmonary** arteries are the two from the heart to the two lungs;

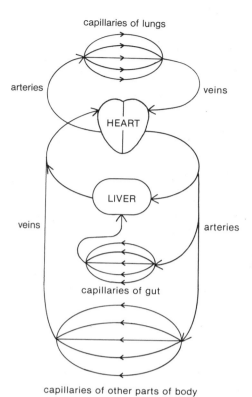

Fig. 24.4 Circulation to lungs, gut, liver and body

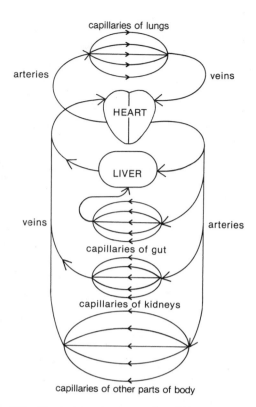

Fig. 24.5 Circulation to lungs, gut, liver, kidneys and body

the **aorta** is the major artery passing to the other parts of the body (it soon divides into other arteries going to the head, arms, parts of the gut, other abdominal organs and the legs – all of which have names, but you need not learn them);

the **pulmonary veins** are those coming back to the heart from the lungs;

the three **venae cavae** (singular **vena cava**) are major veins bringing blood back to the heart from the other parts of the body.

Figures 24.1–24.5 do not show what the circulation looks like or where the arteries, veins, heart, lungs and kidneys really lie in the human body. Figure 24.6 does this.

The heart

Structure

In this unit you have learnt that:
the walls of the heart are of muscle;
its left and right sides are completely separate;
arteries carry blood at pressure from the heart;
veins return blood at near-zero pressure to the heart.

On each side of the heart is a thin-walled chamber that receives blood at near-zero pressure from the veins. These are the two **atria** (singular **atrium**). Each atrium leads by a narrow opening, controlled by a valve, into a thick-walled pumping chamber, the **ventricle**. From the two ventricles blood is pumped out through the arteries. At each of the two exits from the heart is another set of valves. Figure 24.7 shows the main structures of the heart, first as the muscles of the ventricle walls relax and both ventricles and atria fill with blood, then as the muscles of the ventricle walls contract and force blood out of the two major arteries, the aorta and the pulmonary artery.

In the diagrams the left and right sides appear to be reversed because the heart is viewed from the front:
only two of the three venae cavae are shown entering the right atrium;
four pulmonary veins are shown entering the left atrium (the veins do not all join up before they reach the heart);
when one set of valves is open, the other set is closed (this is how the blood is kept flowing in one direction: when the heart is filling up, it draws blood in from the veins; when it is emptying, it pushes blood out of the arteries).

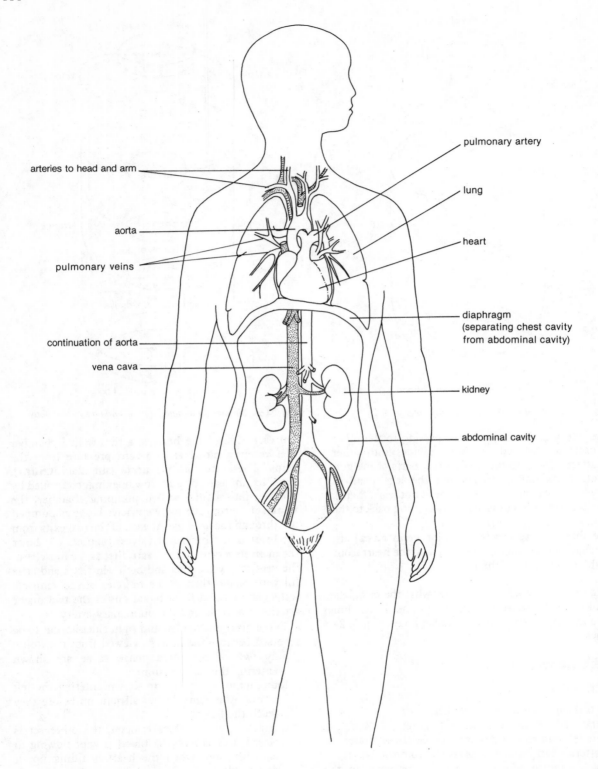

Fig. 24.6 *Major blood vessels and organs in the human body (gut removed)*

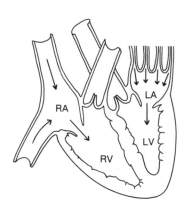
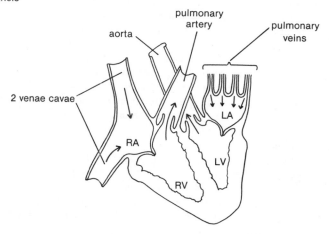

Fig. 24.7 Human heart and blood flow

Valves

Not shown in Figure 24.7 (because it would make the diagrams confusing) are the tendons attached from the ventricle walls to the valves that separate the atria from the ventricles (the **cuspid valves**). You will be able to see them clearly in the heart that you dissect – which will not of course be a human heart but a similar heart of another mammal. The tendons stop the valves, which are only flaps of skin, from turning inside out when the ventricles contract and push blood against them. Think of these valves as swing doors which open only one way (from atria to ventricles). They are stopped from opening the other way by the tendons which hold them closed when the blood is pushed against them from the ventricle side.

The valves (**semi-lunar valves**) at the entrances to the two major arteries are different. They are not flaps of skin but pockets of skin. When the muscles in the ventricle wall relax and the cavity of the ventricle expands, drawing blood into the heart, these pockets fill with blood, blocking the artery tubes. Because blood cannot get back to the heart from the arteries, blood is drawn in from the veins. The action of pocket valves is shown in Figure 24.8.

Figure 24.8 shows how blood-pressure differences control the opening and closing of the valves: when pressure is higher on the left than on the right, the pocket valve opens and blood is pushed towards the right; when pressure is higher on the right than on

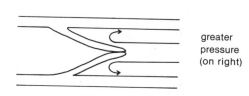

Fig. 24.8 Action of pocket valves

the left, the pocket valve closes and blood cannot flow to the left.

Pocket valves are also found lining the walls of veins. Their function in the veins is the same: they allow only a one-way blood flow, this time back to the heart.

Blood supply to the heart

Shortly after leaving the heart by the aorta, blood passes through two small arteries to the heart wall. These are the two **coronary arteries**, which you should be able to see branching on the outside of the heart itself. A blockage in either of these arteries by a blood clot (coronary thrombosis) or by fatty tissue can cause a 'heart attack'. Such a blockage stops the flow of blood supplying oxygen to the muscles in the heart wall. Being starved of oxygen, the muscles respire anaerobically and collect lactic acid, which causes the pain of the heart attack. The deprived muscles soon stop working and eventually die. The severity of a heart attack depends on where the blockage occurs. If the artery is blocked near its source from the aorta, damage to the heart muscle is severe and causes death. If a small branch is blocked, only a small section of heart wall is damaged – and the heart attack is a useful warning.

Blood circulation: summary

The heart pumps blood out to the lungs and to the other parts of the body via the arteries. Pressure is lower the further blood travels. The arteries branch, getting narrower and narrower until they form capillaries where substances are exchanged between blood and tissues. Capillaries join up to form veins, which join again and again to return blood to the heart. The pressure in the large veins is near zero. Pressure of surrounding body muscles on veins helps push blood back to the heart. The pocket valves prevent it from flowing backwards.

Exercise

Most people have a heartbeat rate at rest of about 75 beats per minute, but there is considerable variation among perfectly normal people. The rate is usually measured by the pulse, the heartbeat transmitted to the arteries. Under stress or during exercise the heartbeat rate increases, with the result that more blood is delivered to the body, particularly to the lungs and muscles. In Experiment 28.4 you will investigate the effect of exercise on pulse rate, i.e. heartbeat rate.

Bleeding

The reason you bleed when you cut yourself is that blood pressure forces blood out of your arteries and veins. Since blood is under far greater pressure in arteries, a cut artery bleeds much more than a cut vein. Bright red blood spurts out of a cut artery in time with each heartbeat or pulse. Blood seeping from a vein flows steadily and is usually a deeper red, because it contains less oxygen than blood from an artery.

Blood clots when it is exposed to air; clotting stops blood escaping. You should study the treatment to stop bleeding in the current *First Aid Manual* (published by Dorling Kindersley for St. John Ambulance, St. Andrew's Ambulance Association and The British Red Cross Society). The purpose of the treatment is to stop or slow the flow of blood to give the blood time to clot.

Questions

Q 24.1 Copy the two tables and complete them to show the differences between (a) arteries, capillaries and veins and (b) atria and ventricles.

(a)

	Arteries	Capillaries	Veins
Thickness of wall			
Direction of blood flow			
Blood pressure			
Presence of valves			

(b)

	Atria	Ventricles
Thickness of wall		
Direction of blood flow		
Blood pressure		

Q 24.2 Figure 24.9 shows blood pressure in the left ventricle, left atrium and aorta.
(a) How long does one complete heartbeat take?
(b) For how long during one heartbeat is pressure greater in the left ventricle than in the left atrium?
(c) What happens to the muscles of the ventricle wall when the pressure in the ventricle is increasing?
(d) What happens to the semilunar valve at the entrance to the aorta when (i) pressure is greater in the ventricle than in the aorta and (ii) pressure is greater in the aorta than in the ventricle?
(e) For how long is (i) the semilunar valve open and (ii) the cuspid valve open? (Remember that valves are open when pressure is greater on one side of the valve and closed when pressure is greater on the other side of the valve.)

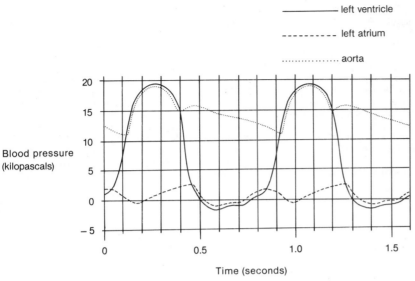

Fig. 24.9

PRACTICAL WORK

Experiment 24.1 To dissect a mammalian heart

Note for teachers
Lambs' hearts from most butchers and supermarkets can be used for dissection even if the major blood vessels and parts of the atria have been removed from them. But complete hearts, which may be obtainable from a local abattoir, are better. A student need not wear gloves, but they should be available because some students do not like to touch uncooked meat.

Materials required by each student
1 dropper pipette
1 dissection dish
1 pair of forceps
1 scalpel
1 pair of scissors
1 pair of disposable gloves
1 heart

Method
1. Lay the heart on the dissection dish with its ventral surface upwards. (An animal's ventral side is the one normally directed downwards or, if the animal stands on two legs like apes and humans, the one normally directed forwards.) If you are not sure which side of the heart is which, pinch both sides between your finger and thumb. The side which feels thicker is the left side and so should be on your right if you have laid it correctly on the tray.
2. Note the small chambers at the top of the heart. These are the atria. The blood vessels which emptied into the atria have probably been removed from the heart you have been given, but you should see smaller blood vessels lying on the surface of the heart muscle. Make a drawing of the heart.

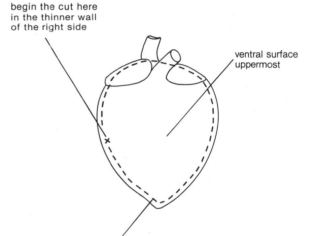

Fig. 24.10 Position of the cuts in the heart wall

3. As indicated in Figure 24.10, cut through the wall of the heart, beginning on the thinner right side, and continue to cut through the heart muscle all around its edge.
4. When you have finished the incision, the top part of the heart should lift away from the bottom part on both left and right. A broad sheet of muscle will still hold the two parts together in the middle. Cut through this sheet of muscle so that the upper part of the heart can be completely removed from the lower part.
5. Examine the lower part of the heart. Identify the atria and ventricles on each side and the septum separating the left and right sides.

6 Locate the position of the bicuspid valve (between the left atrium and left ventricle) and of the tricuspid valve (between the right atrium and right ventricle). Note their structures and the strength of the cords of tissue attaching them to the bottom of the ventricles. (Try to snap one with your fingers.)
7 Locate the positions where the aorta and pulmonary artery leave the left and right ventricle respectively and find the semilunar (pocket) valves at their base. Squirt some water from a dropper pipette against one of the valves in the direction that blood would leave the heart and note what happens to the valve. Squirt the water from the opposite direction and note what happens to the valve.
8 Wash your hands thoroughly and make a drawing of the dissected heart.

Interpretation of results
1 Label your drawing of the external features of the heart.
2 Which tissues are supplied by the small vessels on the outside of the heart? What would happen if one of them became blocked?
3 Label your drawing of the internal features of the heart. Place arrows on the drawing to show the direction in which blood normally flows.
4 Describe what happened to the semilunar valves when you squirted water over them. Use this information to explain how they normally operate.
5 What is the importance of the strength of the cords holding the bicuspid and tricuspid valves to the ventricle walls?

UNIT 25

Blood and its functions

Blood carries substances and heat around the body and protects the body. It is a mixture of a yellow fluid called **plasma**, red and white cells called **blood corpuscles**, and cell fragments called **platelets**. Figure 25.1 shows 100 cm³ of blood separated into its plasma and cells. Plasma is mainly water but contains a number of dissolved substances: inorganic ions, proteins, sugar (glucose), amino acids, fatty acids, gases, urea and antibodies, as well as extremely fine droplets of lipid (which are not dissolved but are 'suspended' in the plasma). These are the substances that plasma carries from one part of the body to another.

Red blood cells (or red corpuscles or **erythrocytes**) are red because their cytoplasm contains a dissolved red pigment called **haemoglobin**. The so-called white blood cells (or white corpuscles or **leucocytes**) are in fact colourless. Red cells outnumber white ones by six hundred to one. Platelets are fragments of colourless cytoplasm, less than half the diameter of red blood cells, without a nucleus. There are about 30 platelets to every one white blood cell. Platelets break up when blood is exposed to air and so are not seen unless special techniques are used.

Red blood cells transport oxygen. Haemoglobin has a special attraction for oxygen and readily combines with it to form **oxyhaemoglobin**. When oxygen is in low supply, especially in acid conditions, bright red oxyhaemoglobin gives up its oxygen and becomes dark red haemoglobin again. In the lungs oxygen is plentiful and diffuses into the blood to form oxyhaemoglobin. Elsewhere in the body, where respiration occurs with the result that oxygen is used and carbon dioxide is given off and forms a weak acid, oxygen is released from haemoglobin. Oxygen diffuses out of the blood to the cells along the concentration gradient.

Red blood cells have no nucleus and consist of cytoplasm containing large amounts of haemoglobin surrounded by a cell membrane. Their shape is unusual: they are squashed flat with each face deeply indented like a biconcave lens or liquorice Pontefract cake. Figure 25.2 shows three views of a red blood

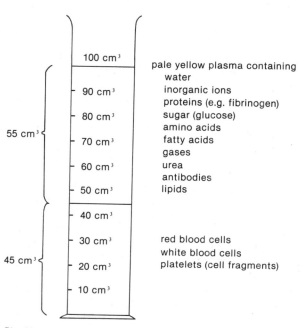

Fig. 25.1 Contents of human blood

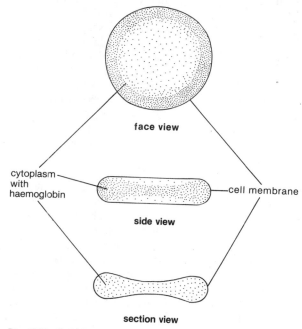

Fig. 25.2 Red blood cells

Fig. 25.3 Red blood cells: scanning electron micrograph

cell. Figure 25.3 is an electron micrograph of red blood cells in several different views. The odd shape of a red blood cell gives it a large surface area in relation to its volume and means that no part of the cytoplasm is far from the surface of the cell. A red blood cell is efficient at taking up and releasing oxygen because the diffusion surface is large and the diffusion distance small. The cells pack closely together like a pile of coins, leaving few plasma spaces between them. The surface area of all the red blood cells in the 100 cm^3 of blood in Figure 25.1 is 60 m^2, the area of a large room. The effect of shape on surface area is described in Unit 37.

White blood cells protect the body. The commonest white blood cells have lobed irregular nuclei and cytoplasm that is able to change its spherical shape (and therefore the shape of the cells). Their cytoplasm is able to surround and destroy harmful bacteria and viruses in the body. By changing shape they push between the cells of the blood capillaries and escape from the blood among the body cells. White blood cells concentrate where there is an infection: white pus at a wound contains large numbers of live and dead white blood cells that have escaped from blood and attacked infecting organisms. Figure 25.4 shows the shape of three different types of white blood cell. You might see any of these through a microscope on a blood slide.

The white blood cell on the left of Figure 25.4 is the common type with an irregular nucleus. The one in the middle also has cytoplasm that can change its shape but is rare. The one on the right with the large central nucleus is a **lymphocyte**: it is not able to capture (engulf) bacteria and viruses, but it can divide and make cells that circulate in the plasma and are able to form protective chemicals called **antibodies**. Different antibodies are able to destroy different bacteria and viruses. Other antibodies act against foreign chemicals, including poisons produced by bacteria, and destroy them (as described in Unit 26).

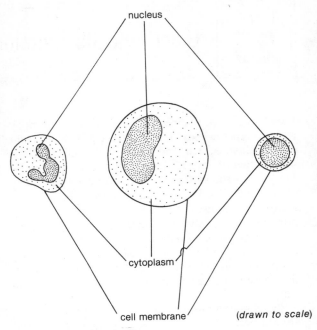

Fig. 25.4 White blood cells

Platelets also protect the body. They form blood clots at wounds and stop bleeding. When the body is cut and blood is exposed to air, the fragments of cytoplasm which form platelets break up and release enzymes. These enzymes set off a chain of reactions which result in the change of **fibrinogen**, a soluble blood protein, into fine strands of **fibrin**. Figure 25.3 shows red blood cells entangled in fibrin strands. Both fibrin strands and red blood cells are far too small to be seen without a microscope: only the electron microscope can show the detail in Figure 25.3.

Unfortunately, platelets may form blood clots not at wounds but inside intact blood vessels. A blood clot that blocks an intact blood vessel is a **thrombosis**. A coronary thrombosis, mentioned in Unit 24, is the formation of a blood clot in a coronary artery. Anti-clotting drugs can be given to people whose blood is clotting too readily, but there is then a risk that their blood will not clot when it needs to at a wound. After an operation, when blood has been exposed to air, there is a greater risk of a thrombosis. Patients are now encouraged to get up as soon as possible after an operation to keep their blood circulating freely and so reduce the risk of clots. Clots seem to form more readily on the inside of blood vessels that are roughened with fatty deposits containing cholesterol: this is the reason for medical advice not to eat much fat.

When blood clots form in vessels where there are alternative blood channels, damage to the surrounding tissues deprived of blood may be slight. But blood clots in the larger vessels of the coronary circulation can be fatal, as can blood clots that block the pulmonary artery leading out of the heart to the lungs. Coronary thrombosis and pulmonary embolism (blockage) are, after accidents, the main causes of sudden death. Clots in blood vessels of the brain are always serious because death of brain cells affects the senses, intelligence or body activities. Damage to parts of the brain caused by a reduced blood supply is called a **stroke**.

Functions of blood

Blood has two major functions: transport and protection.

All living cells in the body lie near a blood supply: by diffusion or active transport they obtain substances from the blood via the tissue fluid that surrounds living cells and dispose of waste products (notably carbon dioxide) via the tissue fluid to the blood. Urea, another waste product, is removed from the liver by the blood. The blood transports:

- oxygen, as oxyhaemoglobin in the red blood cells;
- carbon dioxide, as hydrogencarbonate ions dissolved in plasma and in the red blood cells;
- water, as the major part of both plasma and blood cells;
- food, as dissolved sugar (glucose), amino acids and fatty acids and as small lipid droplets in the plasma;
- inorganic ions, such as calcium and iron, in plasma and red blood cells;
- urea, dissolved in plasma;
- hormones, dissolved in plasma;
- proteins, including antibodies and fibrinogen, suspended in plasma;
- red blood cells, white blood cells and platelets, carried by plasma;
- heat, which warms the blood where the rate of respiration is high, as in the liver and muscles, and passes out of the blood wherever surrounding tissues are cooler.

Blood protects the body by:

- white blood cells which surround and destroy bacteria and viruses;
- white blood cells which produce cells that can make antibodies;
- platelets which cause blood-clotting and prevent excessive bleeding.

Questions

Q 25.1 Copy the table and complete it to summarise the differences between red blood cells, white blood cells and platelets.

	Red blood cells	White blood cells	Platelets
Structure			
Size (small, large or medium)			
Ratio			
Function			

Q 25.2 A few days before a race a long-distance athlete had some of his own blood removed and refrigerated. Shortly before the race he had his refrigerated blood transfused back into his bloodstream. What advantage did this give him?

PRACTICAL WORK

Experiment 25.1 To investigate the structure of blood in a human blood smear

Note for teachers

Before performing class experiments involving blood-letting, teachers must find out if restrictions are imposed by their own Local Education Authorities. Many Authorities have a written code of conduct for such practicals. Teachers are recommended to read the Association for Science Education Committee statement 'Human blood sampling: recommended procedures' in *Education in Science*, April 1979, pages 27-8. No student should be made to participate in bloodletting.

This experiment can be performed, without letting blood, by a microscope study of commercially prepared human blood smears. Alternatively, students can examine photomicrographs of human blood smears.

There is no agreement on which is the best place to puncture the body to obtain drops of blood. The finger tip contains many nerve endings and the risk of subsequent infection is greatest here. The back of the finger near the nail is less painful but produces less blood and there is a risk of damage to the nail bed. The ear lobe requires a second person to collect the drops of blood and there is a risk of damage to the eyes and face if a student jerks his or her head.

The teacher should supervise the issue, use and collection of lancets. After use they should be disposed of aseptically, preferably by autoclaving or burning. If any blood is spilled, the teacher should immediately clear it up wearing gloves and using a suitable disinfectant, e.g. sodium-chlorate(I) (sodium-hypochlorite) solution.

This experiment requires a compound microscope with an objective lens allowing at least ×40 magnification. The experiment can be performed adequately with one microscope for every two students.

Materials required by each student
1 dropper pipette
2 sterile glass slides
1 dish to catch washing fluids and stains
1 microscope
1 stopclock
1 piece of absorbent paper
distilled water
5 cm^3 either Leishman's or Wright's stain (bought ready-made)
1 container of 10% sodium-chlorate(I) (sodium-hypochlorite) solution (a disinfectant) in which the slides can be fully immersed
For blood-letting
1 closable container for used lancets and swabs
1 sterile lancet
2 balls of cotton wool to use as swabs
10 cm^3 ethanol

Materials required by the class
sterile lancets
disposable towels

Method
Do not use your own blood for this experiment if you do not like the idea. If you do use your own blood, you must follow the instructions carefully to minimise any risk of spreading disease.
1. Collect all the apparatus you will need so that once you have washed your hands you need not move away from the bench.
2. Wash your hands *thoroughly* with soap and water. Dry them with a disposable towel or let them dry in the air.
3. Use a swab soaked in ethanol to wipe the part of the skin you have chosen to prick and allow it to dry.
4. Remove the sterile lancet from its packet immediately before use. Do not touch the packet with the part you are going to prick and do not let the sharp end of the lancet touch anything.
5. Using one bold movement, prick the chosen part once with the lancet. Put the lancet in the container and close it. On no account use this lancet again.
6. Let two drops of blood fall on to one of the sterile glass slides. Do not touch the slide with the part you have pricked.
7. Use the second swab to wipe the region of the prick with ethanol. Press slightly as you do so, to stop any more blood coming out. Put the soiled swab in the container and close it.
8. If any blood has spilled on to the bench, inform your teacher immediately.

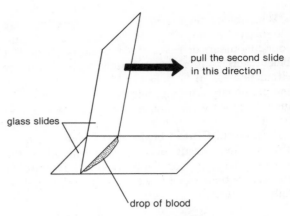

Fig. 25.5 Method of preparing a blood smear

9. Place the end of the second slide on the drop of blood at an angle of about 45° as in Figure 25.5. Quickly pull the second slide across the first so that you make an even film of the drop of blood. Put the second slide into the disinfectant.
10. Dry the blood smear by waving it in the air.
11. Pipette eight drops of the stain on to the dried smear and leave them for about 45 seconds.
12. Add eight drops of distilled water to the stain. Gently rock the slide to and fro to mix the stain and water. Leave the slide for five to ten minutes. While you are waiting, hand in the closed container to your teacher.
13. Wash the stain off the slide in a stream of distilled water. Gently blot the slide dry.
14. Place the slide on the stage of the microscope and examine it to find as many different types of blood cell as you can. Draw an example of each type of cell you find.
15. Put this slide in the container of disinfectant and hand it in to your teacher.
16. Before you leave the laboratory, wash your hands thoroughly with soap and water. Dry them with a disposable towel or let them dry in the air.

Interpretation of results
1. How many different types of cell did you find and draw? Explain the differences between them.
2. Use this unit to help you name the cells you found.
3. What is the function of each cell? Explain how any of the cells are adapted for their functions.
4. The method you followed involved a number of safety precautions. Briefly explain:
   (a) the reason any object contaminated with blood was put in either disinfectant or a sealed container;
   (b) the likely source of any infection to yourself;
   (c) the most likely site of infection.

UNIT 26

Antibodies; allergy; blood groups of the ABO system

Antibodies

We all have antibodies in our blood to protect us. Antibodies form in response to certain chemicals called **antigens**. Among the well-known antigens are those that cause illnesses, allergies and the rejection of transplants. Without antibodies we should have many illnesses and should die young.

Like enzymes, antibodies are globular proteins consisting of chains of amino acids rolled up in an organised way to form complicated molecules. An antibody has a shape that enables it to attach itself to a certain antigen and thereby make it inactive. As with a substrate and enzyme, the shape of the antibody's surface must fit the other chemical exactly or it will not work. Antibodies are produced by **plasma cells**, which form in blood plasma from lymphocytes, the white blood cells that have a large nucleus. In Figure 26.1 the white blood cell on the right is a lymphocyte.

A new-born baby is protected by antibodies that it receives from its mother both before birth and in her milk. After a few months a baby starts to produce its own antibodies.

Antibodies form in response to common illnesses. For example, the virus that causes chicken-pox stimulates the production of antibodies against itself. After a few days these antibodies inactivate the chicken-pox virus and the patient gets well. People who have had chicken-pox are unlikely to get it a second time because their bodies have 'learnt' how to make chicken-pox antibodies. Chicken-pox viruses entering their bodies will usually be destroyed before they can cause illness.

Vaccination

About two hundred years ago Edward Jenner took matter from a sore on the hand of a girl who had caught cowpox from cows she was milking and passed it into the arm of a boy: he thereby protected the boy from the more serious disease of smallpox. Jenner called the matter from the sore a **vaccine** and the process **vaccination** (from the Latin *vacca*, which means *cow*). He successfully vaccinated hundreds of people against smallpox.

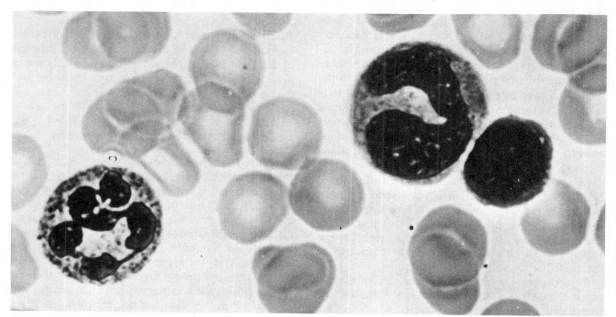

Fig. 26.1 Different forms of white blood cell

At first vaccination meant only the use of cowpox vaccine to protect against smallpox. As a tribute to Jenner its meaning was first extended to the use of any organisms to give protection from more dangerous ones. (By organisms is meant both bacteria and viruses: viruses are on the borderline between life and non-life and it is questionable whether they should be called organisms.) Today vaccination has come to mean the same as **immunisation**: giving immunity or protection against infection by any means. Jenner's method, of vaccinating with live organisms which are closely related to the disease and stimulate the same antibodies, is still used today. Sometimes people are vaccinated with live but weak disease-causing organisms: this is **inoculation**, a method used against smallpox before Jenner's discovery. But most vaccines are now of bacteria and viruses killed in such a way that, though they can no longer cause serious illness, their ability to stimulate antibodies is preserved. Vaccines formed from dead organisms are safer. Examples of diseases that have been almost eliminated in Britain by immunisation are smallpox, diphtheria, poliomyelitis and whooping cough. Smallpox has been almost eliminated throughout the world. Immunisation is not completely safe: the body sometimes reacts against the vaccine.

Most vaccines are injected but some are taken by mouth. Inside the body a vaccine will stimulate the formation of antibodies. If the body has met neither the vaccine nor the disease organism before, it cannot respond immediately: it takes time, sometimes a few days, sometimes a few weeks, for the white blood cells to produce the plasma cells that will form the antibody. Sometimes a second vaccination is necessary before plasma cells and antibodies are formed in sufficient numbers to prevent the disease. Provided there are enough plasma cells or antibodies present in the blood, infection by disease-causing viruses or bacteria is quickly suppressed. Edward Jenner's cowpox vaccine worked against smallpox because the viruses of the two diseases are similar: antibodies against cowpox will also attach to and inactivate smallpox viruses.

German measles is one of the mildest illnesses of childhood. Even in adults it is not a serious illness. But it is extremely serious if caught by a woman early in pregnancy because the Rubella virus that causes German measles can damage a growing embryo. For this reason all girls are now given the chance to be vaccinated while still at school.

In recent years we have discovered that substances other than viruses and bacteria can stimulate the formation of antibodies. Allergic reactions are thought to be the result of antibody formation.

Allergy

An allergy is an exaggerated or unusual reaction by living cells to a substance to which they have been exposed more than once (and to which they showed no exaggerated or unusual reaction on their first exposure). The commonest allergy is **hay fever**. The substance that results in the reactions of sneezing and watering and itching of the eyes is plant **pollen**. Pollen, which is fine dust released from flowers, enters the body through the throat and air passages. It is treated as a foreign substance by the body, which produces both antibodies and a substance called **histamine**. Unlike antibodies, histamine is not shaped to fit the foreign substance and inactivate it. It is histamine that produces sneezing and watering and itching of the eyes. These reactions are much reduced by drugs called **antihistamines**. Since histamine is produced in most allergic reactions, not only in hay fever, antihistamines are widely used in their treatment.

Transplants

The rejection of transplants – for example, of skin, a kidney or a heart – is also known to be the result of antibody formation. The only transplants exempt from this so-called **immune response** are those between identical twins, triplets, etc., and between one part of the body and another (skin and bone are both transplanted in this way). Drugs can suppress antibody formation, but they leave the patient at risk from infections that their antibodies would normally overcome.

Antibody formation must be avoided in blood transfusions. It was only in this century that the existence of different blood groups was discovered. Someone given blood of a different group from his or her own may produce antibodies to destroy it.

Blood groups of the ABO system

Everyone belongs to one of four groups: A; B; AB; O. People whose blood is A have a substance on their red blood cells called antigen A which reacts with antibody A in the plasma of non-group-A people. The red blood cells of group-A people get sticky and clump together (**agglutinate**) if they come in contact with antibody A. Naturally, group-A people do not have antibody A in their plasma or their blood would clump. Group-B people have a substance on their red blood cells called antigen B which reacts with antibody B in the plasma of non-group-B people and clumps the red cells together. Group-AB people have both antigen A and antigen B on their red blood cells: their red cells clump with either antibody A or antibody B. Group-O people have neither antigen A nor antigen B on their red blood cells: their blood does not clump with either antibody A or antibody B. The table in Figure 26.2 shows how the four blood groups react with antibody A and with antibody B.

	Antibody A	Antibody B
Group A	clumping	no clumping
Group B	no clumping	clumping
Group AB	clumping	clumping
Group O	no clumping	no clumping

Fig. 26.2 Blood groups with antibody A and antibody B

A curious feature of antibodies A and B is that we do not know why they are present in the plasma of most people. Young babies do not have these antibodies, yet by the time they are a year old group-A children have antibody B in their plasma, group-B children have antibody A in their plasma, group-O children have both antibodies A and B in their plasma (while group-AB children have neither antibody in their plasma). This is curious because in other examples of antibody formation something (the antigen) has to stimulate the production of the corresponding antibody.

Blood transfusions are safe only in certain combinations. A transfusion is dangerous if the person receiving blood, called the **recipient**, has antibodies that will clump the transfused red blood cells. For example, group-A people have B antibodies that clump transfused group-B or group-AB blood. Because group-O people have both A and B antibodies in their plasma, they must not be given group-A, group-B or group-AB blood: they must be given blood of their own group O. A person giving blood is called a **donor**. Figure 26.3 shows which transfusions are safe and which are not.

Group-AB people can receive blood from anyone: they are called **universal recipients**. Group-O people can give blood to anyone: they are called **universal donors**. Be sure you can reproduce Figure 26.3 for yourself.

	Donor blood group			
Recipient blood group	A	B	AB	O
A	✓	✗	✗	✓
B	✗	✓	✗	✓
AB	✓	✓	✓	✓
O	✗	✗	✗	✓

dangerous ✗
safe ✓

Fig. 26.3 Safe and dangerous blood transfusions

If blood of the wrong group is given, its red blood cells clump together, block the veins, cause pain at the region of injection and put the recipient in a state of shock. Large amounts of incompatible blood disorganise the circulation and cause death.

It does not seem to matter if the transfused blood has antibodies in the plasma which can clump the red blood cells of the recipient. The antibodies in the transfused blood plasma are diluted when they mix with the blood of the recipient.

Clumping of the red blood cells is so obvious that it can be seen with the naked eye. This enables you to determine your own blood group with a very simple test. If you add a drop of your blood to serum that contains antibody A and another drop to serum that contains antibody B, you can soon see whether or not clumping occurs. **Serum** is plasma from which the clotting agents but not the antibodies have been removed. If your blood clumps only with antibody-A serum, you are group A. If it clumps only with antibody-B serum, you are group B. If it clumps with both antibody-A and antibody-B serum, you are group AB. If it clumps with neither antibody-A nor antibody-B serum, you are group O.

Table 26.1 sums up what you need to know about the ABO system and blood transfusions.

Table 26.1 Blood groups, antibodies and transfusion possibilities

Blood group	Antibodies in plasma	Blood safe to receive in transfusion	Blood dangerous to receive in transfusion
A	B	A and O	B and AB
B	A	B and O	A and AB
AB	—	A, B, AB and O	—
O	A and B	O	A, B and AB

We do not know whether it is better to belong to one group rather than another. The proportions of different blood groups vary between countries and between races. Group O is rarer in Asia than in Europe. Group A is frequent among North American Indians yet rare among South American Indians. Group B is almost non-existent in the Polynesian Islands while it is frequent in South East Asia where the Polynesians may have come from. In the United Kingdom the distribution is approximately 42 % group A, 9 % group B, 3 % group AB and 46 % group O.

What we do know is that blood groups are inherited. If your blood group is A, at least one of your parents has red blood cells carrying the A antigen. Blood-group inheritance obeys rules described in Unit 46.

There are other systems of blood grouping besides the ABO. In practice, blood transfusions have to take account of some of these as well as of the ABO system.

Questions

Q 26.1 A man's blood group is B.
(a) What has he on his red blood cells?
(b) What antibodies are present in his plasma?
(c) To whom can he safely give blood?
(d) From whom can he safely receive blood?
Explain your answers to (c) and (d).

Q 26.2 A person wanting vaccination against tuberculosis is first given a surface skin scratch with tuberculosis antigen. If no pus blister (pustule) develops on the skin, the vaccination with weakened bacteria will be given. If a pus blister does develop, the vaccination will not be given.
(a) Suggest why a pus blister may develop.
(b) Suggest why someone whose skin develops a pus blister is not given the vaccination.

PRACTICAL WORK

Experiment 26.1 To determine a human ABO blood group

Note for teachers
Before performing class experiments involving blood-letting, teachers must find out if restrictions are imposed by their own Local Education Authorities. Many Authorities have a written code of conduct for such practicals. Teachers are recommended to read the Association for Science Education Committee statement 'Human blood sampling: recommended procedures' in *Education in Science*, April 1979, pages 27-8. No student should be made to participate in blood-letting.

This experiment can be performed without letting blood. It can be performed with saliva blood-grouping kits, although only about 80% of students secrete blood-group antigens in their saliva, or with pretyped blood cells. At the time of writing both these products are available from Philip Harris.

Though antisera can be purchased, the use of blood-letting cards reduces the possibility of cross-contamination of sera by students and allows each student a permanent record of the results.

There is no agreement on which is the best place to puncture the body to obtain drops of blood. The finger tip contains many nerve endings and the risk of subsequent infection is greatest here. The back of the finger near the nail is less painful but produces less blood and there is a risk of damage to the nail bed. The ear lobe requires a second person to collect the drops of blood and there is a risk of damage to the eyes and face if a student jerks his or her head.

The teacher should supervise the issue, use and collection of lancets. After use they should be disposed of aseptically, preferably by autoclaving or burning. If any blood is spilled, the teacher should immediately clear it up wearing gloves and using a suitable disinfectant, e.g. sodium-chlorate(I) (sodium-hypochlorite) solution.

Materials required by each student
1 dropper pipette
1 blood-grouping card and blood-serum mixing-stick
5 cm^3 distilled water
For blood-letting
1 closable container for used lancets, swabs and serum mixing-stick
1 sterile lancet
2 balls of cotton wool to use as swabs
10 cm^3 ethanol

Materials required by the class
sterile lancets
disposable towels

Method
Do not use your own blood for this experiment if you do not like the idea. If you do use your own blood, you must follow the instructions carefully to minimise any risk of spreading disease.
1 Collect all the apparatus you will need so that once you have washed your hands you need not move away from the bench.
2 Remove the blood-grouping card from its wrapping. Put one drop of distilled water from a dropper pipette on each of the squares on the card which contain the dried antisera. Mix the water and the antiserum in each square together. Be careful not to mix the contents of the different squares together.
3 Leave the prepared blood-grouping card on your bench and wash your hands *thoroughly* with soap and water. Dry them with a disposable towel or let them dry in the air.

4   Use a swab soaked in ethanol to wipe the part of the skin you have chosen to prick and allow it to dry.
5   Remove the sterile lancet from its packet immediately before use. Do not touch the packet with the part you are going to prick and do not let the sharp end of the lancet touch anything.
6   Using one bold movement, prick the chosen part once with the lancet. Put the lancet in the container and close it. On no account use this lancet again.
7   Let one drop of blood fall on to the flat end of the serum mixing-stick and mix this drop with the diluted serum in one of the squares of the grouping card. Do not let the blood mixture run out of that square.
8   Clean the end of the mixing-stick on the swab which you used earlier.
9   Repeat instructions 7 and 8 until you have mixed drops of blood into each square on the card. Put the soiled mixing-stick in the container and close it.
10 Use the second swab to wipe the region of the prick with ethanol. Press slightly as you do so to stop any more blood coming out. Put the soiled swab in the container and close it.
11 If any blood has spilled on the bench or any other apparatus, inform your teacher at once.
12 Gently rock the blood-grouping card to and fro to mix the blood and antiserum within each square for about one minute.
13 Examine the card and compare the contents of each square with those of the control square.
14 Before you leave the laboratory, wash your hands thoroughly with soap and water. Dry them with a disposable towel or let them dry in the air.

Interpretation of results
1   Describe the appearance of the control square.
2   Which squares had the same appearance as the control square?
3   Which squares looked different from the control square? Describe the difference.
4   Explain how the antibody present in the card caused the appearance of your blood which you described in 3.
5   To which blood group do you belong?
6   Would you trust your blood-grouping as a medical diagnosis? Explain your answer.

UNIT 27: Gas exchange in flowering plants and humans

Respiration is the conversion of the energy in food, usually sugar, into a form the cell can make use of. It is described in Unit 5. Most plants and animals, including flowering plants and humans, respire aerobically: their cells use oxygen to release energy by breaking down compounds into carbon dioxide and water.

sugar + oxygen → carbon dioxide + water + energy

Not only must their cells get oxygen from their environment but they must dispose of the waste products of aerobic respiration, carbon dioxide and water, to their environment. Otherwise water would upset the osmotic balance of the cell and carbon dioxide would dissolve in water to form a weak acid. Taking in oxygen from the environment and giving off carbon dioxide to the environment are an example of **gas exchange**. The removal of water is described in Unit 29.

Gas exchange involves diffusion, which is described in Unit 15. Oxygen diffuses into simple plants and animals from the air or water which surrounds them and carbon dioxide diffuses out into the air or water. A single cell satisfies all the conditions for diffusion outlined in Unit 15: its surface is large enough; its radius, the distance a diffusing substance must travel, is short enough; its surface is always moist and its cytoplasm is mainly water, which allows the diffusing substance to move in solution. In larger organisms, in which most cells are not in contact with their environment, gases are diffused from cell to cell. Large animals have a **circulation system**, i.e. the blood system, to transport dissolved gases, as well as a **ventilation system** to take in oxygen and remove carbon dioxide. Ventilation is part of gas exchange.

Gas exchange in flowering plants

In flowering plants gas exchange must provide not only for respiration but also for photosynthesis, which is part of the feeding process and is described in Unit 19. In photosynthesis those of the cells which contain chlorophyll use the energy from sunlight to make sugar from carbon dioxide and water, giving off oxygen as a waste product.

carbon dioxide + water + light energy $\xrightarrow{\text{chlorophyll}}$ sugar + oxygen

Photosynthesis occurs only when it is light, and only then in cells with chlorophyll. Respiration occurs all the time in all living cells. During darkness gas exchange is for respiration and involves taking in oxygen and giving off carbon dioxide. When there is only a little light and plants are photosynthesising only slightly, gas exchange still involves taking in oxygen and giving off carbon dioxide. This is because the small amount of carbon dioxide needed for photosynthesis is given off by respiration and the small amount of oxygen given off by photosynthesis is used up in respiration, while more oxygen is needed for respiration than is provided by photosynthesis and more carbon dioxide is given off in respiration than is used in photosynthesis.

Twice a day between full daylight and darkness there is a short time when there is no gas exchange between green plant cells and their surroundings: all the oxygen needed for respiration is produced by photosynthesis and all the carbon dioxide needed for photosynthesis is produced by respiration.

When the light is good and the plant is photosynthesising strongly, the carbon dioxide needed for photosynthesis is more than that given off by respiration and the oxygen given off by photosynthesis is more than that used in respiration. By day, therefore, gas exchange in flowering plants involves taking in carbon dioxide and giving off oxygen. Photosynthesis is the source of the oxygen in the air which most organisms, including humans, use in respiration. Without photosynthesis we should die. Figure 27.1 shows the gas exchanges of a leaf in darkness and in various light intensities.

Most living cells of flowering plants are in contact with air, whereas most living cells of animals are not. Even flowering plants living in water contain large air spaces. Diffusion through the air spaces between the living cells of the leaf, the stem and the root is sufficient to meet the plant cell's needs for gas exchange. Where there are pores, gases can diffuse in from the air and out into the air. Even where there are no pores, or where the pores are closed, enough gas exchange by diffusion can take place through living

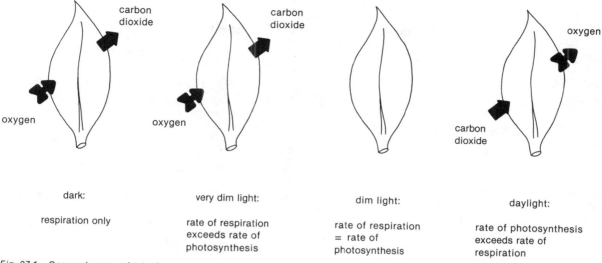

Fig. 27.1 Gas exchanges of a leaf

cells at the surface and from cell to cell or through the air spaces inside the plant.

Unit 19 describes gas exchange in photosynthesis. Gas exchange in aerobic respiration works on the same principle, but *oxygen* is taken in and *carbon dioxide* given out owing to the changed diffusion gradients.

Gas exchange in humans 1

Human cells have no air spaces between them. Moreover, because they are more active than plant cells, they need more oxygen and produce more carbon dioxide, which has to be got rid of. Gas exchange in humans is therefore more complicated than in plants. But the process of diffusion from cell to cell is similar: oxygen diffuses into the cells along a concentration gradient; carbon dioxide diffuses out of the cells along a concentration gradient; the surface of each cell is large enough for diffusion; the distance across each cell is short enough for diffusion; the surfaces of the cells are moist so that oxygen and carbon dioxide can pass in and out in solution. In the human body oxygen is brought from the air to the cells and carbon dioxide is taken from the cells to the air by a combination of the ventilation system and the circulation system.

Gas exchange takes place not at the surface of the cells (as in plants) and not at the surface of the animal (as in small animals such as an earthworm) but at the surface of special organs in the chest, the **lungs**. The ventilation system brings oxygen to the lungs and takes carbon dioxide from them.

Ventilation system

Details of the ventilation system are shown in Figure 27.2. You can see that there is a continuous passage from the air through either the mouth cavity or the nose cavity, down the throat, through the voice-box (**larynx**), down the windpipe (**trachea**), through either the right or left **bronchus** to either the right or left lung. Each bronchus divides immediately it enters the lung into narrower branches called **bronchioles**, which divide further, again and again, until they form the finest microscopic branches of the bronchioles ending in tiny clusters of bubble-like **air sacs** (called **alveoli**, singular **alveolus**). Branching bronchioles are shown in the right lung but the end branches of the bronchioles are microscopic and too small to be shown in Figure 27.2. The end branches are shown, together with the air sacs, in Figure 27.3.

The windpipe, the bronchi and the larger branches of the bronchioles are kept open by rings of gristle-like tissue called **cartilage**. Smaller branches of the bronchioles are kept open by connective tissue. The walls of all the passages in the ventilation system contain muscles. The inner surface of the entire system from mouth and nose to air sacs is kept moist by cells that secrete a watery **mucus**. Fine hairs line all the passages (but not the air sacs): their regular beating wafts the mucus upwards to the throat, where it is swallowed. This keeps the air passages clean and free of dust and bacteria.

Air sacs

The wall of a single air sac is extremely thin, just one layer of flat cells. On the inside it is lined with mucus.

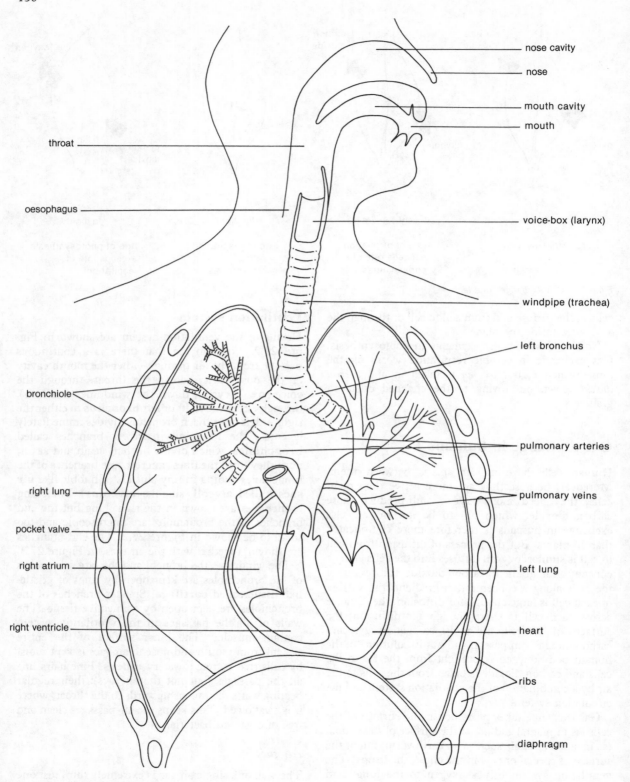

Fig. 27.2 Ventilation system

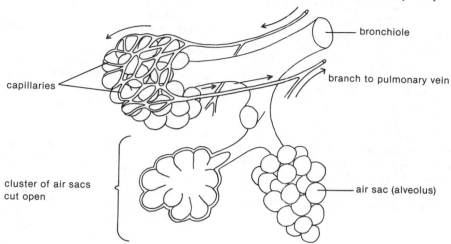

Fig. 27.3 Bronchiole ending

On the outside it is covered with a network of blood capillaries. Capillaries are shown covering a cluster of air sacs in Figure 27.3, while a section of part of a single air sac is shown in Figure 27.4. Since the other name for air sacs is alveoli, the air inside air sacs is sometimes called **alveolar air**.

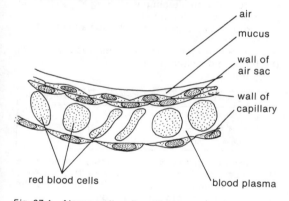

Fig. 27.4 Air-sac wall and capillary

Pulmonary arteries pass to each lung from the heart, branching again and again, getting smaller and smaller, until they form a network of capillaries over the outer surfaces of the clusters of alveoli.

Gas diffusion

Red blood cells move through the capillaries, often in single file, close enough to air in the air sacs for dissolved oxygen to diffuse through the mucus, through the single layer of cells around the air sac, through the single layer of cells forming the wall of the capillary, through the plasma of the blood and through the membrane of the red blood cell, to form the compound oxyhaemoglobin with the red pigment haemoglobin. Dissolved carbon dioxide diffuses in the opposite direction from the red blood cells and plasma via the capillary wall, the air-sac wall and the mucus into the air in the air sac itself.

Capillaries join again and again with one another to form fine branches of veins; vein branches join to form the large pulmonary veins that return blood to the left atrium of the heart. Branching blood vessels passing between the heart and the left lung are shown in Figure 27.2.

All the essentials for gas exchange by diffusion are present: the surface area of the air sacs in the lungs is large (in an adult it is about that of a tennis court); the diffusion distance from air in the air sac to blood is short; the diffusion surface is moist with mucus, allowing the diffusing substance to move in solution; diffusion gradients are present.

Diffusion gradients exist because oxygen is plentiful in the air sac (one fifth of air is oxygen) but scarce in the red blood cells, which have just been around the body giving up their oxygen to the cells. Carbon dioxide is plentiful in the blood, which has just been around the body collecting it as a waste product from the cells, but is scarce in the air sac (only three-ten-thousandths, i.e. 0.03 %, of air is carbon dioxide).

Circulation system

The circulation system is described in Unit 24. Look at Figure 24.5 (page 113), which is a circulation diagram. The capillaries of the lungs are shown. You can follow the path that red blood cells containing oxygen must follow to get to other parts of the body: they must go through the left-hand side of the heart and out to the body via the aorta and other arteries before branching again and again into capillaries. From the capillaries oxygen diffuses from haemoglobin in the red blood cells to the body cells, always dissolved in the fluid that surrounds the cells or in the cytoplasm of the cells, along the concentration gradient. Carbon dioxide, also in solution, diffuses from the body cells to the capillaries along the concentration gradient to be carried away as hydrogencarbonate ions. Carbon dioxide must go through the right-hand side of the heart before passing via the pulmonary arteries to the capillaries surrounding the air sacs.

Questions

Q 27.1 (a) Describe the journey of a molecule of oxygen from the air to a red blood cell.
(b) Describe the journey of a molecule of carbon dioxide from a cell in the liver to the air in an air sac.

PRACTICAL WORK

Experiment 27.1 To investigate the relation between the production of carbon dioxide and its absorption by plants

Note for teachers
Make the hydrogencarbonate indicator solution by dissolving 0.2 g of thymol-blue powder and 0.1 g of cresol-red powder in 20 cm^3 of ethanol. Add this to 0.84 g of 'Analar' sodium hydrogencarbonate dissolved in 900 cm^3 of distilled water in a 1 dm^3 volumetric flask and make up the final volume to 1 dm^3 with distilled water. On the day of its use, remove 2 cm^3 of this solution for each student. Dilute the volume removed with nine times its own volume of distilled water. Using an aquarium or filter pump, aerate this diluted solution for about fifteen minutes to get it into equilibrium with atmospheric air. It should then be red. The leaves in this experiment should be from healthy mesophytic plants.

Materials required by each student
1 2 cm^3 pipette fitted with a pipette filler
1 test-tube rack with five test-tubes
1 pair of forceps
1 plain white tile
1 bench lamp
1 spirit marker
3 rubber bungs to fit the test-tubes
1 pair of safety spectacles
enough aluminium cooking foil to cover one test-tube completely
5 cm^3 dilute hydrochloric acid
20 cm^3 hydrogencarbonate indicator solution
5 cm^3 dilute sodium-hydroxide solution
2 leaves

Materials required by the class
distilled water

Method
1. Wash the five test-tubes in tap water and then in distilled water. Rinse them with a little of the hydrogencarbonate indicator solution. Label them A to E.
2. Pipette 2 cm^3 of hydrogencarbonate indicator solution into tubes A and B.
3. Wearing safety spectacles, add 2 cm^3 of dilute hydrochloric acid to tube A. Record the colour of the indicator solution.
4. Wash the pipette and, wearing safety spectacles, use it to add 2 cm^3 of dilute sodium-hydroxide solution to tube B. Record the colour of the indicator solution.
5. Wash the pipette and use it to put 2 cm^3 of indicator solution into tubes C to E.
6. Roll the leaves lengthwise so that the lower surface is on the outside of the roll. Put one leaf in each of tubes C and D so that it rests on the wall of the tube but does not touch the indicator solution.
7. Cover tube C with aluminium cooking foil so that no light can get through its sides.
8. Put a rubber bung firmly in each tube and stand the tubes in the rack.
9. Put the bench lamp a few centimetres away from the tubes and switch it on.
10. Copy the table.

Tube	Contents	Colour of indicator	Inferred pH
C	Leaf, no light		
D	Leaf in light		
E	No leaf, light		

11. Near the end of the practical session remove the aluminium foil from tube C and hold each tube against the white tile so that you can record the colour of the indicator in your table.

Interpretation of results
1. What colour did the indicator solution turn in (a) tube A with dilute acid and (b) tube B with dilute alkali?
2. Use this information to complete your table to show your conclusions about the pH in tubes C to E.
3. Carbon dioxide is an acidic gas. Assuming that the changes in pH were associated with the presence of this gas, explain your results in tubes C and D.
4. What was the purpose of tube E?

Experiment 27.2 To investigate the structure of the respiratory system of a mammal

Note for teachers
The best source for sets of lungs is an abattoir. If there is no abattoir in your area, you can probably get them from a butcher. A student need not wear gloves, but they should be available because some students do not like to touch uncooked meat.

Materials required by each group of four students
1 dissection dish
1 scalpel
1 hand-lens
1 bowl of water
diluted laboratory disinfectant
1 set of lungs with trachea attached

Materials required by the class
disposable gloves

Method
1. Arrange the lungs and trachea in the dish.
2. Examine the trachea. Note the rings of cartilage along its length and the larynx (voice-box) at its open end.
3. Separate the right and left lungs and note the trachea splitting into the bronchi which enter the lungs.
4. Examine the external appearance of the lungs and note that they consist of a number of lobes. Count these lobes.
5. Make a large labelled drawing of all that you can see of the respiratory system.
6. Cut into one of the lungs at the point where the bronchus enters it. Continue to cut away the lung tissue to expose the bronchus and the smaller bronchioles which emerge from it.
7. Make a large labelled drawing to show the distribution of bronchioles in this side of the lung.
8. Cut off a small piece of lung tissue and squeeze it under water in the bowl. Record what happens when you squeeze it.
9. Examine a small piece of lung tissue with the hand-lens. Describe its appearance.

Interpretation of results
1. What is the function of the cartilage rings in the trachea?
2. Explain why the rings of cartilage do not go all the way round the trachea.
3. Describe the way in which the bronchus you examined branched inside the lung. What is the advantage of this branching?
4. Explain the reaction you observed when you squeezed the piece of lung under water. Use your observations with the hand-lens in your explanation.
5. Briefly explain the significance of the colour of the lungs.

Unit 28: Gas exchange in humans; exercise, smoking and pollution

Gas exchange in humans 2

The lungs of a human adult hold about 6 dm³ of air. When we breathe normally and are at rest, nearly 0.5 dm³ of air passes both in and out of our lungs about fifteen times a minute, a total volume of 7.5 dm³ per minute (15 × 0.5 dm³). As soon as we need more oxygen, or have to get rid of more carbon dioxide, we breathe more deeply and more quickly.

The greatest volume of air which can be breathed out in one breath is the **vital capacity** of the lungs. In an adult this is about 4.5 dm³. This leaves 1.5 dm³ (6 dm³ − 4.5 dm³) of alveolar air (i.e. air in the air sacs) in the lungs which cannot be breathed out. Figure 28.1 shows the total volume of air the lungs can hold, the volume of air that is normally breathed in or out at rest, the greatest volume of air breathed in, the greatest volume of air breathed out (the vital capacity) and the volume of air in the air sacs which we are unable to breathe out. In addition there is always a small amount of air left in the windpipe and throat, but this is not part of the lung volume.

'Ventilation' is a long word for breathing, but it does summarise the function of breathing, which is to take fresh air containing plenty of oxygen into the lungs and to remove stale air containing too much carbon dioxide. Experiments 28.1 and 28.2 will give you some idea of the differences between atmospheric air that you breathe in, called **inhaled** or **inspired** air, and the air you breathe out, called **exhaled** or **expired** air.

Exhaled air has less oxygen and more carbon dioxide than atmospheric (inhaled) air. It is also moister than atmospheric air: the water vapour in exhaled air forms fine droplets if you breathe on a cold shiny surface such as a mirror or a window. Exhaled air is usually also warmer and, if you have a respiratory illness, such as a cold, it will contain viruses.

Because some air in the air sacs cannot be expelled by breathing, fresh air cannot directly reach the mucus in the air sacs. Both oxygen and carbon dioxide diffuse through the air in the air sac because there is a concentration gradient (see Figure 28.2).

Ventilation mechanism

We breathe in and out (ventilate) by moving the chest wall and the diaphragm (the muscles that separate the abdomen from the chest). To breathe in, we increase the volume of the chest cavity, with the outer lung surface held by moisture against the inside of the

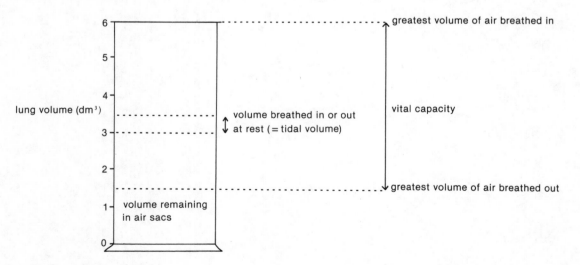

Fig. 28.1 Lung volumes

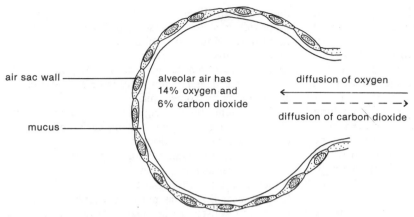

Fig. 28.2 Diffusion through air sac

chest wall. The pressure of air in the atmosphere forces air into the lungs as the space inside them increases. To breathe out, we decrease the volume of the chest cavity, decrease the cavity in the lungs and force air out.

People may stop breathing without being dead – after accidents or heart attacks, for example. It may be possible to restart someone's breathing, even when it has stopped for several minutes, by **mouth-to-mouth ventilation**. If you had stopped breathing and someone else breathed directly into your mouth, air would be forced into your lungs. The exhaled air the other person breathed into your mouth would have only 16 % oxygen, but that would be better than nothing and, together with the mechanical stimulation of your lungs, could start you breathing for yourself. The technique of mouth-to-mouth ventilation is described in the *First Aid Manual* (Dorling Kindersley for St. John Ambulance, St. Andrew's Ambulance Association and The British Red Cross Society). If you learn it, you may one day be able to save someone in need, but it should be shown to you by a qualified person.

Exercise

Although you breathe shallowly and slowly when you are at rest, you breathe deeply and quickly after you have been running. An athlete running 100 metres may take only one breath during the race but will breathe deeply and quickly after it. During a 100-metre race run in about ten seconds there is not enough time for extra oxygen from the air to get down into the lungs, to diffuse through the air sacs into the blood, to pass through the heart and to reach the muscles of the legs which need extra oxygen. The cells of the muscles of the leg therefore use up all their stored oxygen and begin anaerobic respiration. In humans this leads to a build-up of lactic acid which may cause cramp before it is taken away from the cells by the blood. Deep and rapid breathing after a 100-metre race provides oxygen which allows aerobic respiration to remove some of this lactic acid and convert the rest of it back into stored food (glycogen) in the muscles. The build-up of lactic acid creates an **oxygen debt**: breathing is deeper and more rapid until the debt is repaid. More oxygen is needed, not for more muscle contraction, but for the removal of the lactic acid. To a long-distance runner oxygen debt is an even greater problem. Training increases the amount of oxygen that can be taken in and transported to the muscles and hence increases energy production. Judging pace is essentially knowing how fast it is possible to run without creating an oxygen debt that will cause cramp.

Smoking

In breathing we unavoidably take in harmful substances. We breathe in bacteria and viruses, which the white blood cells in our bodies usually make harmless. We breathe in smoke and soot in towns where there are fires. We breathe in petrol fumes where there are cars: the harmful effects of lead in petrol fumes were recognised only in recent years. Asbestos dust is another harmful substance whose danger even in small amounts was recognised only in recent years.

The mucous lining and the wafting hairs in the air passages of the ventilation system constantly remove small particles that we swallow without even noticing them. Larger particles are dangerous because they may completely block the ventilation passages. Many of the dangers we cannot avoid, but some we can.

Cigarette smoke contains hundreds of different chemicals. Even the poison arsenic was once contained in cigarette smoke when arsenic was used to spray tobacco plants against disease. Cigarette manufacturers have managed to remove some of the harmful chemicals they identified, and filter-tipped cigarettes are intended to remove more of them. But cigarette smoke is still an irritant to the inner lung surfaces: it increases the amount of mucus they produce and it slows down or stops the wafting of the hairs. Many smokers therefore have a cough which brings up this excess mucus (phlegm) from the lungs: the severity of a smoker's cough is related to the number of cigarettes smoked.

Coughing and bringing up phlegm are often the first stage of **bronchitis**; at a later stage there is infection by viruses and bacteria, together with fever; eventually the air passages are narrowed when scar tissue forms within them due to constant coughing, to irritation causing the build-up of mucus and to infection by viruses and bacteria. Of course, not all bronchitis is caused by smoking. Though it is much commoner in smokers, non-smokers get it too. Pollutants such as soot and particularly sulphur dioxide also cause bronchitis, which is known in many countries as the 'English disease' because it is more common in England than in other countries.

Chronic, which means long-lasting, bronchitis damages the thin dividing walls between individual air sacs with the result that the area for gas exchange is greatly reduced. A smaller area for gas exchange means that less oxygen can be taken into the blood. Together with the narrowing of the air passages, this causes shortness of breath and difficulty in breathing at the least exertion, the symptoms of chronic bronchitis.

Nicotine, a compound in cigarette smoke, has a number of complex effects on the body: it acts on nerve cells, particularly those in the brain, to increase their activity or, under some conditions, to reduce their activity; it releases hormones, particularly adrenalin, into the blood. Through the release of hormones it increases the heart rate, increases blood pressure and constricts blood vessels: after years of smoking this brings about blood-circulation problems, particularly in the legs. There is also evidence that nicotine increases the deposit of fatty substances inside the arteries and increases the likelihood of blood-clotting inside intact blood vessels (thrombosis).

Carbon monoxide in cigarette smoke combines with haemoglobin in the red blood cells and prevents some of it from combining with oxygen to form oxy-haemoglobin. When a person stops smoking, carbon monoxide is slowly released from the blood, freeing more haemoglobin to combine with oxygen. Though carbon monoxide thus reduces the amount of oxygen that the blood can carry, it does so by only a small amount that does not normally have a drastic effect. But smoking by women during pregnancy significantly lowers the birth weight of their babies: even slightly reduced oxygen in a mother's body significantly reduces respiration in the unborn baby and hence the energy available for its growth.

Cancer-producing substances are present in cigarette smoke and it is widely believed that smoking causes lung cancer. But it has not been shown *how* cancer-producing substances in cigarette smoke cause lung cancer. We can feel confident that smoking causes bronchitis, diseases of the heart and of the blood vessels and reduction in the amount of oxygen carried by the blood, because we can explain how it causes them. That smoking causes lung cancer is uncertain. All we know for sure is that smokers are more likely to get lung cancer than non-smokers. This may be because smoking causes lung cancer, because lung cancer causes smoking, or because something else causes both smoking and lung cancer. People suffering from stress may be more likely than others both to smoke heavily and to get lung cancer.

Pollution

Sulphur dioxide is a gas produced when sulphur is burnt. Sulphur is present in most fuels and is formed at industrial plants, particularly power stations, steelworks, smelters and brickworks. Tall factory chimneys release smoke containing sulphur dioxide high into the air where it may be carried great distances before it is dissolved and falls to earth as **acid rain**. The graph in Figure 28.3 shows the relation between the sulphur-dioxide content of air during a dense London fog and the increased number of deaths. Where there is more sulphur dioxide, more people die – particularly people with breathing problems such as bronchitis.

Carbon-monoxide poisoning is not a serious hazard of smoking. But, when petrol or coal is incompletely burnt, carbon monoxide collects in the atmosphere in sufficient quantities to reduce people's oxygen-carrying haemoglobin to the point where they can become unconscious and even die: running the engine of a car in a closed garage is highly dangerous. Carbon monoxide, like sulphur dioxide, is a harmful air pollutant.

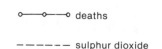
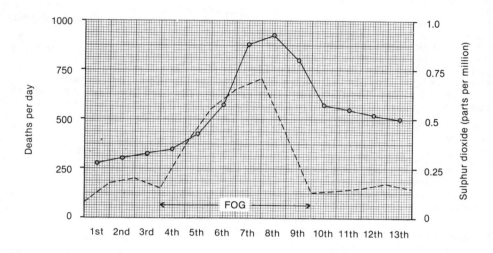

Fig. 28.3 Sulphur dioxide and deaths during a London fog (Source: Committee of the Royal College of Physicians, 1970)

Questions

Q 28.1 The table shows the composition of inspired, alveolar and expired air.

Contents	Percentage composition of dry air		
	Inspired	Alveolar	Expired
Nitrogen	79	80	80
Oxygen	21	14	16
Carbon dioxide	0.03	6	4
Water vapour	variable	saturated	saturated

(a) Explain the difference in the proportions of carbon dioxide in (i) inspired and expired air and (ii) alveolar and expired air.
(b) Explain the difference in the proportions of oxygen in (i) inspired and expired air and (ii) alveolar and expired air.
(c) Explain the difference in the amounts of water vapour in inspired and expired air.

Q 28.2 Look at Figure 28.3 and answer the following questions.
(a) On which day was (i) the death rate highest and (ii) the increase in sulphur dioxide greatest?
(b) Why does the graph suggest that sulphur dioxide may kill people?
(c) Does the graph prove that sulphur dioxide kills people?

PRACTICAL WORK

Note for teachers
There are five experiments involving gas exchange, exercise and smoking. They are grouped at the end of this unit but can be performed whenever there is time available during the course. There are other units that have little or no practical work: Units 33 and 34 are ones during which time will certainly be available.

Experiment 28.1 To compare the carbon-dioxide content of inhaled (atmospheric) air and exhaled air

Note for teachers
The apparatus used in this experiment must be scrupulously clean. Any student with a cold or with a record of respiratory infections should not be allowed to perform the experiment.
   The lime water should be prepared the day before by shaking distilled water with an excess of calcium hydroxide. Close the container to prevent the entry of carbon dioxide and leave the lime water to settle

overnight. Decant the clear solution just before class use.

The method for making hydrogencarbonate indicator solution is described in the 'Note for teachers' in Experiment 27.1.

To avoid accidents to students it is advisable to fit lengths of 5-6 mm diameter glass tubing into the rubber bungs to be used. Select two-hole rubber bungs which securely fit the conical flasks to be used. Cut two 7 cm and two 15 cm lengths of glass tubing for each student and flame the ends. Using a little petroleum jelly as a lubricant, fit one 7 cm and one 15 cm length glass tubing into each rubber bung.

Materials required by each student
2 250 cm^3 conical flasks
1 spirit marker
2 20 cm lengths of 5 mm bore rubber tubing
2 rubber bungs each fitted with one 7 cm and one 15 cm length of 5–6 mm diameter glass tubing
40 cm^3 hydrogencarbonate indicator solution
40 cm^3 clear lime water

Method
1. Set up the apparatus shown in Figure 28.4.
2. Breathe in through the tubing connected to flask A and out through the tubing connected to flask B several times.
3. Note the appearance of the lime water in each flask.
4. Disconnect the bungs from the flasks and discard the lime water. Rinse the flasks with water and reassemble the apparatus with some of the hydrogencarbonate indicator solution in each flask.
5. Breathe in through the tubing connected to flask A and out through the tubing connected to flask B several times.
6. Note the appearance of the indicator solution in each flask.

Interpretation of results
1. What happened to the lime water in the two flasks?
2. What happened to the hydrogencarbonate indicator solution in the two flasks?
3. Explain what these results show about the composition of inhaled and exhaled air.
4. Explain how these differences arise.

Experiment 28.2 To compare the oxygen content of atmospheric air and exhaled air

Note for teachers
Construction of candle holders is described in the 'Note for teachers' in Experiment 5.3.

Materials required by each student
1 clear spoutless 500 cm^3 beaker with glass cover
1 bowl deep enough for the beaker to be covered by water when placed in it on its side
1 stopclock
1 50 cm length of rubber tubing
1 candle holder with attached candle

Method
1. Copy the table.

Contents of beaker	Time for which candle burned (s)
Atmospheric air	
Exhaled air	

2. Light the candle in the candle holder and lower it into the empty beaker until the lid of the candle holder is pressed tightly against the rim of the beaker. Immediately start the stopclock. Record in your table the time for which the candle continues to burn. Remove the candle holder and candle from the beaker.
3. Fill the bowl with water and lay the beaker on its side in the water. When all the air has escaped from the beaker, put one end of the rubber tubing into it and, keeping the open end of the beaker under water, raise the bottom of the beaker until the beaker is vertical (without letting any air get into the beaker).
4. Blow a small volume of air from your lungs into the air and then continue to exhale through the rubber tubing so that your breath collects in the upturned beaker.

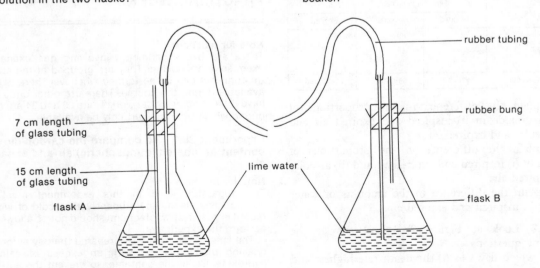

Fig. 28.4 Apparatus used in Experiment 28.1

5 When the beaker is filled with your breath, remove the rubber tubing and place the lid over the submerged end of the beaker.
6 Quickly remove the beaker of exhaled air from the bowl and replace its lid with the lighted candle holder as in instruction 2. Record in your table the time for which the candle continues to burn.

Interpretation of results
1 In which kind of air did the candle burn for longer?
2 What gas must have been present in the air to enable the candle to burn?
3 What can you conclude about the relative concentrations of this gas in the two air samples?
4 Suggest why you were instructed to collect only the latter part of your exhaled breath.

Experiment 28.3 To measure the vital capacity of human lungs

Note for teachers
Students must be supervised during this practical to ensure that they do not hyperventilate before exhaling into the apparatus.

A container of minimum volume $3\,dm^3$ is needed and can be shared by the group. It must be calibrated with marks on the outside and be translucent enough for the water level to be seen through its side. The outside can be calibrated by marking the level of water after successive additions of $500\,cm^3$. The container must have a stopper. A large plastic bottle with a screw-cap is ideal.

Materials required by each student
1 50 cm length of rubber tubing

Materials required by the class
1 large bowl
1 large calibrated container with a stopper

Method
1 Fill the container with water and put in the stopper.
2 Put water in the bowl to a depth of about 5 cm and hold the container upside down with its neck under the water.
3 Remove the stopper from the submerged neck of the container and put one end of the rubber tubing into the container.
4 Record the level of the water in the container using the marks on its side.
5 Take a deep breath and then exhale as much air as you can in one breath down the tubing.
6 Record the final water level in the container.

Interpretation of results
1 What was the volume of your exhaled breath?
2 It is unusual for you to exhale as much air as this in one breath. In what circumstances would you normally breathe so deeply? Explain why.

Experiment 28.4 To investigate the effect of exercise on the pulse rate and on the breathing rate

Note for teachers
This experiment combines two given in the syllabus. Running on the spot is customary, but other forms of exercise may be used provided that they impose no strain on the individual student.

Materials required by each pair of students
1 stopclock
1 chair

Method
1 Copy the table. In the table you will record measurements of pulse rate per minute and breathing rate per minute in successive minutes. To give yourself time to do the recording, you will make the actual measurements during 30 seconds and double the results. For example, pulse and breathing rates of 40 and 10 per 30 seconds become 80 and 20 per minute.

Time (min)	Activity	Pulse rate (per min)	Breathing rate (per min)
1	rest		
2	immediately before exercise		
3	exercise	—	—
4	exercise	—	—
5	immediately after exercise		
6	rest		
7	rest		
8	rest		
9	rest		
10	rest		

2 Work in pairs and follow instructions 3 to 8 in turn.
3 Sit your partner in a chair. Check that you can find your partner's pulse: you do so by placing two fingers on your partner's wrist just behind the thumb. If you cannot find your partner's pulse, ask your teacher for help.
4 Start the stopclock. During 30 seconds of the first minute count your partner's pulse beats while your partner counts his or her own breaths. Double the two measurements and record them in your table.
5 Repeat instruction 4 during the second minute.
6 Your partner should now take two minutes' exercise such as running on the spot. Do not measure your partner's pulse or breathing rates during this time.
7 During each of the fifth, sixth, seventh, eighth, ninth and tenth minutes repeat instruction 4 and record all the doubled measurements in your table.
8 Stop the stopclock, change roles and repeat the experiment.

Interpretation of results
1 Plot two jagged-line graphs of your partner's pulse rate and breathing rate with time on the x-axes and rate per minute on the y-axes. Label the two graphs you draw.
2 Describe the changes that occurred in your partner's breathing rate during the ten-minute period. Explain why these changes occurred. What was the biological value of the change in breathing rate during exercise? Why did the changes after the exercise was stopped occur gradually?
3 Describe the changes that occurred in your partner's pulse rate during the ten-minute period. What effect will the change in pulse rate during exercise have had on the transport of red blood cells? Name three substances which will have moved through your partner's blood more quickly as a result of the exercise.

Experiment 28.5 To investigate whether there is tar in cigarette smoke

Note for teachers
The value of this experiment may be enhanced if those students who admit to smoking are allowed to use one of their own cigarettes in the apparatus. Prepare the apparatus by cutting 5-6 mm diameter glass delivery tubing into two lengths of 15 cm and two lengths of 10 cm per student. Make a right-angled bend in each length about 4 cm from one end. Into each two-hole rubber bung insert the long arms of one 15 cm length of tubing and one 10 cm length of tubing. Cut a further 5 cm length of 5-6 mm diameter glass delivery tubing per student and flute the end of each one so that it will accommodate a cigarette.

Materials required by each student
1 100 cm^3 conical flask
1 flat-bottomed specimen tube
1 U-type absorption tube with side arms
5 cm 5-6 mm diameter glass delivery tubing fluted at one end
1 pair of forceps
1 bowl containing ice
4 lengths of rubber tubing of internal diameter 5-6 mm
2 plain rubber bungs to fit the absorption tube
1 two-hole rubber bung to fit the conical flask, fitted with right-angle lengths of glass delivery tubing
1 two-hole rubber bung to fit the specimen tube, fitted with right-angle lengths of glass delivery tubing
1 small ball of cotton wool
50 cm^3 diluted Universal indicator solution
1 cigarette

Materials required by the class
1 vacuum pump
clamps, stands and boss heads

Method
1. Study Figure 28.5 showing the final appearance of the apparatus.
2. Pour the indicator solution into the conical flask and insert the rubber bung that fits it.
3. Put the ball of cotton wool in the specimen tube and insert the rubber bung that fits it.
4. Insert the plain rubber bungs in the U-tube and put it in the bowl of ice.
5. Connect these three pieces of apparatus with rubber tubing. Use clamps to hold the apparatus straight if necessary.
6. Attach the straight piece of glass tubing to the free side-arm of the U-tube and place a cigarette in its fluted end.
7. Attach the apparatus to the vacuum pump and turn the pump on.
8. Record the appearance of the cotton wool and the colour of the indicator solution.
9. Light the cigarette without removing it from the apparatus.
10. When the apparatus has 'smoked' the cigarette, remove the stub using forceps, extinguish it and dispose of it.
11. Record the appearance of the cotton wool and the colour of the indicator solution.

Interpretation of results
1. Explain the final appearance of the cotton wool.
2. Describe any change in pH which has been shown by the indicator solution and compare this with the pH of the body.
3. Using your answers to the above questions, suggest the chemical changes that may occur in the lining of the lungs of someone who has inhaled cigarette smoke. What effect may these changes have on the normal behaviour of cells in this lining?

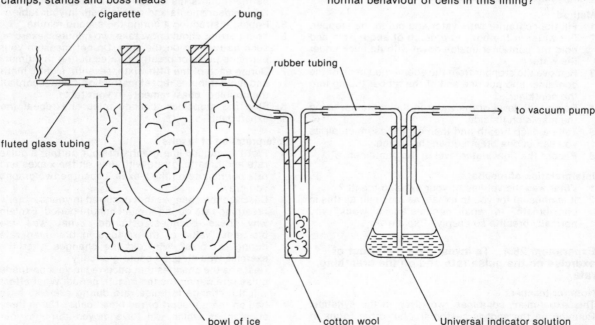

Fig. 28.5 Apparatus used in Experiment 28.5

Unit 29: Excretion in flowering plants and humans

Waste products from chemical activities in living cells have to be got rid of. Otherwise they would not only take up too much room but could poison and kill cells. Getting rid of these waste products is **excretion**.

Excretion in flowering plants

Plants have few waste products because they take in only simple substances: carbon dioxide, water, oxygen and inorganic ions. From carbon dioxide, water and inorganic ions they make the complex substances of which they are composed, e.g. cellulose and proteins. The oxygen may be used in respiration.

Both photosynthesis, a feeding process, and respiration, the energy-releasing process, result in waste products. During photosynthesis the only waste product produced is oxygen: a little of it is used in respiration; the remainder diffuses from the cells where photosynthesis takes place, through the air spaces and out through the stomata. The waste products of respiration are water and carbon dioxide. Waste water from respiration is added to the transpiration stream and removed in transpiration. If carbon dioxide is not immediately used in photosynthesis, it diffuses from the cells into the air spaces and passes out via stomata. If stomata are closed, all the excretory products can diffuse in small amounts through the surface cells.

Plants excrete no nitrogen products because they take in nitrogen, in the form of nitrate and ammonium ions through their roots, only in the amounts needed to build up new compounds.

Excretion in humans

There are two important chemical activities that result in waste products in humans: respiration and the breakdown of excess amino acids.

Respiration

Respiration is the same in humans as it is in plants. Since much more energy is needed by humans than by plants, there is much more waste, but the excretory products are the same: carbon dioxide and water.

Carbon dioxide

The removal of carbon dioxide from the cells by blood plasma and red blood cells, by diffusion at the air sacs and by ventilation from the lungs is described in Units 27 and 28. This is one form of excretion in humans.

Water

If the water formed by respiration in the cells is not needed, it is also removed by the blood plasma. It is added to water that has been absorbed from food and drink in the large intestine, a process described in Unit 22. From this point on excretory water and absorbed water are mixed up. Some of this mixed-up water in the blood plasma will be lost from the body in sweat, in tears, in continuous evaporation from the surface of the skin, and via the ventilation system when moist air is exhaled. Any excess water in the blood plasma is lost via the kidneys in **urine**: this is the only water loss from the body which is carefully regulated. Water from food and drink which is not absorbed in the large intestine is lost from the body in faeces. This water is not excretory water: the contents of faeces have passed straight through the body without being involved in a chemical reaction; they are not excretory products.

Excess amino acids

In our mixed diet we eat many different proteins which yield many different amino acids after digestion. When we are still growing, we need more amino acids per kilogram of body mass than when we are adult; on some days (for example, after an injury) we need more amino acids than others. The body usually takes in more amino acids than are needed but has no means of storing them; it may also take in the wrong ones. Excess amino acids are broken down and excreted.

The body does not waste the whole of the amino-acid molecule. Amino acids contain carbon, hydrogen, oxygen and nitrogen which the liver can break down (**deaminate**) into a nitrogen part and a non-nitrogen part containing carbon, hydrogen and oxygen. The non-nitrogen part can be respired, like sugar, to release energy, carbon dioxide and water. The carbon dioxide and water are disposed of in the same ways as the other carbon dioxide and water produced by respiration of sugar. The nitrogen part can be poisonous and is first made into a fairly harmless excretory product called **urea**. Urea is carried from the liver round the body dissolved in blood plasma to the **kidneys**, where it is excreted. These reactions are summarised in Figure 29.1.

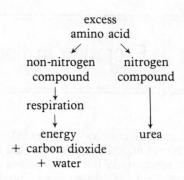

Fig. 29.1 Breakdown of amino acid

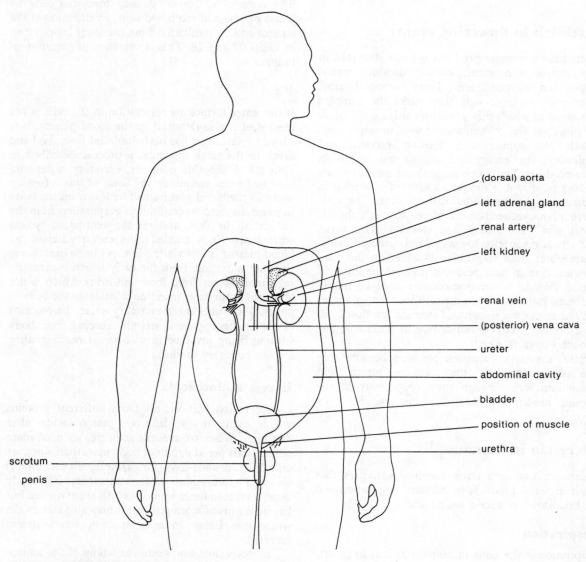

Fig. 29.2 Urinary system: gut removed

The urinary system

The two kidneys, which make urine containing urea, lie one each side on the back wall of the abdominal cavity. Their position is shown in Figure 29.2, though they can be seen only when the stomach and intestine have been removed. They are surrounded by fat which helps to protect them from a blow in the back. A carcass hanging in a butcher's shop often has the kidneys still attached but the stomach and intestine removed.

Figure 29.2 shows the **renal arteries**, which branch from the aorta to take blood to the kidneys, and the **renal veins**, which take blood from the kidneys to one of the venae cavae. A tube, the **ureter**, runs from each kidney down to the **bladder**. Urine, made in the kidney from urea, excess water and excess ions from the plasma, trickles continually down the ureter but is stored in the bladder for several hours. A ring-like muscle keeps the outlet of the bladder closed until, under the control of the brain, the muscle relaxes to let urine out by a narrow tube, the **urethra**. The urethra is shorter in females than in males. In females it opens in front of the vagina while in males it opens at the end of the penis, where there is a single opening from both the urinary and reproductive systems. Otherwise the male and female urinary systems are the same. Attached to each kidney is an adrenal gland, as shown in Figure 29.2, but the adrenal glands are not part of the urinary system: they produce hormones which are described in Unit 34.

The human kidney

You will see a mammal's kidney in the laboratory. The human kidney is shown in Figure 29.3. All mammal kidneys are similar, with an outer **cortex**, an inner **medulla** and a large **pelvis**, which means basin, into which the urine can drain. The pelvis narrows to pass out of the kidney as a thick-walled tube, the ureter. In a human the renal arteries and veins are branched outside the kidney. In a sheep and in many other mammals they join each kidney unbranched as single vessels. In humans the medulla is grouped into about twelve pyramids, each of which passes into a funnel-like extension of the pelvis wall.

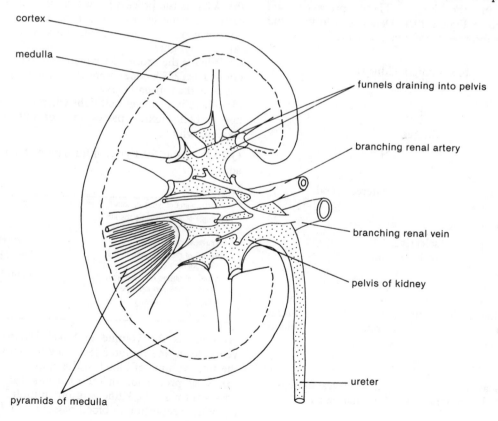

Fig. 29.3 Human kidney cut open

Making urine

In 1 dm³ of urine there are about:
  960 g water
   20 g urea
   12 g salts (mainly sodium chloride)

Urine is made in two stages in the kidney. First, blood is filtered: more than a quarter of the fluid plasma, containing its dissolved substances but not its large protein molecules and not the cells or cell fragments, passes out of the blood. This process is **filtration**. Before the mixture that has been filtered, the **filtrate**, can pass out of the kidney, living kidney cells remove all the substances that the body still needs and return them to the blood. This process is **selective reabsorption**. By the time the filtrate gets to the pelvis of the kidney only excess water, excess salts (mainly common salt, sodium chloride) and excess urea are left in any quantity, though there are small quantities of other substances such as hormones and poisons: all these substances form urine. Excess salts and poisons in urine may have been contained in food or drink or may have been produced by chemical activities in the body. These processes are summarised in Figure 29.4. Details of filtration and selective reabsorption are given in Unit 30.

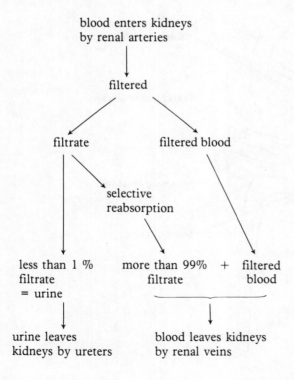

Fig. 29.4 Activities of the kidney

Questions

Q 29.1 Gases passing into and out of a leaf of a plant in daylight and in darkness are shown in Figure 29.5. The size of the arrows is a guide to the amounts involved. More water vapour passes out in daylight than in darkness.

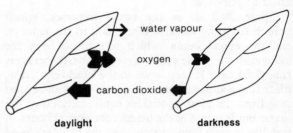

Fig. 29.5

(a) What is the process for which carbon dioxide enters and oxygen leaves a leaf in daylight?
(b) What is the process for which oxygen enters and carbon dioxide leaves a leaf in darkness? Why does not oxygen enter and carbon dioxide leave a leaf during daylight?
(c) What is the process by which water vapour passes out of a leaf? Why does more water vapour pass out in daylight than in darkness?
(d) There is a time at half-light when neither oxygen nor carbon dioxide passes into or out of a leaf. Explain why.

Q 29.2 Table 29.1 gives information about a typical adult.

Table 29.1

Total volume of blood in body	5 dm³
Volume of blood leaving the left ventricle at each heartbeat	0.07 dm³
Heartbeat rate	70 per min
Volume of blood passing to the two kidneys	1.2 dm³ per min
Volume of filtrate removed from blood in the two kidneys	0.12 dm³ per min
Volume of urine leaving the two kidneys	1.5 dm³ per day

(a) What is the volume of blood leaving the left ventricle in one minute? (See how it compares with the total volume of blood in the body.)
(b) What proportion of the total blood in the body passes through the kidneys in one minute?
(c) What proportion of blood passing into the kidney is filtered out?
(d) What is the volume of the filtrate per day? How

does this compare with the amount of urine produced per day?

(e) How much filtrate is reabsorbed in the kidneys per day?

Q 29.3 Explain why the passing of faeces is not excretion.

PRACTICAL WORK

Experiment 29.1 To dissect a mammalian kidney

Note for teachers
Lamb kidneys, obtainable from most butchers, are suitable for this experiment. The fat (suet) should be trimmed away before use. A student need not wear gloves, but they should be available because some students do not like to touch uncooked meat.

Materials required by each student
1 dissection dish
1 pair of forceps
1 scalpel
1 pair of scissors
1 lamb's kidney

Materials required by the class
disposable gloves

Method
1. Examine the outside of the kidney. Look carefully in the 'dimple' at one side of the kidney for the ureter and for any remains of the blood vessels supplying the kidney.
2. With the kidney lying in the dish, slice through horizontally so that the top half lifts off the bottom half.
3. Notice the dark outer cortex region and the lighter medulla region divided into triangular-shaped sections called pyramids.
4. Identify the pelvis of the kidney and look for the cut end of the ureter, which leaves the kidney from the pelvis.
5. Make a large labelled drawing of the dissected kidney.

Interpretation of results
1. Explain why the cortex is darker than the medulla.
2. What major process occurs in (a) the cortex and (b) the medulla? Figure 30.1 will help you answer this question.
3. Kidneys are covered in fat (suet) which is usually trimmed away by the butcher. Suggest a reason for this fat around the kidney.

UNIT 30
Kidney: structure and function; kidneys and health

Filtration and selective reabsorption

As well as branching arteries, veins and capillaries, each kidney contains between one and two million **nephrons**. A nephron is shown, with the blood vessels that surround it, in Figure 30.1.

Each nephron consists of: a cup-shaped capsule, called a **Bowman's capsule** after the man who discovered it; a twisted knot of blood capillaries, called a **glomerulus**; a long thin tube whose walls are only one cell thick, called a **tubule**. Blood enters each glomerulus from a small branch of the renal artery.

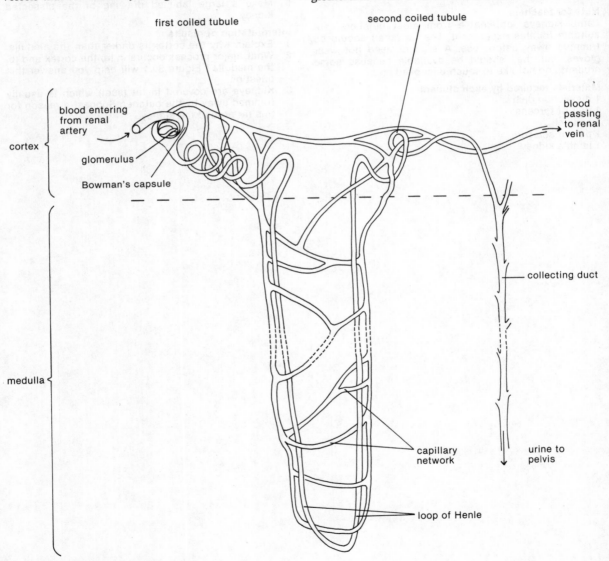

Fig. 30.1 Nephron and blood vessels

The capsules and glomeruli lie in the cortex and can be seen as specks with a hand-lens. The tubule is coiled where it lies in the cortex but is long and U-shaped across the medulla. Since there are more than a million nephrons winding across each kidney, you cannot see one clearly when you dissect a kidney or even when you see sections of a kidney. All the nephrons are jumbled together with their blood vessels: a section cuts across hundreds of different nephrons and hundreds of different blood vessels at different levels and at different angles. Nephrons drain into **collecting ducts** which cross the medulla to open into the pelvis. The branched blood vessels form a capillary network between and around the nephrons.

Blood pressure from the heart, aorta and renal artery forces some of the blood plasma out through the thin wall of the glomerulus and into the cavity of Bowman's capsule. This is **filtration** (often called **ultrafiltration** because it even separates large from small molecules). Blood pressure in the glomerulus is particularly high because the blood vessel carrying blood away from it is narrower than the one carrying blood to it.

Since all the nephrons behave in the same way, once you understand how one nephron works, you understand how the kidney works. After filtration into Bowman's capsule, the filtrate passes down the tubule where three processes are involved in **reabsorption**: diffusion, osmosis (i.e. diffusion of water) and active transport. The extreme length of the tubule increases the surface area over which the processes take place. Substances pass into the cells of the tubule wall, into the cortex, into the medulla and into the capillary network between the tubules to be carried out of the kidney by the renal vein.

Table 30.1 shows the major substances filtered, reabsorbed and excreted in urine.

You can see that only a small proportion of the substances filtered into Bowman's capsule is excreted in urine. Almost all the water, some of the urea and most of the ions are reabsorbed. All the amino acids and glucose, which form the body's food, are reabsorbed. This is the mechanism of selective reabsorption. Only those substances still needed by the body are reabsorbed.

The amounts of water, urea and ions reabsorbed, and therefore the amounts excreted, depend on conditions within the body. If you have drunk a lot of water, more water than usual will be excreted; if you have eaten a lot of protein, more urea than usual will be excreted; if you have eaten a lot of salts, more inorganic ions than usual will be excreted.

Not only does your intake of water, protein and salts in food vary, but the loss of water, urea and inorganic ions varies with the amount you sweat. **Sweating** is one of the major ways by which the body is cooled: if your body is too hot, you sweat; if it is too cold, sweating stops. Even if the body has no excess water, urea or inorganic ions, a proportion of them will be lost when the body gets too hot and sweats. The kidneys are the only organs that can regulate the loss of water, urea and inorganic ions. There is more about sweating in Unit 35.

The process by which the kidneys keep the concentration of substances in the blood more or less the same is controlled by hormones passed into the blood elsewhere in the body and in turn is controlled by a part of the brain. The kidney acts under the direction of these hormones.

The process of keeping conditions inside the body within fine limits is **homeostasis**. Keeping a regulated concentration of inorganic ions and water in the body is **osmoregulation**. Thus osmoregulation is an example of homeostasis. If fluids in the body are too concentrated, more inorganic ions and less water are excreted in urine; if fluids in the body are too dilute, more water and fewer inorganic ions are excreted in urine. Osmoregulation is achieved by selective reabsorption out of the tubule, i.e. by removing controlled amounts of water and ions from the filtrate.

The excretion of urea is another example of homeostasis involving the kidney. Blood always contains a small amount of urea and the amount remains more or less constant: if there is little urea in the blood, the walls of the kidney tubule reabsorb some from the filtrate; if there is a great deal of urea in the blood, not only is it filtered out in the 10 % of blood that leaves the glomerulus but it is secreted *into* the tubule as blood passes through the capillaries around it.

Table 30.1

Substances filtered into Bowman's capsule	Substances reabsorbed from tubule into blood	Substances excreted in urine
water	water (more than 99 %)	water (less than 1 %)
glucose	glucose	—
urea	some urea	some urea
inorganic ions	most inorganic ions	some inorganic ions (especially of sodium chloride)
amino acids	amino acids	—

Kidneys and health

Toxic substances

If toxic substances, i.e. poisons, are taken into the body, they damage it unless they are made harmless or removed. The body deals with harmful substances in three different ways:
- by breaking them down completely into harmless substances;
- by breaking them down partially and excreting the remainder;
- by storing them out of harm's way until they can be got rid of.

Alcohol is a poison that is completely broken down provided that it is not drunk in excess. Since it contains only the elements carbon, hydrogen and oxygen, the body uses it as a food which is broken down in respiration and releases energy. The liver breaks down alcohol but can do so only at a certain rate: if alcohol intake exceeds this rate, liver cells are poisoned and killed by alcohol and are replaced by scar tissue. This condition is **cirrhosis** of the liver, from which heavy drinkers may die.

Barbiturates, drugs described in Unit 33, are broken down so slowly in the liver that their effect in the body lasts for hours. Partly they are excreted unchanged in urine and partly they undergo complex chemical changes which interfere with the normal actions of the liver. This is why barbiturates should not be taken in large quantities.

Lead is a poisonous element which cannot be broken down at all. Either swallowed or inhaled, it collects in bones, where it replaces calcium, or in hairs, where it gradually grows out and is lost. Very small amounts are excreted in urine or secreted in saliva. Lead can damage nerve and brain cells.

Kidney failure

A healthy person may undergo an operation to have a kidney removed so that it may be transplanted to another person who needs it. This is possible because one normal kidney is usually enough to keep a person in good health. But of course it is an advantage to have two kidneys in case one should be damaged or become diseased.

Nowadays a person in whom neither kidney is functioning can lead a fairly normal life by spending two or three nights a week attached to a kidney machine.

Kidney or dialysis machines take over the functions of failing kidneys. What the kidney performs by filtration and selective reabsorption, a kidney machine performs in one step by diffusion. **Dialysis** means selective diffusion through a membrane which allows small but not large molecules (and nothing larger than a molecule) to pass through it. Substances pass from a higher to a lower concentration along a concentration gradient through a dialysis membrane: large molecules cannot pass through. The membrane in the machine separates two solutions: the patient's blood and a solution in the machine itself. The one in the machine is an ideal plasma solution which contains all the small-molecule substances, in the right concentrations, normally found in healthy blood plasma.

Figure 30.2 shows the principles of dialysis. The large protein molecules, the blood cells and the cell fragments (platelets) cannot cross the membrane. The small-molecule substances, such as water, urea, glucose, amino acids and ions do cross the membrane. Any of these in higher concentration in the blood than in the prepared solution will diffuse out of the blood; any in higher concentration in the prepared solution will diffuse into the blood. The success of the machine depends on keeping the prepared solution at ideal concentrations and at the right temperature so that there is no effective loss of the essential substances or of heat in the plasma but there is loss of waste products, especially excess urea, excess water and excess ions.

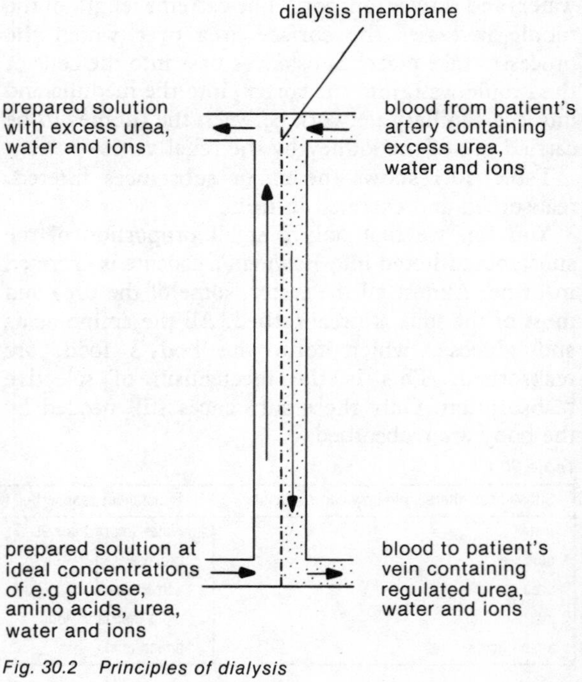

Fig. 30.2 Principles of dialysis

The kidney machine is a complicated and expensive piece of apparatus. The dialysis membrane has to have a large surface area to give effective contact between blood and the prepared solution. The solution must be constantly purified to remove the excess products and maintain it at the ideal concentration. The temperature must be thermostatically controlled. Patients must spend hours on the machine each week, though they can usually spend the time asleep, so that all their blood is regularly purified. Patients may have a tube permanently fitted to an artery and another permanently fitted to a vein in either an arm or a leg so that they can be easily attached to the dialysis machine every few days.

Kidney transplants are the alternative to kidney machines. A donor kidney is usually placed in the groin region of the abdomen while the patient's kidneys are left in their usual position in the body. The surgery in a kidney transplant is not difficult because there are only three tubes to join up: the renal artery must be joined to a major artery in the patient's groin; the renal vein must be joined to a major vein in the patient's groin; the ureter must be joined to one of the patient's ureters before it reaches the bladder. The serious problem in kidney transplants is **tissue rejection**. The white blood cells treat foreign substances, especially the foreign proteins in tissues, as invaders and produce plasma cells that make antibodies to destroy them. It is therefore important that a transplanted kidney is as like the person's original tissues as possible: this is **tissue matching**. An identical twin's kidney is accepted without rejection. If the kidney transplanted comes from someone with different tissue proteins, the antibody production of the patient must be reduced by drugs to allow the patient to accept the transplant. Unfortunately the patient's resistance to disease organisms is also reduced. Tissue rejection is described in more detail in Unit 26 in connection with blood transfusion, which is a form of transplant.

Questions

Q 30.1 The properties and contents of blood leaving the kidney by the renal vein differ from those of blood entering by the renal artery. Make a table like the one illustrated but larger. Enter the properties and contents of blood in the first column. Say whether each one has been decreased, increased or unchanged by the kidney and briefly explain why: (f) has been completed as an example.

Blood	Decreased, increased or unchanged, with explanation
(a) Pressure	
(b) Temperature	
(c) Water content	
(d) Urea content	
(e) Ion content	
(f) Protein content	Unchanged because protein is not filtered out of the blood into Bowman's capsule
(g) Amino-acid content	
(h) Glucose content	
(i) Oxygen content	
(j) Carbon-dioxide content	

Q 30.2 The table shows the sweat, urine and salt (sodium chloride) lost by a person on three different days. The person carried out the same activities each day and ate and drank the same amounts. The only difference was in the temperature.

Conditions	Sweat lost per day (dm^3)	Urine lost per day (dm^3)	Salt (sodium chloride) lost per day (grams)	
			in sweat	in urine
Normal day (slight sweating)	0.5	1.5	1.5	18.0
Cold day (no sweating)	0.0	2.0	0.0	19.5
Warm day (moderate sweating)	1.0	1.0	3.0	16.5

(a) What was the total volume of sweat and urine lost on (i) a normal, (ii) a cold, (iii) a warm day? How is the total volume of sweat and urine lost kept at the same level whatever the external temperature?
(b) What was the total mass of salt lost in sweat and urine on (i) a normal, (ii) a cold, (iii) a warm day? How is the total mass of salt lost in sweat and urine kept at the same level whatever the external temperature?
(c) By taking great exercise on a very hot day, a person could lose as much as $8.0\,dm^3$ of sweat. (i) How much salt would be lost in the sweat if the concentration remained the same as on a normal day? (ii) While the volume of urine excreted would be greatly reduced, it would not stop completely. Suggest why some urine must be excreted. (iii) Apart from water, what must people who sweat a great deal include in their diet?

PRACTICAL WORK

Experiment 30.1 To investigate the microscopic structure of a mammalian kidney

Note for teachers
This experiment requires a compound microscope with an objective lens allowing at least × 40 magnification. The investigation can be performed adequately with one microscope for every two students. If there are not enough microscopes, the teacher can demonstrate the investigation, allowing the students to look down the microscope, or use photographs, or do both. Biological suppliers sell sections of kidneys.

Materials required by each student
1 microscope
1 prepared section through a mammalian kidney

Method
1. Examine the kidney section under the microscope and find the cortex, medulla and pelvis.
2. Look closely in the cortex region for small round structures containing blood capillaries. These are the Bowman's capsules with their glomeruli.
3. Also in the cortex find a number of small cut tubules lined by cells. These are a mixture of the first and second coiled tubules. The first coiled tubules have bigger outside diameters, smaller gaps in the middle, fewer nuclei and thicker walls than the second tubules. They are also more common in the section because they are longer than the second tubules.
4. Make a drawing of a Bowman's capsule, glomerulus, first coiled tubule and second coiled tubule as they appear in the section.
5. Look in the medulla through the high-power lens of the microscope. It contains the loop of Henle and the collecting ducts. The loop of Henle has a much thinner wall than the collecting duct. You may also see a number of small, thin-walled capillaries.
6. Make a labelled drawing of the parts of the tubule which you have found in the medulla.

Interpretation of results
1. What happens in (a) the Bowman's capsule, (b) the first coiled tubule, (c) the second coiled tubule, (d) the loop of Henle?
2. Suggest a function for the numerous capillaries in the medulla region.
3. Suggest one advantage of the thinness of most of the tubule wall.

Experiment 30.2 To test samples of human urine for protein and reducing sugar

Note for teachers
Students often like to perform such tests on samples of their own urine. Any suitable container may be used to collect the sample. If a student is to collect a sample at home, the container should have a few drops of methylbenzene in it to prevent the bacterial breakdown of the urine before it is returned to school or college.

Materials required by each student
1 dropper pipette
1 test-tube rack with 2 test-tubes
1 labstix to test for reducing sugar
1 labstix to test for protein
1 sample of urine

Method
1. Use the dropper pipette to transfer equal volumes of the urine sample into the two tubes.
2. Dip the protein labstix into one of the samples and allow the colour to develop. Compare the colour with the chart on the labstix bottle and record the protein content of the urine sample.
3. Test the second sample with the reducing-sugar labstix. Compare the colour with the chart on the labstix bottle and record the reducing-sugar content of the sample.

Interpretation of results
1. Did the urine contain any protein? From your knowledge of the structure of proteins and the process of ultrafiltration, suggest an explanation of this result.
2. Did the urine sample contain reducing sugar? Suggest an explanation of this result.
3. When may a person's urine contain large amounts of (a) protein and (b) reducing sugar?

UNIT 31 Sensitivity and response 1

The environments of plants and animals are constantly changing: the light intensity changes; the temperature changes; the humidity changes; it rains or stops raining; there is movement of other organisms; there are sounds. The ability of plants and animals to register such events is called **sensitivity**. Altering themselves to take account of such events is **response**. The ability both to sense and to respond is **irritability**.

The event that brings about a response is a **stimulus**. The region of the body which can sense a stimulus is the **receptor**. The part of the body which carries out the response is the **effector**. Suppose that a bang makes you gasp: the bang is the stimulus; the receptor is in your ear; the effector is your ventilation system; the response is your gasping. Try to think of other examples for yourself. (There are some in Question 31.1 at the end of this unit.) Because the receptor is in a different part of the body from the effector, there must be a communication between them. Communication is by nerve impulses or by chemicals called **hormones**. Figure 31.1 summarises the terms that have been used.

Flowering plants

It may surprise you that plants respond. They respond mainly to a few important stimuli that effect their living activities:

light needed for photosynthesis;
water needed for all cell activities including photosynthesis, for transpiration and for transport;
gravity needed to ensure that shoots grow upwards to the light and roots grow downwards to water.

A flowing plant cannot move as an animal can. It responds by growing or bending or both: such a response is a **tropism**. Experiment 31.1 is concerned with a shoot's response to light, a **phototropism**: the stimulus is light; the receptor is the tip of the shoot; the effector is the part of the shoot that bends and grows; the response is bending and growing in the direction of light. Communication between the receptor (the tip) and the effector (the part that bends and grows) is by hormones. There is more about plant hormones in Unit 36.

Humans

Sensitivity and response are more obvious in humans. We are sensitive to at least six different outside stimuli:

light
sound
temperature
pressure
chemicals – which stimulate our sense of taste and smell
gravity

To receive these stimuli we have very specialised receptors: the eye contains the receptor for light, the ear the receptor for sound, the skin the receptors for temperature and for pressure, the nose and mouth the receptors for chemicals and the inner ear the receptor for gravity.

Sense organs

The eye

We rely greatly on our eyes and can distinguish a variety of shades and colours. The receptor is the **retina**, a membrane that lies inside at the back of the eye. Information about images is sent by nerves, in

stimulus —affects→ receptor ----→ communication by nerve impulses or hormones or both ----→ effector —carries out→ response

Fig. 31.1 Irritability

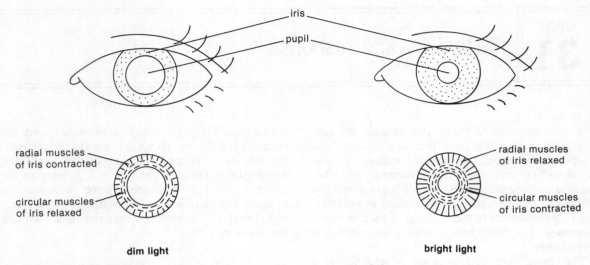

Fig. 31.2 Opening and closing the iris

the form of impulses, from the retina to the brain, which works out what we see. The retina is sensitive to both dim and bright light and can be damaged by very bright light. To protect the delicate retina, the size of the opening that lets light into the eye, the **pupil** in the front of the eye, can be adjusted by the coloured ring, the **iris**, which lies around it. When light is bright, the circular muscles in the iris contract and its radial muscles relax to decrease the size of the pupil and let only a small amount of light into the eye. When light is dim, the circular muscles in the iris relax and its radial muscles contract to increase the size of the pupil and let a large amount of light into the eye. This is shown in Figure 31.2.

The circular and radial muscles of the iris have opposite effects: the contraction of circular muscles closes the iris; the contraction of radial muscles opens it. When one set contracts, the other set relaxes, or they would oppose one another. Sets of muscles with opposite effects are described as **antagonistic**. Antagonistic muscles are involved in movement: they are described more fully in Unit 38.

We have no conscious control over the iris. It opens and closes without conscious thought. Its adjustments are **involuntary** and **automatic**. Bright light closes the iris and dim light opens it. Involuntary automatic actions in which the same stimulus produces the same response are called **reflex actions**. In Unit 32 there is an experiment on the iris reflex.

The ear

Air vibrates to produce sound. The outer ear directs the vibrations down to the ear-drum, which itself vibrates, causing first small bones and then internal fluids to vibrate. The receptors for sound, deep in the inner ear, are stimulated by movements of these fluids. Nerves carry impulses from these receptors to the brain, which automatically works out the nature of the sound.

Impulses are also sent to the brain from receptors for movement and balance in a different part of the inner ear: these are stimulated by the movement of fluid influenced by gravity. If you feel dizzy when you spin around, it is because the fluid in your inner ear is not moving at the same speed as the rest of you. If you still feel dizzy when you stop, it is because the fluid is still moving.

The skin

The skin is the sense organ that contains receptors sensitive to temperature and pressure. It also contains receptors sensitive to pain, but these are little understood. Skin receptors are the endings of single nerve cells. The skin is described in detail in Unit 32. Nerves carry impulses from temperature, pressure and pain receptors in the skin to the brain.

The nose and mouth

The nose and mouth are the sense organs that contain the receptors which give us our sense of smell and taste. In the moist linings of the nose and mouth are scattered cells sensitive to chemicals. Chemicals that stimulate these cells are first dissolved in moisture. Nerves carry impulses from these cells to the brain.

Questions

Q 31.1 Copy the table. Write in the stimulus, the sense organ containing the receptor, the part of the body containing the effector, and the response in each of the seven events described: (a) has been completed as an example.

Stimulus	Sense organ containing the receptor	Effector or part of body containing effector	Response
(a) bang	ear	ventilation system	gasp

(a) A bang makes you gasp.
(b) A speck of dust in your eye makes you blink.
(c) A racehorse gallops when the starting stalls open.
(d) You sweat when the room gets hot.
(e) You swallow a fly and cough.
(f) Steam burns your arm and you move it away.
(g) A cat sees a mouse and pounces.

PRACTICAL WORK

Experiment 31.1 To investigate the effect of unidirectional light on the growth of plant shoots

Note for teachers
This experiment can be linked with Experiment 42.1 investigating the conditions needed for seed germination. If so, students may germinate their own seedlings up to six days before the experiment is begun. If not, the seedlings should be germinated by the teacher, and students should begin at instruction 3.

Materials required by each student
1 pair of scissors
2 petri dishes
1 ruler graduated in cm
1 spirit marker
enough absorbent cotton wool to cover the bottom of the petri dishes
1 cardboard box
cress seeds

Materials required by the class
2 bench lamps
1 dark cupboard

Method
1. Label the bottoms of the two petri dishes with your initials. Line them with cotton wool and soak the cotton wool with water. Spread some cress seeds evenly over the cotton wool in both dishes.
2. Leave the dishes in the dark cupboard. Check that the cotton wool does not dry out while they are there. If necessary add more water to them.
3. After five or six days remove the germinated seedlings from the cupboard and observe the seed leaves (cotyledons) at the tips of the plumules. In each dish cut the tips of the plumules and the cotyledons from half of the plants. Leave the other half uncut.
4. Place one of the petri dishes where it will not be disturbed. Place a bench lamp about 30 cm above it shining down on it.
5. Place the cardboard box on its side where it will not be disturbed. Place the second petri dish of seedlings in this box. Put a bench lamp 30 cm from the box so that it shines on to the seedlings through the open box.
6. After 24 hours examine the way in which the seedlings have grown and record your observations in a table like the one illustrated.

Position of light	Treatment of seedling	Direction of growth of stem
Above petri dish	tip removed	
	tip intact	
To side of petri dish	tip removed	
	tip intact	

Interpretation of results
1. Did the two groups of seedlings with their tips cut off grow differently from each other?
2. Did the two groups of seedlings with their tips intact grow differently from each other?
3. In which petri dish did all the seedlings grow in the same direction?
4. What can you conclude from your experiment about the sensitivity of seedlings to light?
5. Explain the importance of the plant response you have investigated in the normal activity of plants.

Experiment 31.2 To investigate the variety of stimuli to which the human skin responds

Materials required by each pair of students
3 250 cm^3 beakers
3 mounted dissection needles
1 ruler graduated in mm
1 fine spirit marker
2 sheets of graph paper

Materials required by the class
hot water
ice

Method
1. Work in pairs and follow instructions 2 to 4 in turn.
2. Use the fine spirit marker and ruler to draw a square of side 2 cm on the back of your partner's hand and to divide this square into a hundred squares of side 2 mm. Mark three similar grids of a hundred squares on a sheet of the graph paper. Label these grids hot, cold and touch respectively.
3. Put one needle into a beaker of hot (but not boiling) water and a second into a beaker of iced water. Leave the third, which is to test touch only, in the empty beaker.

4 During this part of the experiment your partner must turn away and be unable to see what you are doing. Using the three needles at random, lightly touch your partner's skin twenty times inside the squares of the grid. Touch some squares with more than one needle. Each time your partner correctly identifies a stimulus as hot, cold or touch only, make a tick in the corresponding square of the appropriate grid on the graph paper. Each time your partner incorrectly identifies a stimulus, put a cross in the corresponding square of the appropriate grid on the graph paper.
5 Change roles and repeat the experiment.

Interpretation of results
1 Did you and your partner correctly identify every stimulus as either hot, cold or touch? If not, count the number of correct responses for each stimulus and calculate the frequency of correct responses using the formula:

$$\frac{\text{number of correct responses}}{\text{total number of stimuli}} \times 100 \%$$

2 Was there a difference in the frequency of correct responses to the different stimuli?
3 Were any of the 2 mm^2 areas of skin sensitive to more than one of the stimuli? What do your results suggest about the number of different receptors in the skin?
4 Is heat detected by the same receptor as cold? Explain your answer.

UNIT 32 Sensitivity and response 2

The skin: structure

Figure 32.1 shows a vertical section through the skin. The skin receptors for temperature, pressure and pain are different shapes. The sensitivity of the skin depends on the number of receptors in an area: the more receptors, the greater the sensitivity.

The outer layer of the skin is the **epidermis**: its surface, the part you see, is made up of dead cells that will flake off; underneath are cells that are dying and will replace the surface cells; further underneath are living cells (the **Malpighian layer**, i.e. a layer within the epidermis), which divide to form new cells that replace the dying ones. The middle layer of the skin, the **dermis**, consists of connective tissue containing: the receptor nerve endings; the nerve fibres that lead from them; blood vessels; sweat glands; hairs and the living cells surrounding them; grease (**sebaceous**) glands that produce **sebum** to keep the hairs and skin supple and waterproof; muscle fibres that help to fluff up the hairs in furry animals but do nothing important for our few hairs. The innermost layer of skin, the **hypodermis**, consists of fat-storing cells.

Figure 32.2 is a photograph of a vertical section of the skin on a human scalp. The hairs, sebaceous glands and fat-storing cells of the hypodermis are clearly shown. On the right of the photograph are parts of two sweat ducts. Sections through the skin show only parts of different structures. You are unlikely to be able to identify single receptors. Experiment 32.2 will help you to understand the structure of the skin.

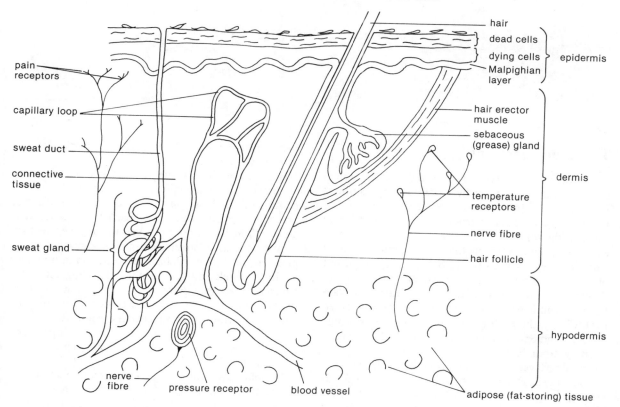

Fig. 32.1 Skin: vertical section

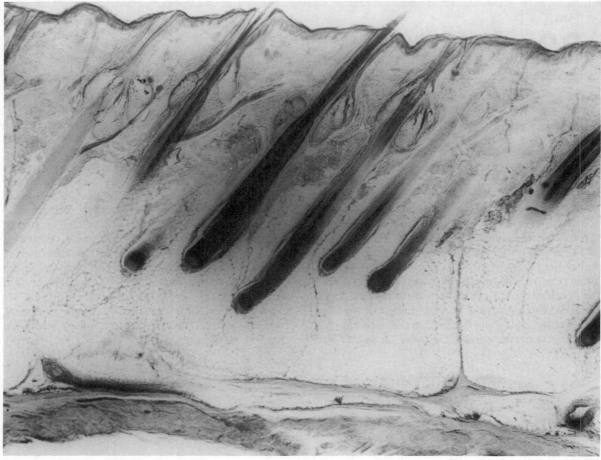

Fig. 32.2 Human skin: vertical section

The skin: functions

The skin is a **protective organ**. The epidermis forms a continuous cover over the body protecting it from infection by bacteria and viruses, from water loss by evaporation, and from mechanical damage by friction: the soles of the feet have a particularly thick epidermis. The waterproofing of the skin by sebum from the grease glands serves both to keep water out of the body and to reduce the rate at which it leaves the body in evaporation. Burning by ultraviolet rays from the sun is prevented by a brown pigment, called **melanin**, in the Malpighian layer: different amounts of this pigment account for different skin shades. Damage from blows and knocks is reduced by the cushioning layer of fat cells in the hypodermis.

The fat cells also make the skin an important **food-storage organ** and **insulation layer** reducing the body's heat loss. Many mammals increase the fat stored in this layer before winter; when mini-skirts were fashionable, the fat layer in women's legs increased in thickness.

Hairs have little value in humans but in most mammals they trap air, forming an air blanket which **reduces heat loss** from the body or, as in camels, prevents excessive heat gain. Each hair has its own muscle. The hair muscle changes the angle of the hair to make the air blanket deeper or shallower. Goose-pimples are all that is left of this function in humans.

Blood vessels in the skin not only bring food and oxygen to all the living cells and remove carbon dioxide and other waste products from them but are vitally important in **temperature regulation**. The sweat glands are also concerned with temperature regulation. There is more about temperature regulation in Unit 35.

As well as performing its important functions of protection, food-storage and temperature regulation, the skin is a sense organ. The function of the skin as a sense organ is described in Unit 31.

Nerve impulses

All the sense organs that have been described send information from their receptors to the brain in the form of electrical impulses or electrical charges. The electrical impulse or charge passes along a fine thread of cytoplasm called a nerve fibre. Individual nerve fibres can be seen only with a microscope, but bundles of nerve fibres called **nerves** can be seen as white strands with the naked eye. The electrical impulses are interpreted by the brain, a process of which we know little except that it involves the passing of more electrical impulses. When the brain has interpreted the electrical impulses, a response may be needed. The brain will then send electrical impulses through a different set of nerve fibres to the effector or effectors, which will carry out the response.

Let us consider what happens when you hear your name called and shout 'Yes' in reply. Air vibrations pass into your ears to vibrate the bones in each ear and the fluid in each inner ear. Certain receptor cells are stimulated. Impulses pass along many nerve fibres in the nerves from your ears to the brain. Not only does the brain recognise that your name was called but it gains a rough idea of where the call came from because it receives impulses from the ear nearer to the call sooner than from the other ear. Impulses are sent from the brain by different nerve fibres to various muscles of the ventilation system, the voice-box and the mouth for you to say 'Yes'. Sound is the stimulus; cells in the inner ear are the receptors; communication is by nerve impulses; the effectors are the muscles involved; the response is saying 'Yes'.

Saying 'Yes' is a **voluntary** action. The decision to say 'Yes' comes from the thinking part of the brain. Talking, writing, walking, singing, playing tennis, playing the guitar – these are all examples of voluntary actions, though, unlike thinking, they also involve the cerebellum, a non-thinking part of the brain. **Involuntary** actions do not involve the thinking part of the brain. The kind of involuntary action you must understand is the **reflex action**, often called a **reflex response** or simply a **reflex**.

Reflex actions

Reflexes are actions we take without deciding to take them. They may involve the non-thinking parts of the brain or they may not involve the brain at all. Reflexes, because they do not involve the thinking part of the brain, are quicker than voluntary actions.

When we move we are continually in danger of overbalancing, but automatically, by reflex actions, we adjust our position so that we do not fall. When we smell food that we like, our gut prepares to digest it and we secrete saliva and digestive juices in the stomach. These are all reflex actions involving the non-thinking parts of the brain. The iris reflex, described in Unit 31, also involves a non-thinking part of the brain: if light is too bright for the retina, impulses are sent by nerve fibres to a non-thinking part of the brain, which sends impulses to the circular muscles of the iris, which contract. Reflex actions are **involuntary** and **automatic**. In a reflex action the same stimulus produces the same response.

An example of a simple reflex, one that does not involve the brain at all until after it has taken place, is a **withdrawal reflex**. Its function is protective: speed is important and reflexes that do not involve the brain are the fastest of all. A withdrawal reflex is the response to an unexpected damaging stimulus. If your arm is unexpectedly pricked by a pin or scalded by steam from a kettle, you immediately pull it away. An impulse travels from the receptors in the skin up the nerve fibres of the arm to the spinal cord, which immediately sends another impulse back down other nerve fibres of the arm to the muscles, which immediately move the arm away. Impulses also pass up the spinal cord to the brain, which interprets what has happened, but only after the response has been made. The response is involuntary and automatic, and it will always be the same if the stimulus is unexpected. If you know someone is going to prick you, you can use your brain to deliver impulses which stop the withdrawal reflex.

Figure 32.3 shows the path taken by impulses that cause the withdrawal reflex of the arm. The spinal cord is shown in transverse section, but the backbone that protects it is not shown. Note that impulses from the receptors pass into the spinal cord by the upper (dorsal) branch of the spinal nerve and pass out to the effector muscles by the lower (ventral) branch. A short distance from the spinal cord **sensory** nerve fibres from the receptors and **motor** nerve fibres to the effectors are bound together in a single **mixed nerve** in which it is not possible to see any difference between them. Sensory impulses pass up the nerve from the receptors to the spinal cord while motor impulses pass down it from the spinal cord to the muscles, though they pass through different fibres. Only one set of nerve fibres is shown: in fact there are thousands running alongside one another in the same nerve. Within the nerve, impulses cannot pass sideways from one fibre to another because each is insulated by its own **fatty sheath**.

A place where nerve fibres meet is called a **synapse**. Here impulses can pass from one fibre to

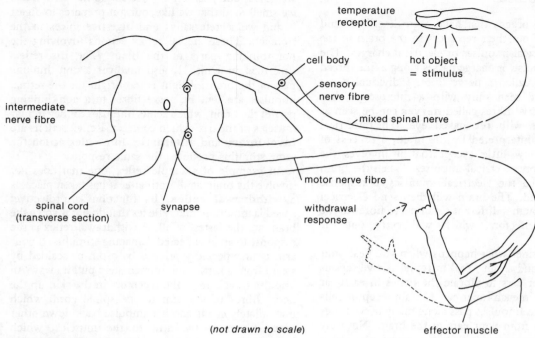

Fig. 32.3 Reflex action (*not drawn to scale*)

another because they lie close together without insulation (though they do not touch). The impulses 'jump' this gap by stimulating secretions of a special chemical.

The path taken by an impulse to produce a reflex action is called a **reflex arc**. It is made up of the nerve fibres and nerve cells through which it passes. Each nerve fibre has a **cell body** containing a nucleus somewhere along its length. The cell body is shown on each nerve fibre in Figure 32.3.

Questions

Q 32.1 Figure 32.4 shows the nerve pathways involved when someone stands barefoot on a sharp object.
What would be the effect of the destruction of nerve tissue at A, at B and at C on this reflex?

Q 32.2 When an object is placed in a young baby's hand, it grasps it tightly. When you grasp a very hot pipe, you immediately let go.
(a) What kind of response is shown by a baby?
(b) Explain your response in the second example.

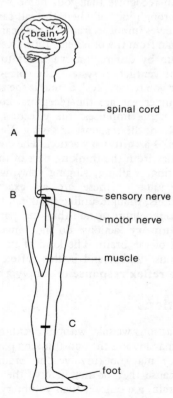

Fig. 32.4

PRACTICAL WORK

Experiment 32.1 To investigate the human skin's ability to distinguish close simultaneous stimuli

Materials required by each pair of students
1 ruler graduated in mm
1 stiff wire 'hair pin' (see Figure 32.5)

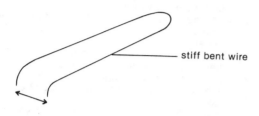

capable of extension to about 7 cm

Fig. 32.5 Wire 'hair pin' used in Experiment 32.1

Method
1. Work in pairs and follow instructions 2 to 4 in turn.
2. With the two points of the wire 'hair pin' 6 cm apart, lightly touch the skin on the back of your partner's hand in different places, sometimes with one point and sometimes with two. Each time ask your partner, who must look away throughout the experiment, whether you are touching with one point or two.
3. Continue to follow instruction 2 but gradually move the points of the 'hair pin' together until your partner cannot correctly identify the number of stimuli. In a table like the one illustrated, record the shortest distance between the points at which your partner can always correctly identify the number of stimuli.

Region of skin	Shortest distance between points when all responses are correct (mm)
Back of hand Finger tip Forearm Neck	

4. Repeat the test on the other regions listed in the table and record your results.
5. Change roles and repeat the experiment.

Interpretation of results
1. The region of skin which resulted in correct responses when the points were closest together is the most sensitive. List the regions of your skin in order of sensitivity, starting with the most sensitive.
2. Relate the sensitivity of the tested areas to their functions.

Experiment 32.2 To investigate the distribution of tissues in human skin

Note for teachers
This experiment requires a compound microscope with an objective lens allowing at least ×10 magnification. The investigation can be performed adequately with one microscope for every two students. If there are not enough microscopes, the teacher can demonstrate the investigation, allowing the students to look down the microscope, or use photographs, or do both. Care must be taken when choosing sections to ensure that hair follicles and sweat glands are present.

Materials required by each student
1 microscope
1 stained slide of a vertical section through human skin

Method
1. Use the microscope to examine the slide. Note that there is a thin outer layer (the epidermis) lying on top of a thicker layer (the dermis). Beneath this is a white layer (the hypodermis).
2. Look at the epidermis. Its lower edge contains living cells (the Malpighian layer) which have been stained during the preparation of the slide.
3. Near the surface the cells of the epidermis are dead and have become scales (of a protein called **keratin**). Find these scales on the slide.
4. Examine the dermis. Most of it is made up of a connective tissue containing other structures. Find:
   (a) blood vessels;
   (b) a hair (contained in a hair follicle);
   (c) a sebaceous (grease) gland;
   (d) nerve endings;
   (e) parts of a coiled sweat gland and its duct going to a pore at the surface of the skin.
5. Note the appearance of the fat (adipose) tissue in the hypodermis. You may also see blood vessels or nerves running through this part of the slide.
6. Make a labelled drawing of exactly what you see. Use your knowledge of the skin to interpret the structures you draw.

Interpretation of results
1. The Malpighian layer contains a pigment called melanin. What is the function of this pigment?
2. Why do the cells of the epidermis die as they are pushed away from the Malpighian layer by new cells?
3. What is the function of the dead cells on the outside of the epidermis?
4. In the dermis you probably could not find a hair follicle or sweat duct which you could see along the whole of its length. Explain this, bearing in mind how the slide has been produced.
5. Make a list of the structures you found in the dermis. Refer to it in making another list of the functions the dermis performs.
6. Explain why the cell structure of the white cells in the hypodermis could not be seen.
7. Give two functions which the hypodermis performs.
8. Which layer of the skin contains the sense organs which were investigated in Experiment 31.2 and Experiment 32.1?

Experiment 32.3 To demonstrate human reflexes

Note for teachers
Two reflexes are used in this experiment: withdrawal of the hand from a hot object and the iris reflex. It is better, though not essential, to demonstrate the iris reflex in a darkened room.

Materials required by each pair of students
1 250 cm^3 beaker
1 glass rod
1 torch
1 disposable towel
hot water

Method
1. Work in pairs. Follow instructions 2 to 4 and 6 to 8 in turn.
2. Put the glass rod in the beaker of hot water.
3. Your partner's eyes must be closed during this part of the experiment. While your partner rests an arm on the bench, quickly remove the glass rod from the hot water, dry it and touch your partner's hand with it. Record your partner's response.
4. Repeat instruction 3 but say when you are about to touch your partner's hand with the rod. Again record your partner's response.
5. Change roles and repeat this part of the experiment.
6. Sit opposite your partner so that you are face to face in a darkened room. Without warning, shine the torch into your partner's face. Record the response in your partner's eyes. If you do not notice any change, switch the torch off and try again after about one minute.
7. Repeat instruction 6 but say when you are about to shine the torch into your partner's eyes. Again record your partner's response.
8. Change roles and repeat this part of the experiment.

Interpretation of results
1. Describe what happened when you touched your partner's hand with the hot glass rod.
2. Did your partner respond any differently when warned of what you were about to do? Explain your answer.
3. Describe what happened to your partner's pupils when you shone the torch on them.
4. Did your partner respond any differently when warned of what you were about to do? Explain your answer.
5. What is the value of the two reflexes you have just investigated?
6. Is the brain involved in either of these reflexes? Explain your answer.

UNIT 33: The brain and drugs

The brain: structure and function

Figure 33.1 shows the positions of the brain inside the skull and of the spinal cord inside the backbone (vertebral column). The skull and backbone protect the important nervous tissue within them from damage by blows and knocks.

The brain in side view shows three important regions: the largest part, the **cerebrum**, hangs over the rest of the brain, except for the **cerebellum** at the back and the **medulla** at the bottom, where it narrows and joins the spinal cord. The **grey matter** where brain activity takes place is made up of nerve cells lying just below the surface of the brain. Within the brain are fluid-filled spaces and **white matter** consisting of bands of nerve fibres.

The cerebrum is deeply folded and divided into two halves, the **cerebral hemispheres**. Thinking is done in the cerebrum, which is relatively larger in humans than in any other animal. Most of the sensory information from the receptors throughout the body goes via other parts of the brain to the grey matter of the cerebrum, where it is interpreted. The grey matter of the cerebrum co-ordinates all the voluntary activities of the body, i.e. the things we do consciously, sending impulses mainly via the spinal cord to the effectors.

In the cerebrum, grey matter forms only a 3 mm surface layer, the **cerebral cortex**. Deep folds enlarge the grey matter's area to about 2200 cm^2.

The cerebellum is also deeply folded. As in the cerebrum, this greatly increases the number of cells able to lie in the grey matter just below the surface. Although much smaller than the cerebrum, the cerebellum has a surface area of over 1500 cm^2. The cerebellum co-ordinates balance, posture and movement of the body at an unconscious level. It receives impulses from the receptors concerned with the position of the body and transmits impulses to the appropriate effectors, i.e. the muscles attached to the skeleton, to keep the body continually balanced. The cerebellum is like a control box: it receives information relating to balance and orders necessary corrections.

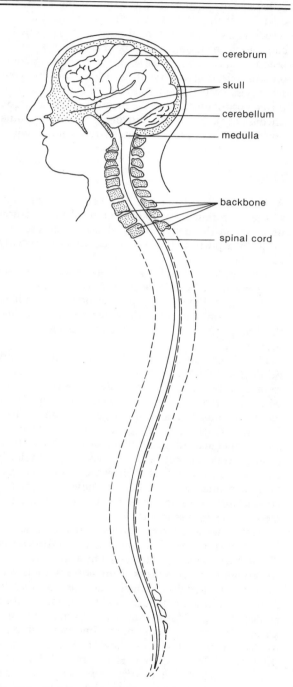

Fig. 33.1 Brain and spinal cord: side view

The medulla joins the spinal cord to the rest of the brain. Impulses pass into it from the receptors, either directly from the head or indirectly through the spinal cord. Although it is only 3 cm long, it is one of the most important parts of the brain. By continuous reflex actions it controls the action of the heart, the diameter of the arteries and the rate of breathing. In recent years it has been realised that it is the major controlling mechanism of the nervous system: it monitors incoming sensory impulses, passing some on but rejecting others; it is responsible for consciousness even though conscious thoughts occur in the cerebrum (damage to the medulla makes a person unconscious); it is also involved in maintaining muscle tone, i.e. keeping muscles ready for activity.

Drugs

A drug is 'a substance taken to help recovery from sickness, to relieve symptoms or to change natural processes in the body'. 'Drug abuse' is the use of a drug for a purpose that is not medically or socially acceptable.

Not everyone would agree on 'abuse'. Nearly everyone has an occasional aspirin to relieve headache or other pain. But some people take dozens of aspirins a day, thereby damaging their nervous systems and the linings of their stomachs and causing irregular heartbeats.

The drugs described in this unit are usually **addictive**, which means that people who take them cannot easily give them up. **Tolerance** is occurring when the *physical* effects are such that people need more and more of a drug to get the same effect. Examples of addictive drugs are alcohol, morphine and heroin (an artificial form of morphine). But many people drink alcohol without becoming physically addicted to it. The World Health Organisation uses 'drug dependence' to include any physical or mental craving for a substance whether it gives pleasure or merely relieves discomfort.

Most drugs affect the nervous system. Those that increase mental and physical activity are **stimulants**. Those that reduce mental and physical activity are **sedatives** or **depressants**. **Pain-killers** are treated as a third category, though they are a form of depressant. You are not expected to remember lots of names of drugs. Many drugs have alternative names. Drug slang changes from time to time and from place to place. You need to understand the broad effects of the three different categories of drug, the detailed effects of alcohol and the personal and social problems that drug dependence can cause.

Stimulants

Stimulants increase activity, alertness and brain power. **Caffeine** is a stimulant in coffee, tea, cocoa and cola. In the small amounts present in these drinks it is usually harmless, but large amounts cause trembling and increased heartbeat and breathing rates. **Amphetamines**, known as pep-pills, speed or uppers, produce effects very like the body's own stimulant **adrenaline** described in Unit 34. The mental stimulation of amphetamines may result from an increased blood flow to the brain. Amphetamines increase heartbeat, blood pressure and blood sugar, as does adrenalin.

Doctors popularised amphetamines as slimming pills and as a treatment for depression and tiredness. The first people to become dependent on them were middle-aged, overweight women. Now it is known that large doses of amphetamines cause dangerous delusions' and hallucinations. The slogan 'Speed Kills' refers not only to reckless driving but also to amphetamines. The prescribing of stimulants has decreased since their dangers have been recognised.

Sedatives

Sedatives, which include **tranquillisers**, decrease body, especially brain, activity. The cerebellum is particularly affected, which results in unsteadiness and loss of balance. **Barbiturates**, known as barbs, candy or goof balls, and **alcohol** are both sedatives. They can cause slowing of reactions, confusion, difficulty in speaking, unsteadiness, poor memory, faulty judgement, inability to concentrate, irritability, excessive emotion, hostility, suspiciousness and suicidal feelings. Heavy drinking may overwork the liver and cause cirrhosis, described in Unit 30. Barbiturates encourage sleep and were commonly prescribed as sleeping tablets. They are habit-forming and people easily become dependent on them. The first sedative-addicts, like the first stimulant-addicts, were respectable people, mainly the middle-aged and elderly. Fewer barbiturates are prescribed by doctors these days.

Alcohol and barbiturates are calming at first but lead to confusion and unconsciousness as the dose is increased. Though a sedative, alcohol is often thought of as a stimulant. This is because its calming effect reduces anxiety and social restraint, enabling people to 'let themselves go' at a party or in a social gathering.

Pain-killers

Pain-killers, more technically **pain relievers**, can be mild and fairly harmless, such as **aspirin**, or powerful and extremely habit-forming, such as **morphine**, formed in the seeds of opium poppies, and **heroin**, an artificial form of morphine. Morphine relieves pain and suppresses coughing but stimulates vomiting.

Other drugs

Cannabis, known as marihuana, hash, pot, dope or grass, which comes from the hemp plant, *Cannabis sativa*, and **solvents**, such as glue, are examples of drugs that cannot be classified as stimulants, sedatives or pain-killers.

Cannabis is not physically addictive, though people may become psychologically dependent on it. Smoking cannabis can give a feeling of well-being, often accompanied by hallucinations, followed by lack of energy and effort. It can also, like drinking alcohol, make a bad frame of mind worse, turning depression to despair or anxiety to panic. Some people think cannabis should be legalised, on the ground that it is no more harmful than alcohol.

Glue-sniffing, technically **solvent abuse**, also produces effects similar to alcohol and other sedatives: in small doses it gives a feeling of well-being and leads to drowsiness. Though glue is freely available in the shops, glue-sniffing is more dangerous than smoking cannabis: glue-sniffers can die from suffocation, choking on vomit, falling unconscious in dangerous places, and heart failure due to sudden shock.

Personal and social effects of drug dependence

Some drugs are less harmful than others, but none is harmless and all of them can kill in large enough doses. Barbiturates, the respectable sleeping tablets, have killed tens of thousands. Barbiturates and alcohol are particularly dangerous when taken together. Alcohol addiction kills more people than heroin and costs the nation millions of pounds a year because addicts cannot work or are in prison. Many people admitted to mental hospitals are alcoholics. Of course heroin is a much more serious drug than alcohol. The figures are explained by the fact that alcohol is freely available while heroin is illegal and difficult to come by.

Illegal drugs are expensive. Those dependent on them may mix with criminals to get supplies and may commit crimes themselves to get money to pay for them. Because illegal drugs are often contaminated and impure, those who take them risk being poisoned. Injection of drugs by contaminated needles and syringes causes infections. The health of those dependent on illegal drugs often suffers because they have no proper homes and do not eat enough.

All drugs, whether taken for kicks or prescribed by doctors, interfere with processes in the body. A doctor should prescribe a drug only if its likely good effects outweigh its possible harmful effects. For example, antihistamines are prescribed for hay fever and other allergies despite the fact that they produce drowsiness, dizziness, hallucinations and about a dozen other side-effects: it is not worth taking antihistamines for hay fever unless you suffer from it badly. Moreover drugs may have side-effects which neither you nor doctors know about but which will reveal themselves years later. A good rule is never to take any drug if you think you can manage without it. A pregnant woman should go to great lengths to avoid drugs since they may harm her more vulnerable baby.

You can find more information about drugs in *What Everyone Should Know About Drugs* by Kenneth Leech (Sheldon Press, 1983) and *Drugs: Medical, Psychological and Social Facts* by Peter Laurie (Penguin, 1978).

Questions

Q 33.1 List the functions of (a) the cerebrum, (b) the cerebellum, (c) the medulla.

Q 33.2 Copy and complete the table. Give two examples of each group of drugs.

Drug group	Examples	Effects
Stimulants		
Sedatives		
Pain-killers		

PRACTICAL WORK

Note for teachers
There is no practical work either in this unit or the next. This leaves time both to discuss the subject-matter of this unit and to complete the extensive practical work on breathing and smoking in Unit 28.

UNIT 34

Hormones; human hormones

Communication between receptors and effectors, i.e. between sensitive regions and the regions that respond, is by nerve impulses or hormones or both. Units 31 and 32 deal mainly with nerve impulses.

Hormones are chemicals produced in one part of a plant or animal and passed to other parts, **target organs**, where they have their effect. Plants have no nerve impulses: all the communication between receptors and effectors in plants is by hormones. Plant hormones are described in Unit 36.

Hormones in humans

Hormones are complex chemical compounds, though not as complex as proteins. (The molecules of some are no bigger than a single, rearranged amino acid; others consist of several amino acids joined together; yet others are different chemical compounds called **steroids**.) All hormones are made and secreted by groups of cells, called **glands**, straight into the blood capillaries that pass through them. Blood transports the hormones around the body to the target organs. The hormone-secreting glands are called **ductless glands** (or **endocrine glands**) because they differ from other glands which have tubes or ducts through which they secrete their products. The adrenal glands, which secrete adrenaline, are an example of ductless glands. Sweat glands and salivary glands are examples of glands that do have ducts.

The two communication systems, nervous and hormonal, are very different and well suited to their different tasks. Nervous communication is used when the response must be rapid but does not need to be kept up for long; the response is usually localised and the nerve carrying the impulse may go to only one effector. A typical nervous response is turning the head at the sound of a sudden noise. Though hormone communication may also produce a short-term response, it is used mainly when the response needed is long-term and when many different effectors throughout the body need to be reached.

The differences between hormonal and nervous control of responses are shown in Table 34.1.

There are dozens of human hormones but you need to know only a few of them: insulin, adrenaline and the main sex hormones.

Insulin

Insulin is a hormone secreted by the **pancreas**. The pancreas lies under the stomach and is joined by a tube to the small intestine. Its position (with the stomach and liver pulled up) is shown in Figure 34.1.

The pancreas is a complex organ because it is both kinds of gland in one: it has a duct passing enzymes into the small intestine; it also secretes insulin into the blood like a ductless gland. The cells in the pancreas which secrete enzymes are different from those which secrete insulin.

Insulin regulates the sugar (glucose) content of the blood, always bringing it down to about $0.6\,g$ per dm^3 of blood. The pancreas cells which make insulin are able to monitor the sugar level of the blood. If it rises above $0.6\,g$ per dm^3, the pancreas secretes insulin, which passes round the body in the blood. When insulin reaches the cells of the liver and muscles, it makes them remove sugar from the blood and store it as glycogen. As the level of sugar in the blood falls, less insulin is secreted; when it reaches a level of $0.6\,g$ per dm^3, no insulin is secreted. It can take a few hours for insulin to bring a raised sugar level down to normal. Control of the sugar level of the blood by insulin is a simple example of homeostasis, which means keeping something in an organism constant or with narrow limits.

More than $0.6\,g$ sugar per dm^3 blood $\longrightarrow$ leads to more insulin produced by the pancreas $\longrightarrow$ leads to only $0.6\,g$ sugar per dm^3 blood

Diabetes mellitus

If the pancreas produces too little insulin, or if the body cannot use insulin efficiently, there is soon too much sugar for the kidneys to reabsorb after

Table 34.1

Hormonal control of responses	Nervous control of responses
by chemicals	by electrical impulses
through blood vessels	through nerves
slow response	quick response
widespread response	localised response
short-term and long-term effect	short-term effect
starts at ductless glands	starts at brain or spinal cord

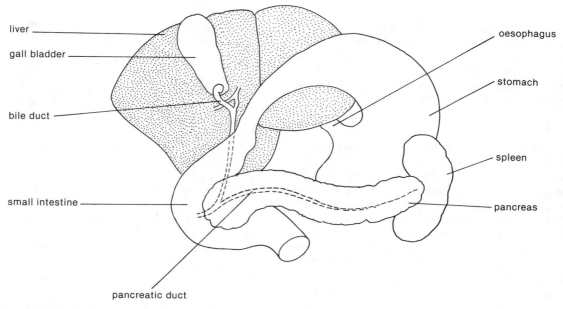

Fig. 34.1 Position of pancreas

filtration. Sugar then passes into the urine. People whose bodies cannot control their blood-sugar level are suffering from *diabetes mellitus* or *diabetes* for short. Diabetes is easily diagnosed from sugar in the urine. With a low-sugar diet and with controlled doses of insulin to help the liver and muscles convert excess sugar to glycogen, diabetics can lead perfectly normal lives.

Adrenaline

Adrenaline is a hormone that stimulates the conversion of glycogen back to sugar. Its secretion is more complicated than insulin secretion; it also has more obvious effects. Adrenaline is secreted by two ductless glands, the **adrenals**, which lie above each kidney. Figure 34.2 shows the position of the adrenals. Adrenaline is known as the 'fight-or-flight' hormone because it increases the efficiency of fighting or escaping. At times of danger, nerve impulses from the brain stimulate the adrenals to pour out large amounts of adrenaline. Small amounts of adrenaline are always present in the blood.

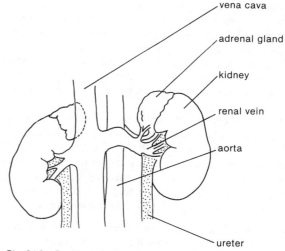

Fig. 34.2 Position of adrenal glands

Adrenaline, together with nerve impulses directly from the medulla of the brain:

    increases the blood-sugar level (by encouraging the release of sugar from stored glycogen in the liver);
    increases the diameter of small arteries in the muscles (**vasodilation**);
    increases the rate of heartbeat;
    increases blood pressure;
    increases the breathing rate;
    decreases the diameter of small arteries in the skin and gut (**vasoconstriction**).

Since blood carries adrenaline around the body, the effects are widespread. All the effects increase the efficiency of fighting or escaping: blood is withdrawn from unessential organs, such as the gut and skin, so that more is available for the muscles; more blood flows through the capillaries of the muscles and blood travels more quickly; more oxygen arrives at the lung surfaces for the blood to take up; more sugar and more oxygen are carried to the muscles for increased energy. When the danger has passed, the nerve impulses from the medulla cease and the adrenaline level slowly falls back to normal. If excess sugar remains in the blood, it stimulates the pancreas to secrete insulin: the liver and muscles remove the sugar and convert it to glycogen.

Sex hormones

Humans have a long childhood before they mature and are able to reproduce. During childhood they grow and learn, and the development of the reproductive system is delayed. The start of sexual maturity, when eggs and sperms can be produced, is called **puberty**. The age of puberty varies greatly but is usually in the early teens. Puberty involves many changes apart from the first production of eggs and sperms. All the changes are controlled by hormones, most of them by the sex hormones, i.e. hormones produced by the sex organs. The sex organs are described in detail in Unit 43.

There are three major sex hormones: **testosterone**, **oestradiol** and **progesterone**. The sex organs of both sexes produce all three but the proportions vary. The two testes in the male produce mainly testosterone, while the two ovaries in the female produce mainly oestradiol and progesterone. Testosterone is the male sex hormone; oestradiol and progesterone are the female sex hormones. Progesterone is concerned mainly with pregnancy and is dealt with in Unit 43.

A gland below the brain (the pituitary gland) controls the time of puberty as well as the production of the sex hormones. Once having begun at puberty, the secretion of sex hormones continues throughout life, though in women it is greatly reduced at about fifty when eggs cease to develop and they cease to have periods and to be fertile (the **menopause**).

Testosterone is secreted in males by cells within the testes. It has some once-for-all effects at puberty:
 the penis, scrotum and testes become larger;
 the voice-box and vocal cords change to give a deeper voice;
 the skin becomes more greasy and sweaty;
 a growth spurt increases height, breadth and mass, developing muscle in particular.
Testosterone has other effects from puberty throughout life:
 the testes manufacture sperm;
 fluid is produced by glands in the reproductive system to form semen;
 hair grows around the external genital organs and up from the lower abdomen to the navel, in the armpits, over the chest and face and elsewhere;
 hair may be lost from the temples and scalp (baldness).

Oestradiol is secreted in females by cells in the ovary which surround a developing egg. Since usually only one egg is produced in a four-week **menstrual cycle**, oestradiol is usually produced by only one of the two ovaries. The once-for-all effects of oestradiol at puberty are:
 the uterus and vagina (parts of the internal female reproductive system) become larger;
 the breasts become larger;
 more fat is deposited under the skin of the thighs and elsewhere.
One effect of oestadiol lasts from puberty until the menopause:
 the lining of the uterus thickens, is shed and is replaced in every four-week menstrual cycle.
Other effects of oestradiol last from puberty throughout life:
 hair grows around the genital organs and in the armpits.
The manufacture of eggs is not controlled by oestradiol (as is the manufacture of sperms by testosterone). Undeveloped eggs, to last a woman's lifetime, are present in her ovaries at birth.

All the changes that come at puberty are known as **secondary sexual characteristics**. The **primary sexual characteristics** are those we are born with. They include the penis, scrotum and testes in boys, the vagina, uterus and ovaries in girls.

Questions

Q 34.1 Copy and complete the table.

Hormone	Where produced	How production is stimulated	Hormone effects
Insulin Adrenaline Testosterone Oestradiol			

PRACTICAL WORK

Note for teachers
There is no practical work in this unit. This leaves time to complete the extensive practical work on breathing and smoking in Unit 28.

UNIT 35
Temperature regulation; the liver; homeostasis

Mammals are warm-blooded. The human body's temperature of about 37°C is high compared with the temperature of air. We associate air temperatures above 37°C with hot deserts. It is rare for the air temperature in Britain to be above 37°C. We are likely to complain that it is too hot when the air temperature is 30°C.

Over a 24-hour period, 95% of the energy released in respiration by humans is in the form of heat. By respiration the body can generate more heat than it needs in normal circumstances. This enables the body to keep its temperature at or near 37°C by controlling the amount of heat lost to the air. We also keep warm by reducing the heat lost to the air in artificial ways: we wear clothes and have fires and central heating. You often feel that your body is gaining heat from a fire. Usually all that is happening is that, because the air is warmed, your body is losing less heat than it would otherwise do. Only if you were very close to a fire would the air temperature be above 37°C, i.e. above your body temperature. The body needs heat from its surroundings only if its own temperature has fallen significantly below 37°C: people suffering from **hypothermia** should be moved to a warm room and their bodies should be covered with insulating material and hot water bottles.

Temperature regulation

The body's cells work most efficiently at temperatures of about 37°C. It is therefore important that the temperature anywhere in the body should be near to 37°C. Temperature regulation involves not only controlling the heat lost to the environment but also distributing heat throughout the body.

Because heat is released in respiration, it forms everywhere in the body where there are living cells. Brown-fat tissues (which lie below the skin at the back of the neck and between the shoulder blades) and organs such as the liver and muscles, in which respiration is vigorous and a lot of energy is released, are the hottest regions of the body. Blood passing through these organs is heated and carries heat away with it. When this blood reaches cooler regions, it gives up some of its heat to them by conduction.

Evaporation

Wet regions exposed to air are cool because evaporation is taking place there. In order to evaporate from liquid to gas a substance needs energy. This energy can be supplied as heat: when the liquid turns to gas, heat is used and the part providing the heat gets cooler. You may have noticed this when a quickly evaporating substance, such as a deodorant or scent, touches your skin. Your skin feels cool because it has given up heat for evaporation. Your skin does not stay cool for long because it regains heat from the warmer blood flowing through it.

Sweating is a way of increasing evaporation. It takes place from the sweat glands in the skin. When we are hot, the brain sends nerve impulses to increase sweating. Sweat, consisting of water, inorganic ions and urea, is extracted from the surrounding blood capillaries and flows up the sweat duct and out of the sweat pore. It spreads over the surface of the skin: when it evaporates, the skin is cooled. Sweating from a sweat gland is shown in Figure 35.1.

The hotter we are, the more we sweat. With an air temperature above 37°C we tend to sweat profusely. In hot dry climates, such as hot deserts have, sweating is an efficient way of cooling the body. The most uncomfortable climates are hot humid ones because the moisture already in the air slows down evaporation.

The coolest regions of the body are those that are exposed to air and are wet: the skin after sweating and the lining of the ventilation system.

Vasodilation and vasoconstriction

The cooling of the body is further adjusted by allowing more or less blood into the skin. The more blood that flows in the skin, the cooler the blood will become because the more heat will be lost to the air. To keep as much heat in the body as possible, blood must be withdrawn from the skin. Your skin is darker when you are hot and losing heat; it is paler when your body is retaining heat. The small arteries in the skin carry blood to the capillaries near the surface. When the muscles in the walls of these small arteries relax, the diameters of the artery cavities increase,

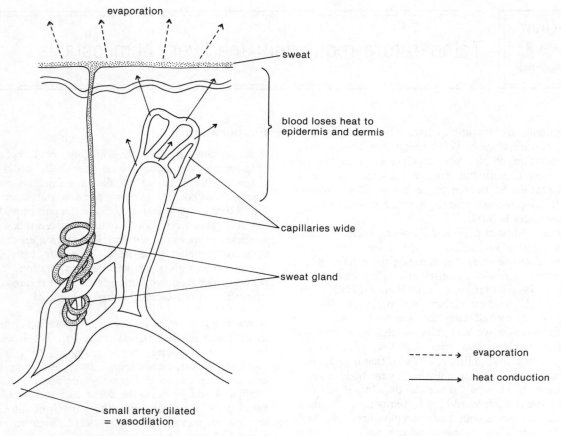

Fig. 35.1 Heat loss by sweating and vasodilation

more blood flows through them into the surface blood capillaries, and more blood is cooled. This is called **vasodilation**. Sweating is usually accompanied by vasodilation in the skin. When the muscles in the walls of these small arteries contract, the diameters of their cavities decrease, less blood flows through them into the surface blood capillaries, and less blood is cooled. This is called **vasoconstriction**. The muscles in these arteries are controlled by the medulla of the brain.

Reduced heat loss

In very cold weather it is important for the body to retain its heat, but some is always lost from the ventilation system and the skin: the air you breathe out is warmer than the air you breathe in; though sweating can be stopped, there is always slight evaporation from the cells of the epidermis.

Heat loss is reduced by the layer of fat below the dermis. Fat is an insulator: the thicker the layer, the less the heat lost from the body. Fat performs the same function as the layer of air trapped by fur in other mammals.

Other mammals reduce heat loss by fluffing up their fur to trap air within it and so insulate themselves more efficiently. Our hair does not insulate us, but we get the same effect by wearing suitable clothes. We call clothes that provide good insulation 'warm clothes' though the clothes themselves are not warm. Have you ever wondered why Arabs in hot deserts cover themselves in clothes? At night of course, when the desert is very cold, the clothes keep the body's heat in. But the reason they wear the same clothes during the day is that the insulation they provide keeps the heat out. Their camels' fur acts on the same principle: in hot weather it usually lies flat to allow heat to escape from the body; in very hot weather, however, the fur fluffs up, just as it does in cold weather, to trap air within it and so provide insulation. Whereas in cold weather the insulation keeps heat in, in very hot weather it keeps it out.

Heat gain

The only method of generating heat in the body is respiration. When we are cold, respiration is increased. **Shivering** is one way of making the muscle cells respire more vigorously and generate more heat. A more efficient way is running about. When we are cold, liver and brown-fat cells also become more active and generate heat: we need more food in winter simply to generate more heat.

The liver

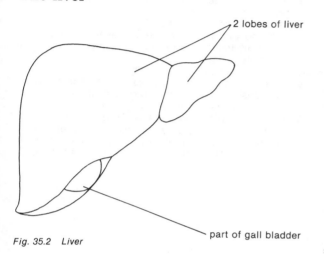

Fig. 35.2 Liver

The liver lies below the diaphragm with its lobes partly covering the stomach and the first parts of the small intestine. Figure 22.1 shows its position in the abdomen and Figure 35.2 shows its shape. The gall bladder, where bile collects, is embedded in the underside of the liver. The liver plays a central role in keeping internal conditions in the body constant or within narrow limits, i.e. in homeostasis. Because it has many different functions, the liver is mentioned in many different units. Its important homeostatic functions are:

- conversion of sugar to glycogen under the influence of insulin, storage of glycogen, and conversion of glycogen to sugar under the influence of adrenalin;
- regulation of the amino-acid and lipid content of the blood, deamination of excess amino acids and formation of urea;
- breakdown of alcohol and barbiturates;
- production of heat;
- production of bile;
- storage of iron;
- breakdown of red blood cells.

Homeostasis

Homeostasis, the process of keeping conditions in an organism constant or within narrow limits, takes many forms. Temperature regulation is a good example of it. Temperature receptors in the brain monitor the temperature of the blood as it passes through: if the temperature is different from 37°C, they send nerve impulses to start correction processes. When the temperature is above 37°C, impulses put into operation mechanisms for losing heat (such as sweating); when the temperature is below 37°C, impulses put into operation mechanisms for gaining heat (such as shivering).

Other examples of homeostasis are: control of the sugar, amino-acid and lipid content of the blood by the liver, described in Unit 24; removal of urea by the kidney, described in Unit 30; osmoregulation, control of the inorganic-ion and water concentrations, by the kidney, described in Unit 30.

Homeostatic mechanisms often involve hormones.

Questions

Q. 35.1 List the methods of (a) gaining heat in the body and (b) losing heat from the body.

Q. 35.2 Explain why:
(a) we lose less heat when air is damp than when it is dry even if the air temperature is the same;
(b) a fan helps to keep us cool;
(c) we can lose heat at an air temperature of 45°C;
(d) a hot dry climate is more comfortable than a hot moist climate.

Q. 35.3 Draw a section through the skin showing only the epidermis, the dermis and a set of capillaries leading from a small *dilated* artery to a small vein. Make a second drawing to show the same set of blood vessels after vasoconstriction of the artery. Add arrows to both drawings to show the direction of blood flow.

Q. 35.4 Explain why it is important that the temperature inside the body should not vary very much.

PRACTICAL WORK

Experiment 35.1 To investigate the skin and muscle changes that occur during temperature regulation

Note for teachers
Running on the spot is customary in this experiment but other forms of exercise may be used provided that they impose no strain on the individual student.

Materials required by each pair of students
1 stopclock
1 bucket
ice

Method
1. Copy the table.

Activity	Face darker or paler	Presence of sweat on face	Skin on arm smoother or less smooth	Shivering
Both arms in iced water				
After exercise				

2. Work in pairs and follow instructions 3 to 5 in turn.
3. Ensure that your partner's sleeves are rolled up as high as possible. Immerse your partner's arms in a bucket of ice and water for as long as your partner feels comfortable. Record your partner's reactions in your table.
4. Remove your partner's arms from the water and dry them. Your partner should now take about a minute's exercise such as running on the spot.*
5. Record in the table your partner's reactions to the exercise.
6. Change roles and repeat the experiment.

Interpretation of results
1. Did your partner have a darker skin with both arms in water or after exercise?
2. What caused the darkness?
3. How do the events in the skin which cause such colour changes help to keep the temperature inside the body constant?
4. What is the value of shivering?
5. When did your partner sweat more? Explain what happens to sweat and how sweating helps to regulate the temperature inside the body.
6. If ice in a bag had been wrapped round your partner's neck, the changes you observed would have occurred even faster. What does this suggest about the site of temperature receptors inside the body?

UNIT 36

Hormones in plants; artifical use of hormones

Plant hormones

Hormones are chemicals which are produced in one part of the body and have an effect elsewhere. All hormones are complex chemical compounds and plant hormones are chemically different from animal hormones. Plant hormones also differ from animal hormones in that they are not produced in glands and are not transported by blood. They are produced only in certain regions of the plant but from cells that look the same as other plant cells. They are produced in tiny quantities: concentrations of a thousand millionth (0.000 000 001) of a gram can have visible effects on a plant. Because they control the growth of plants, plant hormones are often called growth substances: they may encourage growth, slow it down, stop it altogether or change its pattern in some way. Tropisms (responses to such stimuli as light, gravity and water, described in Unit 31) are controlled by plant hormones. Produced in the growing tips of shoots and roots, these hormones diffuse to regions where cells are taking in water and expanding: these are the regions that show the bending response.

A simple experiment showing the production and effect of plant hormones is summarised in Figure 36.1. The tips of the shoots of maize seedlings are cut off and placed on agar. Later, pieces of this agar are placed on one side only of a fresh set of maize shoots whose tips have been cut off. Pieces of pure agar (i.e. agar which has not been in touch with the tips of the shoots of maize seedlings) are placed on one side only of another fresh set of maize shoots whose tips have been cut off. The first set of shoots grow more on the side with the agar, and bend, while the second set continue to grow straight. The explanation is that a hormone has diffused from the maize tips into the agar and then into one side only of the first set of shoots: greater growth on the side with the hormone has caused bending.

The shoots with the pure agar on them are a **control**: only the factor being investigated, the hormone, is different in the control; all other conditions of the experiment must be the same. One example only of each stage is shown in Figure 36.1: to be valid the experiment must have at least ten sets of each procedure because single results could always be freakish. Maize is used in the experiment because it has no leaves around the tips of the seedling shoots.

The experiment outlined in Figure 36.1 shows all the important features of plant hormones: they are chemicals produced in one region which have an effect elsewhere; they move by diffusion; they can remain effective even when removed from the plant. In this experiment they have increased the rate of growth of the shoot. Most plant hormones influence growth in one way or another.

New research will add to, and perhaps revise, our ideas about plant hormones.

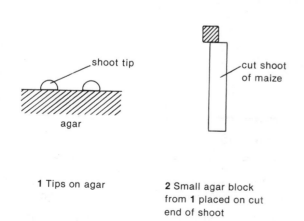
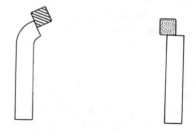

1 Tips on agar

2 Small agar block from 1 placed on cut end of shoot

3 Bending resulting from hormone in agar

4 No bending with pure agar

Fig. 36.1 Effect of hormone from shoot tip

Artificial use of hormones

Plant hormones

Plant hormones are widely used in agriculture and horticulture. Plants produce hormones in such small amounts that their extraction and use would be uneconomical. Since their complex structures are known, they are made in factories. Successful plant-hormone products include those that increase fruit production, selective weedkillers (herbicides) and rooting compounds.

Increased fruit production is achieved in several different ways. There are different compounds which can be sprayed on plants to encourage fruits to form, to prevent fruits dropping off before they are mature enough to pick, and to ripen fruits quickly. Other compounds can be sprayed after harvesting to reduce the number of fruit buds the following year: too many fruit buds result in small undersized fruits.

Selective weedkillers, those that kill some plants but not others, are used in dozens of different forms by farmers, market gardeners and nurserymen producing food and flower crops. Many are extremely dangerous and are not on sale to the general public. Even those on sale to amateur gardeners can be dangerous and must be used according to the manufacturers' instructions. Selective weedkillers are used on lawns where they kill broad-leaved plants but not narrow-leaved ones: the broad leaves of the weeds absorb more of the chemical than the narrow fairly erect leaves of the grass. To ensure that the weeds are killed but not the grass, the exact concentration recommended by the manufacturer must be used at exactly the right time. Many other weedkillers also depend for their effect on the plant hormones they contain. But many weedkillers contain chemicals other than plant hormones. Weedkillers and other chemical products should be applied to food plants only when and where they are essential: like drugs, they may have harmful effects, including some that are not yet known.

Rooting compounds are applied to shoot cuttings to encourage roots to grow from them so that they become new plants. They are often mixed with substances that kill fungi (**fungicides**) to prevent infection and decay of the cut ends of the shoots before they grow roots.

Plant hormones are used to produce seedless grapes, to speed the germination of seeds, and to reduce the stem length of cereal crops so that the heavy seeds are less likely to be knocked down by winds and rain.

Undoubtedly there are better and more economical ways of increasing food production than spraying plants each year with hormones. A method practised for thousands of years is **artificial selection**: humans have chosen to grow crops from varieties that give them the characters they want. In recent years cross-breeding experiments have speeded the development of new varieties. Now it is hoped that **genetic engineering** will speed the process further. Artificial selection is described in Unit 48 and genetic engineering in Unit 47.

Animal hormones

Factory-produced hormones have been fed to farm animals to speed their growth and increase meat production. Hormones extracted from other animals are used in the treatment of hormone deficiencies in humans. Diabetics have been treated with injections of animal insulin since 1922. Infertility in men and women can be similarly treated by giving them extra hormones. Supplementing insufficient hormones in the human body to bring them up to their natural level is not an artificial use of hormones. It is different when athletes take a modified male sex hormone to increase their muscle and bone, because they are not deficient in hormone and end up with more than is natural.

The most common artificial use is in the female contraceptive pill. There are several different varieties but all contain factory-made (synthetic) versions of female sex hormones: some contain only oestrogens (the group that include oestradiol) or progesterone, while others contain both. The quantities of these hormones in the pill prevent eggs from maturing or from implanting in the uterus wall or both. The contraceptive pill is described in Unit 44.

Questions

Q 36.1 List (a) the similarities and (b) the differences between plant and animal hormones.

Q 36.2 Describe one artificial use of plant hormones. Give one advantage and one disadvantage of this use.

PRACTICAL WORK

Experiment 36.1 To investigate the effect of rooting powder on the production of roots by stem cuttings

Note for teachers
Further details of taking stem cuttings can be found in Experiment 40.1 and the two experiments can be performed at the same time. Various brands of hormone rooting powder are available from garden centres and biological suppliers: all are suitable for this experiment. Stem cuttings can be grown in pots of vermiculite instead of in compressed peat pellets. But the pellets, consisting of dried peat which expands to form a small pot after soaking in water, have proved more successful, especially with students who have little experience of cultivating plants.

Materials required by each student
1 scalpel
1 seed tray
6 compressed peat pellets
6 plastic labels
1 bowl
1 ruler graduated in mm
1 spirit marker
several *Pelargonium* (geranium) plants with non-blooming side branches

Materials required by the class
hormone rooting powder
powdered talc

Method
1. Soak the compressed peat pellets in a bowl of water until they expand fully.
2. Choose six well-developed side shoots and use the scalpel to cut them from the *Pelargonium* plant at an angle to their direction of growth.
3. Cut off the lower leaves and the portion of stem below the lower node of each cutting.
4. Dip the cut ends of three of the cuttings in rooting powder. Lightly tap them to remove excess powder so that only a thin coating is left on each.
5. Plant these cuttings in three of the soaked peat pellets. Label them A, B and C and add your initials and the date. Ensure that the stems rest against the bottoms of the holes in the pellets with no air gap.
6. Repeat instructions 4 and 5 using powdered talc instead of rooting powder and labelling the peat pellets D, E and F.
7. Put all the peat pellets in a seed tray and put the seed tray in a warm place away from direct sunlight. Whenever the pellets appear to be drying out, water the cuttings, giving each one the same volume of water.
8. After eight weeks carefully remove the cuttings from the pellets and wash the roots.
9. Record the number of roots on each stem and the length of each root.

Interpretation of results
1. Why were the cuttings labelled D, E and F dipped into talc?
2. Why were the cuttings (a) left in a warm place, (b) left away from direct sunlight, (c) watered?
3. Calculate the mean number of roots grown by the cuttings dipped in (a) rooting powder and (b) talc.
4. Calculate the mean length of the roots produced by cuttings which had been dipped in (a) rooting powder and (b) talc.
5. What do your results indicate about the importance of rooting powder to plant growers?

UNIT 37: Living on land: support in flowering plants and arthropods; surface area and volume

Animals and plants were living in water on planet Earth over 500 million years ago. Land-living animals and plants appeared about 300 million years ago. The plants did not have flowers but were as large as trees; the animals included millipedes, mites, spiders, wingless insects and animals like today's woodlice. These animals belonged to a group still successful today: the arthropods. Living on land poses problems not faced by organisms living in water.

Problems of living on land

Protoplasm is mainly water. Unless air is saturated with water, which sometimes happens in a mist, anything wet will dry out in air as water evaporates. Protoplasm dries out and dies in air unless something prevents it.

Air is not as dense as water. Water is much more supportive than air: you can float on water but you cannot float on air – not without a parachute.

Reproduction presents problems on land: plants and animals living in water can release sperms and eggs for external fertilisation and external development of an embryo. This is not possible in land-living organisms. External fertilisation and embryo development in the trout are described in Unit 44.

The supply of oxygen is *not* a problem on land because air contains much more oxygen than water does. But gas-exchange surfaces, since they must always be moist, must be protected from contact with dry air.

Overcoming the problems of living on land

To prevent excessive evaporation, land-living organisms have waterproof surfaces and enclosed gas-exchange surfaces. Gas-exchange surfaces of plant cells are enclosed in the leaves, stems and roots; gas-exchange surfaces of the air sacs in humans are enclosed in the chest cavity. Both are described in Unit 27.

A firm framework provides support for land-living organisms.

Instead of external fertilisation, land-living organisms have internal fertilisation: fertilisation in flowering plants is described in Unit 42 and in humans in Unit 43.

Preventing evaporation

Little water evaporates through the waxy cuticle on the leaves and young stems of plants or through the bark on old stems. Water evaporates more easily through the pores that are present for gas exchange.

Arthropods, which have been so successful on land, are also protected from drying out by the waxy surface of their outer cuticle. Like plants, they lose some water through the pores that are present for gas exchange.

The epidermis of the skin of humans is also fairly waterproof. The oily secretion from the glands surrounding the hairs helps to waterproof the skin. But some water still evaporates through the epidermis.

Thus none of these surfaces is completely waterproof.

Support

All except a few very small and shapeless organisms need to be supported in air. Protoplasm has a texture like half-set jelly: without support in air it collapses. The problem is less acute in plants, in which every cell has a cellulose cell wall supporting it. But a cellulose framework is not strong enough to support large plants.

Arthropods are supported by their outer cuticle, which can be both hard and thick. The outer cuticle of crabs' legs can be several millimetres thick. The cuticle is its skeleton. Because it is on the outside of the body it is called an **exoskeleton**. The soft living tissues hang on the inside of the skeleton.

Flowering plants

Young stems are first supported entirely by turgid plant cells with cellulose walls: they are full to bursting with water in their central vacuoles. These cells are described in Units 13 and 16. The whole stem is under tension and held together by its outer layer, the epidermis. The stem is like a blown-up balloon except that it is made up of separate units (cells) and that they are full of water. Just as a balloon without internal air pressure collapses, so stems without enough water droop and wilt.

Older stems are also supported by turgid cells but in addition have lengthwise supporting struts of thick-walled dead cells, especially in their outer regions. The walls are thickened with deposits of **lignin** on the original cellulose, forming the substance we know as wood. Figure 37.1 is a photograph of a transverse section of a stem showing the arrangement of about 23 supporting struts near the outside.

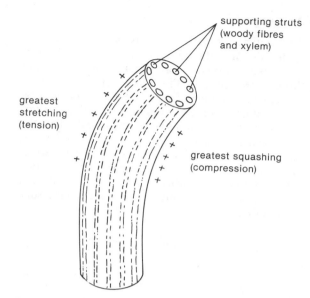

Fig. 37.2 Bent stem

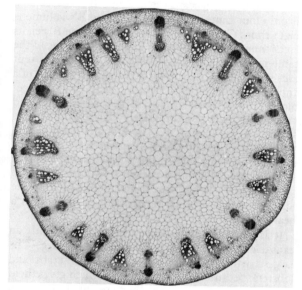

Fig. 37.1 Stem: transverse section

The outer cells and some of the inner cells of each strut are small in transverse section, but they are thick-walled, long and woody. They are called **fibres** and are only for support; in the photograph they are very dark. The other inner cells of the struts, which are wider but also thick-walled, long and woody, are xylem cells, called **vessels**, which transport water as well as giving support. Lengthwise struts separated by turgid cells are mechanically much more efficient than a single central strut would be: there can be some sliding movement between the struts; they give support at the outer regions where there is greatest stress when the wind blows and the stems are bent. Figure 37.2 shows a bent stem: the greatest squashing (compression) is on the inside of the curve; the greatest stretching (tension) is on the outside of the curve. These are just the places where the supporting struts are.

Tree trunks are supported by almost solid wood.

Leaves are thin and flat. This is the shape that gives maximum support in air. It is the shape of a bird's wing and of a hang-glider. Leaves are held out flat by networks of supporting veins, shown in Figure 37.3.

Fig. 37.3 Leaf: supporting veins

Arthropods

The cuticle of arthropods is secreted by their epidermis. All arthropods have a cuticle which forms their exoskeleton. Figure 37.4 shows a section through the cuticle. The outer surface is waxed, reducing evaporation of water from the animal. Most of the cuticle consists of protein, which gives it slight flexibility. The cuticle also contains strands of **chitin**, laid down at all angles, which give it strength, and calcium-carbonate salts, in the outer layers, which give it hardness. Figure 37.4 shows that the cuticle varies in thickness: where it is thin, it is flexible enough to form the joints between the thicker solid pieces of cuticle.

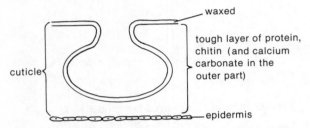

Fig. 37.4   Arthropod cuticle: section

The exoskeleton is easy to see in a large arthropod such as a crab. Like a suit of armour, it protects and supports the soft tissues inside. But the exoskeleton cannot grow and from time to time has to be replaced. When the tissues inside have grown to fill it, the exoskeleton splits, the soft-bodied arthropod squeezes out, quickly expands in size and secretes a new and larger cuticle. Until the cuticle has hardened, which may take hours and even days, the arthropod is vulnerable to attack by predators. Shedding its cuticle in this way is called **moulting** (or **ecdysis**). It is wasteful of the substances it contains: wax, protein, chitin and calcium carbonate. An exoskeleton is mechanically sound, giving support at the surface, where it is most needed, but there are limitations on its size.

Size

Although arthropods have been living on land for 300 million years, they have not grown very big. They have changed hardly at all from those found as fossils in rocks 250 million years old. One limit on their size is mechanical: the exoskeleton needed to support a larger animal would be too heavy for it to carry.

Look at the cubes in Figure 37.5 and at the calculations underneath them. Cube B's surface area is only four times Cube A's. But Cube B's volume is eight times Cube A's. This means that cube B weighs eight times as much as cube A (if they are made of the same material). Question 37.1 at the end of this unit asks you to calculate the surface area and volume of a cube of side 3 cm. You will see that, as cubes get bigger, the difference between their surface area and volume also gets proportionately bigger. Similarly, as arthropods get bigger, they have bigger volumes and weigh more in proportion to their surface areas. They therefore need proportionately thicker exoskeletons to support them. Beyond a certain size, an exoskeleton has to be so thick that there is not enough room inside for the amount of muscle necessary to move such a heavy structure.

It is therefore impossible for an animal with an exoskeleton to be greater than a certain size. The largest arthropods living today are the crabs and lobsters: because they live in water, which gives them more support than air can, they are able to be much larger and heavier than arthropods on land. Only small crabs are ever seen running about in air on the seashore.

The relation between volume and surface area has implications for mammals too. A mammal is warm-

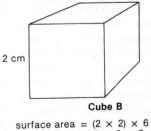

Cube A
surface area = (1 × 1) × 6 = 6 cm²
volume = 1 × 1 × 1 = 1 cm³

Cube B
surface area = (2 × 2) × 6 = 24 cm²
volume = 2 × 2 × 2 = 8 cm³

surface area cube A : cube B = 6 : 24 = 1 : 4
volume cube A : cube B = 1 : 8   1 : 8

If cube A weighs 10 g, cube B will weigh 80 g

Fig. 37.5   Size: surface area and volume

blooded and maintains its temperature at a constant level. If its surroundings are cold, it loses heat from its surface: a small mammal such as a shrew or mouse has a proportionately large surface area from which it loses heat. If it is very cold, a small mammal may lose heat faster than it can find food and eat it to release energy to maintain its heat: small mammals often hibernate, i.e. spend the winter in an inactive state in which it does not matter that their temperature drops; few large mammals hibernate.

At the other extreme, a large mammal in hot weather has difficulty losing enough heat to keep cool: its body cells continue to produce heat which, because of its proportionately small surface area, cannot be lost quickly enough. In hot weather elephants are lethargic and do not exert much energy, with the result that they make less heat; they seek shade and flap their huge ears which, because they are thin and have a large surface area, give off heat and help to cool the blood. Elephants also douse themselves with water in hot weather: the water evaporates and helps to cool them.

There are two possible reasons why there is no land animal bigger than an elephant: a bigger animal might not be able to lose enough heat; its skeleton might not be able to support it in air. A whale is bigger than an elephant, but it is supported by water and it loses heat to water quicker than a land animal loses heat to air.

Size and shape

If a cube is converted to a flat structure with the same volume, its surface area becomes larger. Figure 37.6 shows A and B flattened out so that they are only half as deep. Though the volume of each has remained the same, cube A's surface area has increased from 6 to 7 cm² and cube B's from 24 to 28 cm². Thus the surface areas of both have increased by a sixth. (As in Figure 37.5, B's surface area is only four times A's, yet its volume is eight times A's and it therefore weighs eight times as much.)

Flat structures have a proportionately larger surface area than cubes of the same volume and mass. This mechanical principle explains the shape of leaves. Their large surface area enables the air to support them: if they were more compact in shape, they would weigh the same without having the advantage of some support from air. Flat leaves also provide a large surface area for the absorption of light in photosynthesis.

Cube A

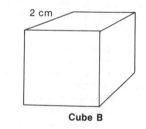

Cube B

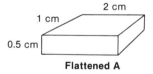

Flattened A

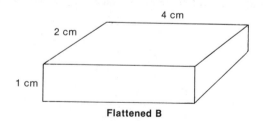

Flattened B

volume of cube A = 1 × 1 × 1 = 1 cm³
volume of flattened A = 1 × 0.5 × 2 = 1 cm³
surface area of cube A = (1 × 1) × 6 = 6 cm²
surface area of flattened A = (1 × 1) × 2 = 1 cm²
  + (2 × 0.5) × 2 = 2 cm²
  + (1 × 2) × 2 = 4 cm²
  ─────
  7 cm²

volume of cube B = 2 × 2 × 2 = 8 cm³
volume of flattened B = 2 × 1 × 4 = 8 cm³
surface area of cube B = (2 × 2) × 6 = 24 cm²
surface area of flattened B = (2 × 1) × 2 = 4 cm²
  + (2 × 4) × 2 = 16 cm²
  + (1 × 4) × 2 = 8 cm²
  ─────
  28 cm²

Fig. 37.6 Surface area and shape

Questions

Q 37.1 (a) What is the surface area of a cube of side 3 cm?
(b) What is the volume of a cube of side 3 cm?
(c) What is the mass of a cube of side 3 cm if a cube of the same material of side 1 cm weighs 10 grams?

Q 37.2 (a) Explain how a stem is supported with woody struts.
(b) Explain how a stem without woody tissue is supported.

PRACTICAL WORK

Experiment 37.1   To investigate the distribution of supporting tissues in a plant stem and a mammalian limb bone

Note for teachers
The first part of this experiment requires a compound microscope with an objective lens allowing at least ×10 magnification. The experiment can be performed adequately with one microscope for every two students. If there are not enough microscopes, the teacher can demonstrate the investigation, allowing the students to look down the microscope, or use photographs, or do both.

Unstained sections of plant stems can be obtained from biological suppliers. If teachers intend to prepare their own, they should use a sharp razor blade or hollow-ground cut-throat razor to take thin transverse sections through the stem of a herbaceous plant such as a geranium. The sections should be stored in 95% ethanol until needed.

Phloroglucin stain should be made by dissolving 3 g of phloroglucinol in 100 cm^3 of 95% ethanol. Just before use the stain solution should be acidified with 2 cm^3 of concentrated hydrochloric acid.

Fresh bones can be obtained from butchers and should be sawn in thick transverse sections across the diaphyses.

Materials required by each student
1 dropper pipette
2 glass microscope slides and coverslips
1 pair of forceps
1 mounted dissection needle
1 microscope
5 cm^3 acidified phloroglucin solution
2 transverse sections of a geranium stem in 95% ethanol
1 transverse section of a mammalian limb bone

Method
1. Use the forceps to put one of the transverse sections of a geranium stem on to a microscope slide.
2. Add a few drops of the phloroglucin stain to the plant section and lower a coverslip on to it, using the mounted needle to avoid trapping air bubbles.
3. Leave the mounted section for two to three minutes to allow the stain to take, and then examine the section under the microscope. Make a labelled drawing to show in which regions of the stem cells have stained red. Do not attempt to draw the cells.
4. If none of the cells in your mounted section have stained red, repeat instructions 1 to 3 using another plant section.
5. Examine the cut end of the bone. Make a labelled drawing to show the position of the hard bone tissue in the bone.

Interpretation of results
1. The phloroglucin made cell walls containing lignin turn red. Lignin is found in supporting tissues in plants. Describe the positions of the supporting tissues in the plant-stem section which you examined.
2. Was the distribution of hard bone tissue in the limb bone similar to that of lignified tissue in the plant stem? If so, suggest why the supporting tissue of these structures is best placed where it is. (Think of the forces acting on the stem and bone.)

UNIT 38
Support and size in mammals; movement of bones

The shape of an arthropod is entirely that of its exoskeleton. A mammal gets its basic shape from its internal skeleton or **endoskeleton**. The skeleton in Figure 38.1 is unmistakably that of a human being. All mammals are recognisable from their skeletons. One function of both exoskeletons and endoskeletons is to give the body shape. More important is the support they give to the soft parts of the body, which in air would collapse without a skeleton.

Bone structure

Look at the separate bones of the limbs in Figure 38.1. They are all long and narrow at the **shaft** but enlarged at the **head**. The hollow in the bone shaft is filled with a light connective substance called **marrow**, which makes some of the white blood cells in living bone. The hollowness makes bone lighter without sacrificing the strength provided by its cylindrical shape. The enlargement of bones at their heads (i.e. where bones meet) increases stability and provides space for the attachment of tendons and ligaments.

Figure 38.2 is a long limb bone cut in half longitudinally. In Experiment 37.1 you saw a transverse section through the cylindrical shaft of a similar bone. The cylindrical shaft can take enormous stresses when the animal is running. At the ends of the bone there is white smooth **cartilage**, which moves with little friction against the limb girdle and against the lower leg bones at the joints.

Bone does not look like living tissue, but living cells are spread throughout it. All the non-living part of bone is secreted by these living cells. The non-living part consists of protein strands and, surrounding them, inorganic salts: the protein strands are extremely tough but allow slight flexibility; the inorganic salts, mainly calcium phosphate, are hard.

In the swollen heads of limb bones, bone tissue is far less dense and hard; indeed it often has a spongy appearance. Whereas the stresses on the shaft are vertical, those on the heads are often in different directions. In the heads small bars of bone lie in all directions, forming a network with fluid-filled spaces between.

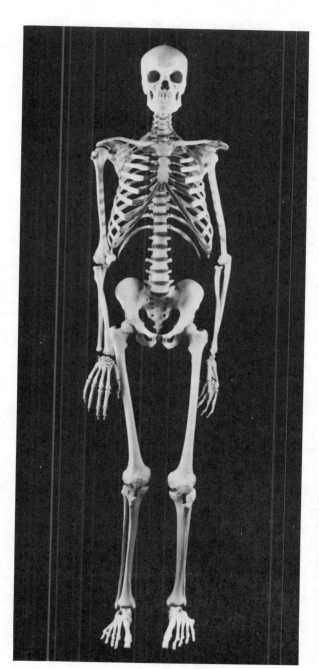

Fig. 38.1 Human skeleton

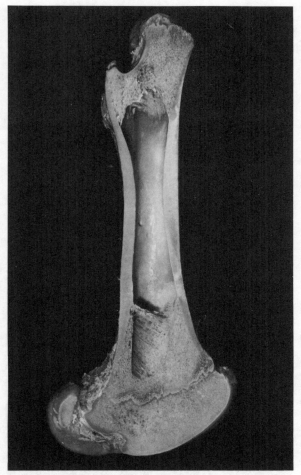

Fig. 38.2 Femur bone of sheep: cut in half longitudinally

Size

Because they are supported internally, mammals can grow much bigger than arthropods. But there are upper limits to the size of a mammal: one limit is that the mass of a very large animal, when supported by only four legs, would be more than the bone could bear. The only land-living mammals much taller than elephants are giraffes, which are thinner and weigh less. Some of the four-legged reptiles living 100 million years ago must have weighed 30 tonnes and been 7 metres high and 25 metres long. Some used a large tail as a fifth supporting 'leg' and some lived partly in water. But *Brontosaurus*, shown in Figure 38.3, is one of the largest known dinosaurs and is now thought to have lived on land. It is possible, therefore, that gravity was weaker when the dinosaurs lived: if gravity was weaker, they will have weighed less than they would today.

Fig. 38.3 Brontosaurus

Land animals have their weight pressing down on the bones of their legs and feet. It is the problem of supporting themselves in air which limits their size. Mammals in water grow much bigger than mammals on land: the whale is now the biggest animal alive and it never leaves the water. Whales washed up on beaches die, crushed by their own enormous mass.

Movement

Movement in both arthropods and mammals is brought about by the power of muscles attached to the skeleton. The power of muscles comes from their ability to shorten in length, i.e. contract. By contracting, muscles pull bones: they cannot push them. The skeleton is free to move because, between the rigid supporting sections, there are **joints**. In arthropods these take the form of thin pliable regions of cuticle, shown in Figure 37.4. In mammals, joints are formed at the ends of bones where they are covered by smooth cartilage. Smooth cartilage of one bone moves against smooth cartilage of its neighbour: a lubricating fluid is secreted between the two surfaces, giving almost frictionless movement. When there is no movement, for example when a person is standing still, tension in the muscles holds the bones in place.

In arthropods the muscles lie inside the skeleton, whereas in mammals they lie outside it. In both arthropods and mammals, muscles are attached to the skeleton by **tendons** across the joints.

Movement at a joint in a mammal

Movement of a single limb, let alone of the whole body, is complex. There are many blocks of muscles attached at their ends to different regions of different bones, bringing about small and subtle movements in different planes. But all movements of the skeleton are governed by the same principles. Muscles work in pairs. When one muscle of a pair contracts, it pulls a bone in one direction; when the other contracts, it pulls the same bone in the opposite direction. Because they produce opposite movements, the two muscles are called **antagonistic muscles**. When one muscle is fully contracted, the other is fully relaxed.

The biceps and triceps muscles are an antagonistic pair of muscles in the upper arm. They move the lower arm at the elbow joint. They control movement *only* at the elbow joint and *only* in one plane. Other pairs of antagonistic muscles surround them, attached to different regions of the bones, producing twisting and sideways movements of great precision. Yet other pairs of antagonistic muscles control movement at the shoulder between the scapula and the humerus.

The biceps and triceps bring about bending and straightening of the elbow. The four different bones involved in this movement are shown in Figure 38.4. The upper arm bone is the **humerus**, the shoulder blade is the **scapula**, and the two parallel bones in the forearm are the **ulna** and **radius**. The elbow joint, where movement is produced by the biceps and triceps, is between the bottom of the humerus, the radius on the thumb side of the forearm and the ulna on the little-finger side of the forearm.

Bones are attached to one another and held firmly in place by tough white **ligaments** which can 'give' a little to allow movement. Ligaments are not shown in Figure 38.4. Muscles narrow where they are attached to bones, or there would not be enough space for all the different attachments. Tendons, which join muscles to bones, must be strong and unstretchable because they have to deliver the full power of muscle contraction to the bone. Remember that ligaments join bones to bones, tendons join muscles to bones.

When both the biceps and the triceps are relaxed, the arm hangs down loosely at the side of the body. When the biceps contracts and the triceps is relaxed, the lower arm moves upwards. When the triceps contracts and the biceps is relaxed, the lower arm moves downwards.

The **biceps** (so-called because it has two points of attachment at one end) joins the radius to the scapula. Because the radius fits into a groove at the bottom of the humerus, contraction of the biceps pivots the two bones around the elbow joint and brings the radius upwards towards the humerus. The arm is bent or flexed at the elbow: the biceps is often called the **flexor** muscle.

The **triceps** (so-called because it has three points of attachment at one end) joins the ulna to the scapula by one attachment and the ulna to the humerus by the other two attachments. This is shown clearly in Figure 38.4. When the triceps contracts, it pivots the ulna around the bottom of the humerus and straightens or extends the arm: the triceps is often called the **extensor** muscle.

Movement produced around the elbow joint by the biceps and triceps muscles is in one plane: the joint is

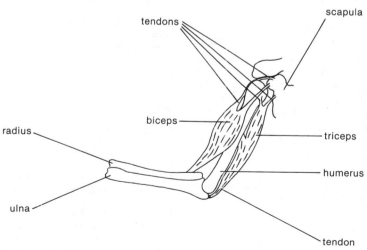

Fig. 38.4 Human elbow: bones and muscles

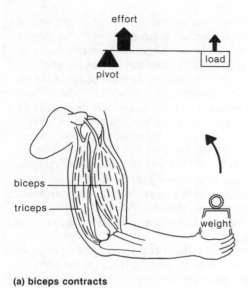
(a) biceps contracts

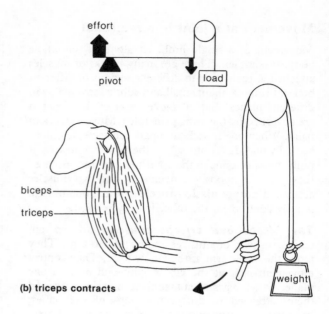
(b) triceps contracts

Fig. 38.5 Movement by levers

described as a **hinge joint**. Movement of segments of skeleton in arthropods is usually by hinge joints.

Figure 38.5 shows the work that the biceps and triceps muscles do when they move the forearm at the elbow. In mechanical terms the forearm is a bar or **lever**; the elbow is a **pivot**; the muscles provide the **effort** shown by the wide arrows; the work to be done is the movement of the **load**, i.e. the weight, by the movements of the hand shown by the narrow arrows. In Figure 38.5(a) the hand moves upwards as the result of the contraction of the biceps; in Figure 38.5(b) the hand moves downwards as the result of the contraction of the triceps.

On a see-saw the lighter of the two people sits further from the pivot in order to balance it. This shows that it is easier to exert a force on a load as you get further away from it. The structure of the arm is such that moving a weight from the elbow is mechanically inefficient. In both examples in Figure 38.5 the distance from the load to the pivot is greater than the distance from the effort to the pivot. To move the weight at the other end the muscles must exert a greater force than that of the weight itself. Fortunately the inefficiency of the muscle attachment is compensated for by the efficiency of the muscles. The biceps is the larger of the two muscles because it usually acts against gravity: the triceps can be smaller because, in moving the forearm down, it is usually helped by gravity.

The joint at the elbow illustrates the principle of levers. Movement of the whole body is carried out by a series of levers.

Muscles

Muscle contraction is controlled by nerve impulses. In the reflex action by which the forearm is quickly withdrawn from a painful stimulus, the impulses pass from the receptors to the spinal cord by the sensory nerve fibres and out to the effector muscles via the motor nerve fibres. Impulses pass to both muscles: the biceps contracts at the same time as the triceps relaxes.

Muscle contraction uses energy. The energy is supplied by respiration. Muscles are well supplied with blood bringing sugar and oxygen and removing carbon dioxide. When muscles are very active for some time, or if many muscles are used at once, the oxygen supply runs out and anaerobic respiration takes place in the muscles. Lactic acid forms, accumulates in the muscles and causes cramp. This is described in Unit 28.

Questions

Q 38.1 Explain why a pair of muscles is needed to control each movement at a joint.

Q 38.2 (a) What prevents displacement of the bones at the elbow joint?
(b) Explain how bone movement at the elbow joint takes place with minimum friction.

(c) (i) Holding on to an overhead crossbar with both hands, you are able to raise your body off the ground by bending your arms at the elbow. What muscles are involved? Which muscles do most work? What is the load being lifted by the muscles? (ii) With your forearms each resting on one of two parallel bars, you are able to raise your body by straightening your arms at the elbow. What muscles are involved? Which muscles do most work? What is the load being lifted by the muscles? (iii) Which of (i) and (ii) is the more difficult activity? Suggest a reason.

PRACTICAL WORK

Experiment 38.1 To investigate the role of calcium salts and protein in producing strength in bones

Materials required by each student
1 250 cm^3 beaker
1 Bunsen burner
1 pair of tongs
1 heat-resistant mat
1 ruler graduated in mm
1 spirit marker
1 pair of safety spectacles
1 piece of red litmus paper
100 cm^3 dilute hydrochloric acid
3 rib bones from a cooked chicken carcass

Materials required by the class
balances to weigh up to 100 g (1 per 4 students)

Method
1. Copy the table at the foot of the page.
2. Taking care not to mix up the bones, weigh and measure the lengths of two of them. Call these two A and B and record these measurements in your table. Call the third bone C.
3. Wearing safety spectacles, pour hydrochloric acid into the beaker.
4. Put bone A into the beaker and ensure that it is completely immersed.
5. Use the tongs to hold bone B in a Bunsen flame.
6. As bone B burns, hold the moistened litmus paper in the gas given off from the bone and record in your table any colour change that occurs.
7. When the bone has stopped burning, put it on to the heat-resistant mat. When it has cooled, measure its length and weigh it. Record these measurements in your table.
8. When bone A has been in acid for not less than one hour and for longer if possible, put on safety spectacles and, using the tongs, remove bone A from the beaker of acid, wash it in water and dry it. Measure its length and weigh it. Record these measurements in your table. Also record any change in bone A's appearance.
9. Bend bone C steadily until it breaks, making a mental note of how far it bends before it breaks and of how much effort you have to exert to break it. How far it bends is its bendability or pliability. How easily it breaks is its brittleness. This bone is a control that enables you to judge how much the experimental treatments have affected the pliability and brittleness of the other two bones. It can be assumed that all three bones had roughly the same pliability and brittleness to start with.
10. Bend each of the other two bones steadily until it breaks and complete your table by recording whether the pliability and brittleness of each have increased or decreased (relative to the pliability and brittleness of bone C).

Interpretation of results
1. The non-living material of the bone contains inorganic salts which are mainly calcium phosphate. These are dissolved by acids. What proportion of the mass of the bone was accounted for by these salts? Suggest how you could find the mass of these salts more accurately.
2. What effect did the hydrochloric acid have on the strength of the bone?
3. The bone also contains a protein which is removed by burning. What proportion of the mass of the bone was accounted for by protein?
4. Using the result of your litmus test, suggest the gas which is produced when proteins burn. (Remember that proteins contain nitrogen.)
5. What effect did burning have on the strength of the bone?
6. Explain any change in the appearance of bone A and bone B.
7. What do your results suggest about the roles of inorganic (mainly calcium) salts and protein in normal bones?

Bone	Treatment	Mass (g)		Length (mm)		Change observed in bone's			Change in colour of litmus paper
		at start	at end	at start	at end	appearance	pliability	brittleness	
A	Soaked in hydrochloric acid								
B	Burned in Bunsen flame								

UNIT 39 Growth

Most plants and animals start life as single cells. As they grow, they usually both make more cells and get bigger. But getting bigger is not necessarily growth: a python gets bigger when it eats a calf and a snail gets bigger when it comes out of its shell to move, but neither is growing.

Growth is an increase in body parts such as protoplasm, skeleton and fur. The mere replacement of body parts is not growth. All increases in body parts take the form either of new protoplasm or of new matter dependent on protoplasm; for example, skeleton is a secretion from protoplasm. Making more protoplasm, i.e. more than is needed for replacement, is therefore crucial to growth.

To make more protoplasm, plants and animals need food. Amino acids provide the materials to make the protein of protoplasm. Sugar and lipids provide energy to put the different chemicals together to make protoplasm. Water is always needed because protoplasm contains a lot of water. As they grow, plants and animals need more supporting material to prevent them from collapsing. As well as making more protoplasm, plants make more cellulose and lignin, mammals make more bone, and arthropods make more exoskeleton.

Plants and animals grow in different ways: only parts of plants grow, whereas animals usually grow all over.

Plants

The stems and roots of flowering plants grow at their tips. These growing regions are well protected, the stem tip by a bud and the root tip by a root cap. Growing regions are delicate places: if one of them is damaged, growth stops there. Leaves have scattered growing regions until they are fully formed, when they stop growing.

Seeds are suitable for growth studies because they grow quickly. In potting compost kept moist and warm seedlings grow noticeably in a day. When a seed first gets bigger, it has not made more protoplasm but has merely absorbed water: this is not growth. As the root and shoot appear, growth is taking place. A growing seedling of French bean is

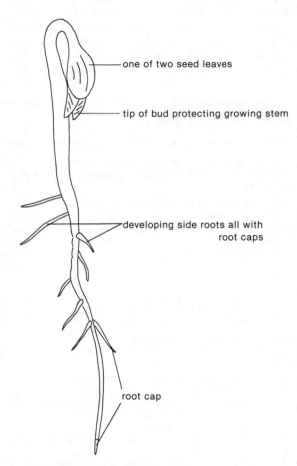

Fig. 39.1 Seedling of French bean

shown in Figure 39.1. Its stem tip is protected not only by a bud but by the two seed leaves which enclose it and by the fact that it is hooked down as it pushes through the soil. All the root tips are protected by root caps. At first all the food for growth comes from the digested food stores in the seed leaves: amino acids, digested from protein, pass from the seed leaves to the growing points to form protoplasm; sugars, digested from starch, pass from the seed leaves to the growing points to provide energy and to form the new cellulose walls; water passes from the root hairs on the roots to the growing points to form protoplasm and to enlarge the new cells.

Later the roots absorb water and inorganic ions from the salts in the soil and, when green leaves have grown, the young plant makes sugars by photosynthesis. Now the plant has a continuous supply of all that it needs for growth: sugars for energy; sugars to make cellulose; sugars and inorganic ions to make amino acids and then the proteins of protoplasm; water to go into protoplasm and enlarge the cells.

Plants usually go on growing until they die: stems and roots get longer and may get wider; new stems, roots and leaves form all the time. Trees may live hundreds or even thousands of years, growing bigger every year.

Animals

Animals grow all over, adding more protoplasm wherever it is needed. But animals stop growing when they reach a certain size. They still make protoplasm and new cells, but only for replacement of parts that are being worn out: the epidermis of the skin forms new cells all the time as the outer surface gets worn away.

Animals need the same food materials as plants for growth: amino acids to make the protein of protoplasm; sugars and lipids to provide the energy to put the chemicals together; water to add to the protoplasm; salts to supply inorganic ions. Because animals cannot make their own food as plants can, they need supplies of far more complicated substances: vitamins in small quantities are particularly needed during growth.

Look at Table 39.1, which lists some of the foods needed each day by a two-year-old child weighing 13.3 kg and by an active adult man weighing 65 kg.

Table 39.1 Some daily food needs

	Child aged 2 years of mass 13.3 kg	Active adult man of mass 65.0 kg
Energy foods	5.7 MJ	12.6 MJ
Protein	16.0 g	37.0 g
Vitamin D	10.0 μg	2.5 μg
Vitamin C	20.0 g	30.0 g
Calcium	0.5 g	0.5 g
Iron	10.0 mg	9.0 mg

The man weighs about five times as much as the child, but he does not need five times as much of any of the foods. In fact the child needs more vitamin D and iron than the man: vitamin D is important for the growth of bone and iron is needed to make red blood cells. The child needs the same amount of calcium as the man: calcium is important for the growth of bones and teeth.

Measuring growth

If we could measure the amount of protoplasm or count the number of cells in a plant or animal, we could easily work out if it was growing. There is no way we can do this. Instead we rely on approximate methods of measuring growth.

Weighing a baby regularly is only a rough guide to how well it is growing: its mass will depend not only on how much it has grown but also on when it was last fed and when it last passed urine and faeces. This is why, as a child gets older and can stand up straight, its height is a better guide to how well it is growing. But this too is only a rough guide because some children are tall and thin while others who have grown by the same amount are short and fat.

We can use height or mass to measure the growth of plants. But plants too can be tall and narrow or short and wide, and increase in mass may be due to taking in water. There are further problems in measuring the mass of plants: they have to be removed from soil and the soil has to be washed off them, after which they cannot be replaced in soil and be expected to grow normally. When measuring the mass of plants, it is best to use samples and throw them away afterwards.

A special difficulty arises in measuring the growth of an arthropod. As the animal grows, its exoskeleton prevents it from increasing in size: after a time its growth results in the shedding of its exoskeleton, and its tissues expand before the new exoskeleton hardens. With the loss of its exoskeleton the arthropod experiences negative growth, i.e. a reduction in body parts. At the same time as the arthropod is 'ungrowing' by shedding its exoskeleton, it is increasing in size by expanding its tissues. If we measure an arthropod's growth by its length, we shall fail to record its growth inside its exoskeleton. Moreover, when it sheds its exoskeleton and expands its tissues, we shall record it as growing when in fact it is 'ungrowing'. If we measure its growth by its mass, we shall correctly record growth as occurring inside its exoskeleton and we shall correctly record negative growth when it sheds its exoskeleton. But, as we have seen, there are other objections to measuring growth by mass. Growth of arthropods is described as **discontinuous**. In all other animals, including humans, growth is described as **continuous**.

Unicellular organisms such as bacteria and yeast grow and divide as soon as they reach a certain size. The growth of colonies of bacteria or yeast can be measured by counting under a microscope the number of cells in small sample volumes. Table 39.2 shows the number of yeast cells seen in the field of view of a microscope when a yeast colony was grown in a sterile solution of food in a flask. Look at the table and you will see that after 24 hours growth slows down: the food supply may be running out or waste products from the yeast may be poisoning it. After 40 hours the number of yeast cells is reduced: yeast cells are beginning to die and growth is said to be negative. If you graph the data in Table 39.2, you will get a typical growth curve: when we are elderly, we too experience negative growth, i.e. our bodies have less protoplasm, (because we have less muscle), less skeleton, less hair, etc.

Table 39.2 Growth of a yeast colony

Age of colony (hours)	Number of yeast cells in microscope field	Increase in number of yeast cells
0	2	
4	5	3
8	8	3
12	15	7
16	29	14
20	52	23
24	86	34
28	116	30
32	134	18
36	142	8
40	143	1
44	142	−1
48	141	−1
52	130	−11

Growth rates

The rate of growth, i.e. the amount of growth that takes place in a certain time, depends on several factors.

The kind of organism: sunflowers grow faster than pine trees; rabbits grow faster than tortoises.

The variety of the organism: climbing French beans grow faster than dwarf French beans.

Inheritance: children of tall parents tend to grow faster than children of short parents.

External factors: growth is faster at warm than at cold temperatures; plants grow faster in good light; animals grow faster if they have plenty of food.

The age of the organism: animals, and even plants, grow more slowly as they get older; eventually animals stop growing altogether.

Hormones: both plants and animals need growth hormones in the right quantities or they become either too small or too big.

Questions

Q 39.1 Provided it has warmth, a seed needs only water and oxygen for growth.
(a) Where do the foods to make protoplasm, cellulose walls and energy come from?
(b) Why does the seed need water to grow?
(c) Why does the seed need oxygen to grow?

Q 39.2 Table 39.1 shows that, although the man is nearly five times heavier than the child, he needs less than five times as much food. Explain why this is true of each of the foods listed in the table.

Q 39.3 Why is mass used to measure a baby's growth while height is used to measure an older child's growth?

Q 39.4 The graph in Figure 39.2 shows the dry mass and fresh mass of samples of ten seedlings during germination.

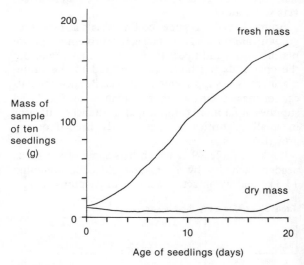

Fig. 39.2

(a) Suggest reasons for the difference between the fresh mass and dry mass of the seedlings.
(b) Explain the slight fall in dry mass during the first sixteen days.
(c) Explain the rise in dry mass during the last four days.
(d) Explain the frequent small variations in both fresh and dry mass throughout the twenty days.

Q 39.5 (a) (i) Draw a graph of the data in Table 39.2 with age of the colony on the horizontal axis and number of yeast cells on the vertical axis.

(ii) Approximately how many yeast cells are there in the microscope field after 21.5 hours?

(b) What further experiments could you carry out to investigate whether growth has slowed down because of lack of food or because of too great a concentration of waste products?

Q 39.6 Table 39.3 shows the mean mass and the mean height of girls and boys from birth to the age of eighteen.

Table 39.3

Age (years)	Mean mass (kg)		Mean height (cm)	
	Girls	Boys	Girls	Boys
Birth	3.2	3.4	50.8	50.8
1	9.5	10.2	76.2	76.2
2	11.8	13.2	83.8	83.8
3	14.1	15.0	88.9	88.9
4	16.3	16.3	96.5	96.5
5	17.7	18.6	104.1	104.1
6	19.1	20.0	111.8	111.8
7	22.7	22.7	119.4	119.4
8	25.0	25.0	127.0	127.0
9	27.7	27.7	132.1	132.1
10	31.8	31.8	137.2	137.2
11	33.6	35.0	139.7	144.8
12	39.0	38.6	147.3	147.3
13	45.8	40.4	152.4	154.9
14	50.8	49.0	157.5	160.0
15	56.7	54.5	162.6	167.6
16	58.1	60.8	165.1	172.3
17	58.6	64.9	167.6	175.3
18	59.0	69.0	170.2	177.8

(a) (i) Draw a graph with age on the horizontal axis and mass on the vertical axis. Plot the data for girls and the data for boys on the single set of axes.
(ii) Explain the differences between the growth of girls and boys as far as you can.
(b) (i) Draw a graph with age on the horizontal axis and rate of growth (cm per year) on the vertical axis. Calculate the growth in height each year for both girls and boys. Plot these data on a single set of axes.
(ii) Explain the changes in the rates of growth as far as you can. (iii) Different parts of the body do not grow at the same rate. Which parts would you expect to grow faster in the early years? Which parts account for the increase in growth rate in the early teens?

PRACTICAL WORK

Note for teachers
Investigating the growth of organisms may mean taking measurements over a period of several weeks. This has to be planned for.

Experiment 39.1 To investigate the growth of an insect

Note for teachers
Any common laboratory insect will do. A stick insect such as *Carausius morosus*, obtainable from biological suppliers, is recommended because:
(i) stick insects show incomplete metamorphosis, which means that measurements of the nymph and adult can be directly compared;
(ii) stick insects usually keep still when they are handled, which means they can be measured without being anaesthetised or killed.
Mature female stick insects lay eggs over a number of weeks without the need for fertilisation (parthenogenesis). An egg is about 2 mm long with a dark brown shell which is rounded at one end and has a pale cap at the other. To incubate the eggs, place them on damp cotton wool or filter paper in a suitable container at room temperature two to three months before the nymphs are required.

Materials required by each student
1 small jar to hold privet or ivy
1 glass container large enough to house the nymph, the small jar and the privet or ivy, covered in such a way that air can get in and out but the nymph cannot escape
1 pair of dividers
1 ruler graduated in mm
cotton wool
1 freshly hatched nymph of *Carausius morosus*
fresh stems of privet or ivy

Method
1 Using the dividers and ruler, measure the length of a freshly hatched nymph of *Carausius morosus* from the tip of its head (do not include its antennae) to the far end of its abdomen. Record your measurement and the date.
2 Place the twig of privet or ivy in a little water in the small jar and plug the top with cotton wool. Place the jar inside the glass container, add the nymph you have measured and cover the top of the container. Leave the covered container in a safe place.
3 Every seven days carefully remove the nymph, measure it in the same way and replace it inside the container. Make at least six weekly measurements and more if you can. Record each measurement and the date on which you make it.

Interpretation of results
1 At the end of your experiment draw a graph to show time from the start of your measurements on the x-axis and length of the insect's body on the y-axis.
2 Explain what your graph shows about the growth of this insect.

Experiment 39.2 To investigate the growth of a plant stem

Note for teachers
Any growing plant stem will do provided that it is not a twining stem. Dwarf varieties of the French bean are suitable. The plumules and radicles should be well developed. Since the experiment makes use of freshly germinated seeds, it may be timed to follow Experiment 42.1 to investigate seed germination.

Materials required by each student
1 small plant pot
potting compost
1 ruler graduated in mm
1 spirit marker
1 freshly germinated seedling

Method
1. About half fill a plant pot with moist potting compost and make a small well in the compost.
2. Taking care not to damage the seedling, put the radicle (developing root) in the well in the compost and, holding the seedling in place, sprinkle more potting compost around it. Lightly press the compost around the base of the exposed seedling.
3. Using the spirit marker, lightly draw a line on the seedling just above the compost. Place a ruler against the plant and measure the height of the stem between the spirit mark and its growing tip. Do not include the length of any leaves that may surround the tip. Record your measurement and the date.
4. Leave your potted plant in a well-lit place where it can be watered as required.
5. Every seven days measure the height of the stem between the spirit mark and its growing tip. Make at least six weekly measurements and more if you can. Record each measurement and the date on which you make it.

Interpretation of results
1. At the end of your experiment draw a graph to show time from the start of your measurements on the x-axis and height of the stem you measured on the y-axis.
2. Using the same axes, draw a second graph to show mean height of the stems measured by the whole group against time from the start of your measurements.
3. Compare the graph of your own results with that of the group results. Why is the group graph more reliable?
4. Compare the group graph of the plant stem's growth with your own graph of the stick insect's growth.

UNIT 40 Cell division and mitosis; asexual reproduction

Cell division

Growth involves making more protoplasm and eventually more cells. Examples of plant and animal cells are described in Unit 13. New cells are made by the division, usually into two, of an existing cell. The new cells are known as **daughter cells**. Cells that divide usually consist only of a nucleus and cytoplasm, i.e. they are unspecialised. Specialised cells, ones adapted for different functions, do not usually divide. Dividing animal cells are not easy to find because animals grow all over. A growing region with dividing cells which is easy to get at is the root tip.

Mitosis

Mitosis is the process whereby a cell divides to form two new cells, each with a nucleus exactly like its own. It is important because the nucleus contains the coded instructions for making enzymes and thus for controlling all the activities in the cell. The coded instructions are in the form of long narrow DNA strands: each long narrow strand is only one molecule. Mitosis ensures that every cell of a plant or animal (except cells resulting from reduction division, explained in Unit 41) contains the same DNA in its nucleus.

Cell division may occur twice an hour or less than once a week. During the 'resting stage' between the completion of one mitosis and the beginning of the next, the DNA in the nucleus is in the form of **chromosomes**. Each chromosome is a strand, i.e. a single molecule, of DNA inside its own protein coat with a small structure on it called a **centromere** (which is not necessarily at the centre). As the time for mitosis approaches, each chromosome **replicates** (duplicates itself) to form two identical **chromatids** which lie side by side joined only by the centromere. In flowering plants and humans, chromosomes are usually in pairs. Figure 40.1 shows two pairs of chromosomes with a total of eight chromatids. You can recognise the pairs of chromosomes by their shapes.

During mitosis the nuclear membrane disappears, each centromere divides and the chromatids of each chromosome separate and go to opposite ends of the cell as shown in Figure 40.2. Now that they have separated, what were the chromatids have become new chromosomes.

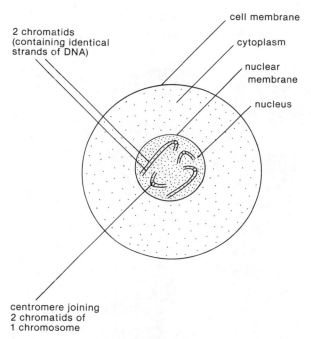

Fig. 40.1 Start of mitosis: two pairs of chromosomes

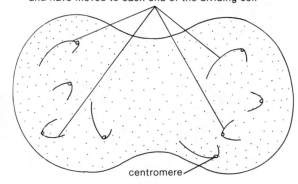

Fig. 40.2 Chromosomes (formerly chromatids) at each end of the cell

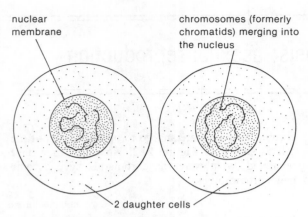

Fig. 40.3 Chromosomes (formerly chromatids) merging into new nuclei

The cell divides into two as the new chromosomes at each end merge to form two new nuclei in new nuclear membranes. This is shown in Figure 40.3.

Now mitosis is complete and each cell goes into another 'resting stage' during which each chromosome replicates (duplicates itself) to become two chromatids joined by a centromere. By the time the two new cells are ready to begin mitosis, each has four chromosomes containing two identical chromatids, or eight chromatids in all, like the original cell in Figure 40.1.

Figure 40.4 is a series of photographs of root-tip cells taken at different stages of mitosis. Individual chromosomes are not clear because they are crowded together in the cell. A photograph of individual human chromosomes is shown in Unit 41.

In photograph A chromosomes are beginning to appear. Though each chromosome contains two chromatids, it is impossible to distinguish them in the photograph. The darkly stained blob is a **nucleolus**, a reserve of chromosome-like material. Present only in some cells, a nucleolus disappears as the chromosomes form.

In photograph B chromosomes are lying across the centre of the cell. The nuclear membrane has disintegrated.

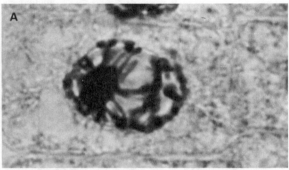

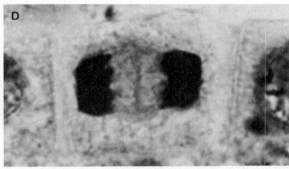

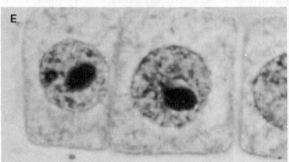

Fig. 40.4 Root-tip cells: mitosis

In photograph C the centromeres have divided and separated. The two chromatids of each original chromosome have become two new chromosomes and have gone to opposite ends of the cell. The two ends of the cell now contain identical combinations of chromosomes and therefore identical sets of DNA.

In photograph D the two sets of new chromosomes are merging to form two new nuclei, one at each end of the cell.

In photograph E newly formed nuclei are in the resting stage. Each again contains a nucleolus. A new cell wall has begun to form between the two cells.

'Resting stage' is an unsuitable name because a lot goes on during it. The DNA is replicated inside the nucleus, which continues to exert control over the cell's other activities. By the time the chromosomes re-emerge, before the nucleus divides again, each consists of two chromatids joined by a centromere as in photograph A.

Reproduction

In single-celled organisms such as bacteria, making more cells is the same as making new organisms. Making new organisms is **reproduction**. The division of a single-celled organism into two is a simple example of **asexual reproduction**. Asexual means not sexual.

Sexual reproduction, described in detail in Unit 41, involves the fusion of two cells, such as an egg and a sperm in humans. It introduces variety among offspring. Usually sexual reproduction involves two parents, though some organisms, including the French bean, can fuse their own cells. Having two parents greatly increases variety among the offspring: one of each pair of chromosomes is inherited from one parent and the other from the other parent. You have 23 pairs, i.e. 46 chromosomes, 23 from your father and 23 from your mother.

Asexual reproduction always involves only one parent and results in no variety at all.

Asexual reproduction in flowering plants

Many plants spread outwards through the soil as they grow. So long as the parts of a plant are still attached to one another, it is one plant, but, when the parts are separated and are able to live independently, asexual reproduction has taken place.

Clumps of bamboo, mint and nettles have underground spreading stems from which new shoots grow. When the underground stems die and decay, many new, i.e. separate, plants are formed.

Some plants have parts which, if they are accidentally separated from the parent plant, quickly develop roots and grow. Stonecrops shed their fleshy leaves as soon as they are touched: many of these develop roots and grow.

Some plants develop special structures which can carry out asexual reproduction. The small sections of a garlic 'bulb' are each able to grow and form independent plants. Potatoes, which are enlarged stem tips, are special structures which can carry out asexual reproduction. About fifteen potatoes develop in the soil around a potato plant. In winter all the other parts of the plant die. In spring about fifteen independent potato plants grow from the potatoes in the soil.

Advantages of asexual reproduction

Plant offspring of asexual reproduction are given a good start in life. They receive food from the parent during the time they are attached to it. They may be large and well developed before they become independent. They usually become independent close to the parent: since the conditions have suited the parent, they should also suit the offspring, which by mitosis inherit the exact features of the parent. Identical offspring constitute a **clone**: they can be produced only by asexual reproduction. Once one plant has established itself in a new area, it can quickly form a large population by asexual reproduction, i.e. it can be a successful coloniser.

Disadvantages of asexual reproduction

Lack of dispersal can cause overcrowding. Lack of variation in the offspring's DNA means that, if the parent has an inherited defect, all the offspring have it too: for example, if the parent is sensitive to frost, all the offspring will be sensitive to frost.

Asexual reproduction in saprophytes

In saprophytes such as bacteria and fungi asexual reproduction is common. After growth and mitosis, a bacterium forms two bacteria. After growth and mitosis in fungi, special reproductive bodies called **spores** are produced. Spores are small and are enclosed by a protective wall. Spores are produced in thousands, sometimes millions, as a result of asexual reproduction and float everywhere, invisible in air. Those that land where the conditions are suitable grow into new fungi.

Asexual reproduction in animals

Only very small animals reproduce mainly asexually: single-celled animals divide and separate; slightly larger animals such as freshwater *Hydra* bud off small versions of themselves. Occasionally a fertilised egg or an embryo divides to form two or more separate cells or embryos. This accounts for *identical* twins, triplets, etc. Though the fertilised egg is the result of sexual reproduction, any later division resulting in two or more separate individuals is asexual reproduction. Being the result of mitosis, identical twins, triplets, etc., have identical DNA and constitute a clone.

Artificial use of asexual reproduction: vegetative propagation

Gardeners and horticulturalists make much use of the readiness of flowering plants to carry out asexual reproduction and form clones. When they have a plant with a desirable feature, such as beautiful flowers, a strong scent, resistance to frost or a fast growth rate, they increase its numbers asexually. Because all the cells divide by mitosis, the offspring have exactly the same DNA as the parent.

Growers do not rely on a natural method of asexual reproduction. They take pieces of stems or leaves, called **cuttings**, and keep them in moist conditions to encourage them to produce roots and grow. Though not all cuttings will take root, they can get many offspring from one parent plant. The artificial multiplication of plants from parts of their stems, roots and leaves (*not* from their flowers and seeds) is **vegetative propagation**. Growers use rooting powders containing plant hormones to stimulate root growth and fungicides to prevent infection and decay. Plant hormones are described in Unit 36. Some plants do not easily form roots from cuttings. These can often be **grafted** on to the stems of different plants with vigorous roots. The shoots that grow from the grafted portion will have the DNA and the characteristics that are wanted even though the roots have the DNA of an inferior plant. Vines and many fruit trees are grown in this way. Figure 40.5 shows a rooted stem cutting and a stem graft.

Plants can now be grown by a method known as **tissue culture**. It is expensive and is used only for commercially valuable plants such as orchids, palm and vines. A few cells from the parent plant can be kept alive on a nutritive mixture containing plant hormones while the cells divide continuously by mitosis. The result is a disorganised mass of tissue in which every cell's nucleus contains an exact copy of the parent's DNA. Groups of a mere few cells can be

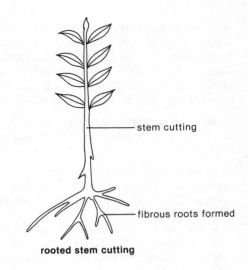
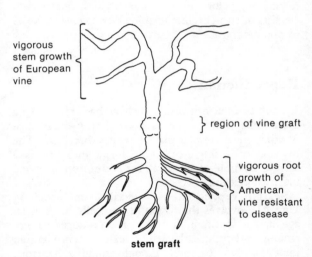

Fig. 40.5 Vegetative propagation

grown into offspring identical to the parent plant. These offspring, formed by mitosis from a single parent, also constitute a clone.

Questions

Q 40.1 Figure 40.6 shows one pair of chromosomes in a nucleus.

Explain why the two chromatids (A and B) of a single chromosome are identical but two similar chromosomes (I and II) in a nucleus are not identical.

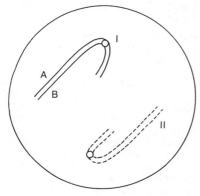

Fig. 40.6

Q 40.2 What is asexual reproduction? What are its advantages and disadvantages in plants?

Q 40.3 Stem cuttings were divided into two batches. Each cutting in batch I had a ring of phloem removed from its stem just above its cut end. Both batches of cuttings were planted in moist potting compost. Figure 40.7 shows where after a few weeks roots developed on both batches of cuttings.

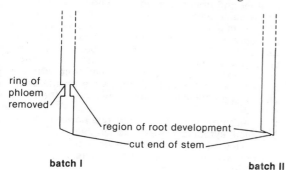

Fig. 40.7

(a) What is the function of phloem?
(b) The development of roots is encouraged by growth hormones produced in the stem tip. Suggest why roots grew in the different positions in the two batches.

PRACTICAL WORK

Experiment 40.1 To demonstrate a number of plant propagation methods

Note for teachers
This experiment requires plants to be kept permanently in the school or college. In the absence of greenhouse facilities all the plants may be grown in a warm laboratory throughout the year.

Although the propagation methods may be performed at any time of year, the best results are obtained in September-October or late spring.

Plans must be made for the propagated plants to be stored in a warm well-lit place for several weeks after the experimental session. If storage space is limited, the number of plants can be reduced by letting the students work in groups.

All the horticultural items can be obtained from good garden centres.

Materials required by each student
1 250 cm^3 beaker
1 scalpel
12 stainless dissecting pins
2 seed trays
6 peat pots
3 compressed peat pellets
1 bowl
1 ruler graduated in cm
2 clear polythene bags to enclose the seed trays

Materials required by the class
moss peat
sharp sand
garden loam
rooting powder
enough *Begonia* plants to provide 2 large leaves for each student
enough *Bryophyllum* (air-plant) plants to provide 3 plantlets for each student
enough *Chlorophytum* (spider-plant) plants to provide 3 offsets for each student
enough *Pelargonium* (geranium) plants to provide 3 side shoots for each student

Method
(a) *Stem cuttings*
1. Soak the three compressed peat pellets in the bowl of water until they are fully expanded.
2. Use the scalpel to cut from the geranium plant three side shoots 8 to 10 cm long containing at least two leaf nodes (regions of the stem where leaves attach). Choose well-developed side shoots which are not blooming, ignoring those that are either old and brown or young and light green.
3. Cut off the lower leaves and the portion of stem below the lower node and put the cuttings into a beaker of water.
4. Dip the cut ends of the cuttings in rooting powder. Lightly tap them to remove excess powder so that only a thin coat is left on each.
5. Plant these cuttings in the three soaked peat pellets. Ensure that the stems rest against the bottoms of the holes in the pellets with no air gap.

6   Put the peat pellets in a seed tray.

(b) *Offset plants*

An offset is a branch that takes root at its tip.

1   Use the beaker to put equal parts of garden loam, sharp sand and moss peat into the bowl. The contents of the bowl must be enough to fill six peat pots. Mix them thoroughly.
2   Put the filled peat pots in water and let the potting mixture become moist.
3   Cut three plants from the tips of the *Chlorophytum* (spider-plant) offsets.
4   Make a small hole in the potting mixture in each of three peat pots and plant one *Chlorophytum* in each. Lightly press some of the potting mixture around each plant so that it is firmly held. Put the three peat pots in the seed tray with the geranium cuttings.

(c) *Young plantlets*

1   Remove three young plantlets from the edge of a *Bryophyllum* (air-plant) leaf.
2   Make a small hole in the moist potting mixture in each of the remaining three peat pots and plant one plantlet in each. Lightly press some of the potting mixture around each plantlet so that it is firmly held. Put the three peat pots in the seed tray with the geraniums and spider plants.
3   Put the seed tray inside a large polythene bag and fasten the end. Leave the seed tray inside the bag in a warm place away from direct sunlight.
4   After two to three weeks open the polythene bag for one hour on the first day and for increasingly longer on each successive day for a week. After a week remove the tray from the bag, water the pots and leave the tray in bright light for another week.

(d) *Leaf cuttings*

1   Mix equal parts of garden loam, sharp sand and moss peat to make enough potting mixture to fill a seed tray. Put the filled tray into water and let the potting mixture become moist.
2   Cut two large leaves from the *Begonia* plant.
3   With both leaves lying upper surface downwards, make a number of cuts in the larger veins on the lower surface.
4   Sprinkle a little rooting powder over the lower-leaf surface and lightly tap the leaves to remove excess powder.
5   Use the dissecting pins to fix the two leaves, lower face downwards, on the surface of the potting mixture in the seed tray. Put the seed tray inside a large polythene bag and fasten the end. Leave the seed tray inside the bag in a warm place away from direct sunlight.

Interpretation of results

1   By which type of cell division have the new plants been produced?
2   How will the plants produced by this method of cell division be different from plants produced by sexual reproduction?
3   Suggest why horticulturalists and gardeners may prefer to reproduce some of their plants by these methods rather than to allow sexual reproduction.

Experiment 40.2   To demonstrate mitosis with the use of bead models

Materials required by each student
60 poppit beads (30 of each of two colours)
a small piece of modelling clay

Method

1   Make two fifteen-bead chains, one of each colour, and stick a piece of modelling clay on each. The chains represent a pair of chromosomes in the nucleus of a cell. The pieces of modelling clay represent the centromeres of the chromosomes. The colours represent the origins of the chromosomes, one from the egg cell, the other from the sperm cell.
2   Make an exact copy of each of the two chains, lay each copy alongside the original and stick it to the modelling clay on the original. This represents the chromosome replication that occurs inside the nucleus before it divides: each chromosome becomes two chromatids joined only by a centromere.
3   Lay the two pairs of chains on the bench so that one chain in each pair lies on each side of an imaginary line and the other chain in each pair on the other side of the imaginary line. This represents the way in which chromatids line up across the cell just before the nucleus divides. (See photograph B in Figure 40.4).
4   Split the modelling clay on the two pairs of chains. Move the chains in each pair further away from one another and from the imaginary line between them. At the same time move the chains on the same side of the imaginary line towards one another. (See photograph C in Figure 40.4). This represents the formation of the nuclei of what will become two new cells: each chromatid has now become a new chromosome; the new chromosomes from different pairs of chromatids (those that were on the same side of the imaginary line) now belong to the same nucleus.
5   In a real organism each of the two new nuclei will organise cytoplasm from the original cell round itself. This will result in the division of the cytoplasm to form two daughter cells each with its own nucleus and its own cytoplasm.

Interpretation of results

1   Each chromosome in your model produced a pair of joined chromatids. This is chromosome replication.
    (a) When does chromosome replication occur in living cells?
    (b) Is there a difference between (i) the chromatids in each pair and (ii) a chromatid and the chromosome from which it was produced?
2   Was the number of each kind of chromosome the same in the model nuclei of the parent and the two daughter cells?

UNIT 41: Life cycles of humans and flowering plants; sexual reproduction

Life cycles

Which came first, the chicken or the egg? This unanswerable question makes the point that the lives of organisms are a continuous cycle. Offspring develop from mature adults, grow, become mature and produce offspring to carry on the cycle. This unit examines a number of organisms' life cycles. The life cycle of the French bean is described in Unit 42.

Asexual reproduction

In asexual reproduction offspring are produced from a single parent: each offspring's DNA is identical to its parent's DNA and each offspring therefore has the same inherited characteristics as its parent. The identical offspring of a single parent are a clone. Figure 41.1 shows two examples of asexual life cycles.

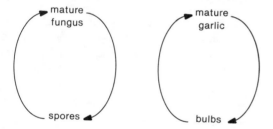

Fig. 41.1 Life cycle of a fungus and garlic

Sexual reproduction

Most organisms reproduce not asexually but sexually. This is the process in which two cells, called **gametes**, fuse together to form a single cell, called a **zygote**. The fusion is **fertilisation**. In very simple organisms the two gametes that fuse look alike, but in most organisms they are different.

The **female gamete** or **ovum**, often called an egg, is the larger of the two, is usually stationary, and has some food material in its cytoplasm. The smaller **male gamete** can either move on its own or be moved. There are more male gametes than female gametes. If a male gamete can move on its own, it is called a **spermatozoon**, or simply a **sperm**.

Humans

In humans the male gamete is a sperm with a long tail which it uses to swim with. The tail is made of cytoplasm; the rest of the sperm's cytoplasm surrounds its nucleus.

The sperm swims to the egg and fuses with it to form a zygote. This is fertilisation. The food in the cytoplasm of the egg will provide the raw materials and energy for the early growth of the zygote. During the next 38 weeks ('nine months') the zygote develops inside the mother's womb (uterus) into an **embryo** and then into a baby. Human reproduction is described in detail in Unit 43.

Figure 41.2 shows a human egg and a human sperm. They are not drawn to the same scale: their approximate sizes are given. Even with its long tail the sperm is less than half the size of the egg.

The nuclei of the egg and the sperm each have only one set of chromosomes, i.e. one set of DNA. These nuclei are called **haploid**. In the zygote formed when they fuse, the nucleus has two sets of chromosomes, i.e. a pair of each kind of chromosome. This nucleus, which is the usual kind in the cells of the body, is called **diploid**. In humans the haploid nucleus of an egg has 23 chromosomes and the haploid nucleus of a sperm has 23 chromosomes. When the two fuse, the diploid nucleus of the zygote has 46 chromosomes,

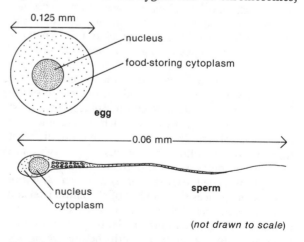

Fig. 41.2 Human egg and sperm

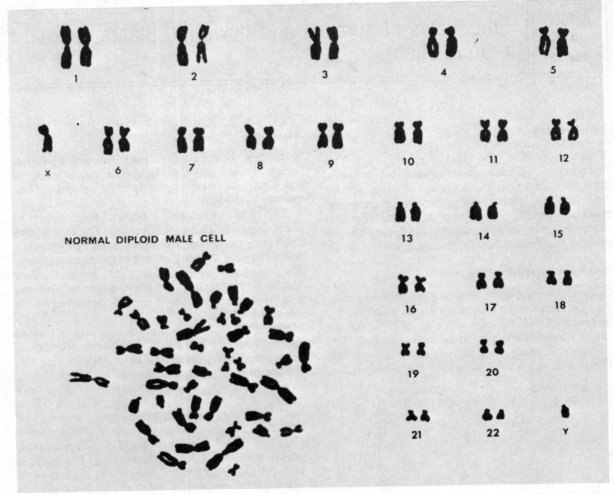

Fig. 41.3 *Human chromosomes after replication: each shows two chromatids joined at the centromere*

two of each kind, i.e. 23 pairs. Twenty-three chromosomes have come from the father, 23 from the mother. Figure 41.3 is a photograph of the 23 pairs of human chromosomes arranged artifically in order of size.

Since the human egg and human sperm contain only 23 chromosomes each, and the other cells of the human body all contain 46 chromosomes, there must at some stage be a process by which the chromosome number is halved. This is a special division, different from mitosis, in which the nucleus divides twice but the chromosomes split into their chromatids only once. It is called **meiosis** or **reduction division** (because it reduces the number of chromosomes) and takes place in every life cycle in which there is sexual reproduction. In the human male, reduction division takes place during the two months before sperms are mature. In the human female, reduction division is a remarkable process. Eggs are not formed in the adult female. Every baby girl is born with ovaries containing more than enough eggs to last her lifetime. Reduction division actually begins before she is born, as the eggs form in her ovaries. A reduction division which begins in the womb is not completed until the egg is shed from her own ovary, which may be 50 years later.

Life-cycle diagrams are a conventional way of showing the chromosome number of the nuclei at each stage. When the nuclei have only a single set of chromosomes, they are haploid: this is shown on the diagram as n where n stands for the number of different chromosomes in the nucleus. When the nuclei contain a double set of chromosomes, they are diploid: this is shown on the diagram as $2n$. In humans only the nuclei of the gametes are n (haploid); all the other cells have $2n$ (diploid) nuclei. In humans

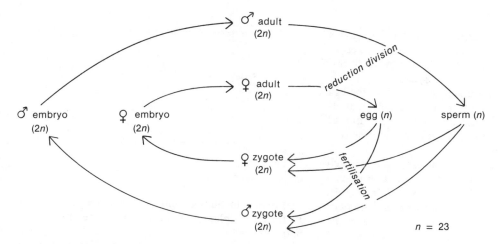

Fig. 41.4 Human life cycle

$n = 23$ and $2n = 46$. Forty-six chromosomes are present in Figure 41.3. Figure 41.4 is a conventional life-cycle diagram for humans: ♂ means male and ♀ means female.

Flowering plants

In flowering plants sexual reproduction takes place within flowers. Flowers are extremely varied but the details of sexual reproduction are similar in all of them.

The female gamete or egg is similar to that in humans: it is a haploid nucleus surrounded by cytoplasm without the cellulose wall and large vacuole usual in a plant cell. Most egg cells of plants and animals have this simple form. Each plant egg lies inside a larger structure called an **embryo sac**, which lies inside a yet larger structure called an **ovule**. Ovules are small oval bodies, difficult to see with the naked eye; when they develop into seeds and get bigger, they can be seen more easily. The typical form of an ovule is shown in Figure 41.5. At one end there is a small hole, the micropyle, through which the pollen tube can grow bringing the male gamete to the egg for fertilisation.

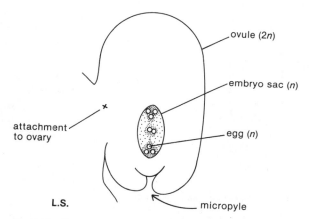

Fig. 41.5 Egg cell in embryo sac in ovule: section

Usually in French beans between three and six ovules lie inside a single **ovary** in each flower. Other species of flowers may have hundreds of ovules and even tens of ovaries. The ovaries form part of a **carpel**, which also includes a **stigma**, a sticky part to receive the pollen, and a **style**, which supports the stigma. These parts of the French bean are shown in Figure 41.6. Carpels and their contents are the female

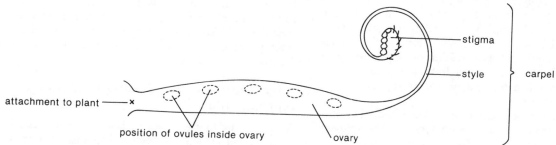

Fig. 41.6 Carpel of French bean

parts of a flower. Carpels, usually found at the centres of flowers, are very different in different flowering plants. The French-bean carpel grows into the 'bean' or pod that we eat.

In flowering plants the male gamete is like the egg in having no cellulose wall or large vacuole, but its haploid nucleus is surrounded only by a thin film of cytoplasm. The male gamete is not a sperm because it cannot move on its own.

Male gametes develop from **pollen grains**, which form the fine dust called **pollen**. A pollen grain is a single cell formed inside the **anthers** of a flower. An anther is the swollen head of a **stamen** and stamens surround the carpel. A stamen with an anther of the French bean is shown in Figure 41.7. Stamens and their contents are the male parts of a flower.

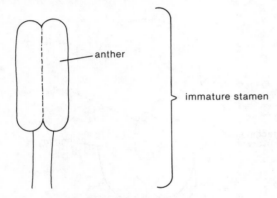

Fig. 41.7 Immature stamen of French bean

Pollen is released from the anthers and most is wasted, as are most male gametes in other organisms. But some pollen may be transferred to the stigma of a plant of the same kind, or even of the same plant, by wind, by insects or by birds. This is **pollination**. **Self-pollination** in the French bean is described in detail in Unit 42.

When the pollen grain lands on the stigma, it still has to reach the egg cell in the embryo sac. It grows a tube, called the **pollen tube**, down through the style to the micropyle of the ovule. Food for growth is supplied to it from the cells of the style and ovary wall. The tip of the pollen tube, shown in Figure 41.8, disintegrates and discharges the male gametes into the embryo sac. One of the male gametes fuses with the egg cell to form a zygote. This is fertilisation. The zygote develops into a plant embryo and then into a seedling.

In a flowering plant fertilisation is possible only after pollination. The development of a French-bean zygote into an embryo and a seedling is described in Unit 42.

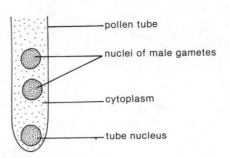

Fig. 41.8 Tip of pollen tube

Whereas in humans only the gametes are haploid, in flowering plants all the nuclei of the embryo sac and the pollen tube are also haploid. Since the haploid chromosome number in the French bean is eleven, the nuclei in all its gametes, embryo sacs, pollen tubes and pollen grains have only one set of eleven different chromosomes.

Reduction division occurs before the embryo sac and pollen grains are formed. When the zygote is formed after fertilisation, there are again two sets, i.e. 22 chromosomes, in the diploid nucleus: the nuclei of the embryo of the seedling and of the mature plant of French bean all contain 22 chromosomes. The life cycle of the French bean is shown in Figure 41.9.

Chromosomes

Chromosome numbers vary greatly in different organisms: for example, there are two pairs in roundworms, four pairs in fruit flies, eleven pairs in French beans, twelve pairs in pine trees and 23 pairs in humans. The number of chromosomes is constant in any one species, is always halved at reduction division and is always doubled at fertilisation. Gametes have the haploid number, zygotes the diploid number. Whenever there is fertilisation in a life cycle, there is also reduction division to keep the chromosome number constant. Reduction division begins in the eggs inside the human female embryo and is completed many years later when an egg is ready for fertilisation, but it is exceptional for it to take so long. In human males it occurs as the sperms are maturing. In flowering plants it occurs at pollen and embryo-sac formation. There are other organisms, especially among simple plants and fungi, in which it occurs at different stages in the life cycle.

Advantages and disadvantages of sexual reproduction

Almost every living organism has a form of sexual reproduction in which gamete nuclei fuse to form a zygote. In **cross-fertilisation** the gametes come

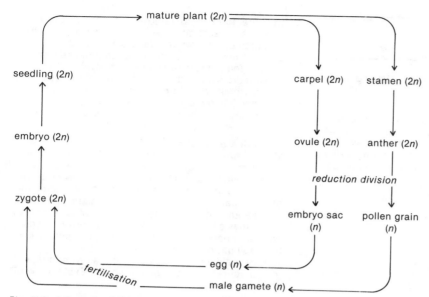

Fig. 41.9 Life cycle of French bean

from different organisms, in **self-fertilisation** from the same organism. In humans there is always cross-fertilisation. In the French bean there can be either cross-fertilisation (after cross-pollination) or self-fertilisation (after self-pollination). After cross-fertilisation there are chromosomes with different DNA and there is greater variety in the offspring. After self-fertilisation the DNA, although it comes from the same parent, is reshuffled and offspring are different from their parent, but there is less variety among them than after cross-fertilisation because the supply of DNA in the gametes is not so varied. The greatest advantage of sexual reproduction, and in particular of cross-fertilisation, is the variation it produces in offspring.

If there are changes where organisms are living, or if they move to a new area, some offspring may be better suited to the new conditions than either of their parents. Thus variety gives the species a better chance of surviving in the new conditions.

The major disadvantage of sexual reproduction is that it is always slow because every offspring must grow from a single cell, the zygote.

In flowering plants sexual reproduction leads to the formation of seeds, which are a dormant stage in the life cycle: as seeds, plants can survive unfavourable conditions. Another advantage of seeds is that they enable plants to be dispersed over a wide area, which increases their long-term chance of survival. The disadvantage is that many seeds do not end up where there are suitable conditions for growth. There is therefore a great wastage of seeds.

Questions

Q 41.1 Figure 41.10 shows the life cycle of a single-celled plant called *Chlamydomonas*.

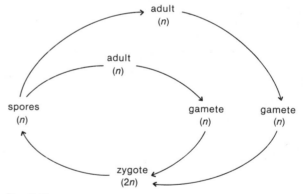

Fig. 41.10

(a) Make a copy of the diagram and add the points in the life cycle at which (i) fertilisation, (ii) mitosis, (iii) reduction division occur.
(b) How is the life cycle of *Chlamydomonas* different from that of humans?

Q 41.2 Explain why self-pollination is possible if the anthers of a flower mature some time before the stigma of the flower but is not possible if the stigma matures some time before the anthers.

PRACTICAL WORK

Experiment 41.1 To investigate the structure of a flower

Note for teachers
The flower of the French bean is small and its structure more difficult to interpret than that of other commonly used flowers such as the buttercup. Teachers may wish to use a simpler flower to introduce students to flower structure and let them study the structure of the French-bean flower later.

Materials required by each student
1 pair of forceps
1 hand-lens
1 plain white tile
1 French-bean flower

Method
1. Use the hand-lens to examine one of the flowers and make a drawing of it from one side.
2. Use the forceps to remove the sepals (the outer green scales) from the flower and place them together on the white tile. Do the same with the petals, taking care to remove each petal separately. Record the number of sepals and petals.
3. Make a drawing of the male and female parts of the flower which you can now see.
4. Remove the male and female parts of the flower separately and place them on the white tile. Record the number of stamens and styles.

Interpretation of results
1. Many flowers have a symmetrical structure in which the number of sepals, the number of petals, etc., form a simple ratio. Is this true of the French bean?
2. Use your observations of the size, shape and colour of the sepals to suggest two functions they perform.
3. The colour and size of the petals will attract insects which may carry pollen from another flower. Suggest how the shape and arrangement of the petals may help pollination by insects.
4. Are the male and female parts of the flower adapted to encourage cross-pollination? Explain your answer.

UNIT 42 — Life cycle of the French bean

This unit describes how a French-bean seed develops into a French-bean plant which produces new French-bean seeds. This is the 'life cycle' of the French bean. It explains how the French bean can go on reproducing itself indefinitely.

The seed

A French-bean seed is shown in several views in Figure 42.1. Inside its dry protective coat, the **testa**, is an **embryo** consisting of two large seed leaves or **cotyledons**, an embryo shoot or **plumule** and an embryo root or **radicle**. The cells of the cotyledons contain food stores of protein and starch. In this form the seed can stay dormant until conditions for growth are favourable, which means it usually stays dormant throughout winter. The testa has a large scar where the seed was attached to the parent plant and a small hole, the **micropyle**, through which the pollen tube containing the male gametes entered for fertilisation.

Germination

Germination is the development that occurs between the time a seed first takes in water and the time the seedling is able to photosynthesise. Favourable conditions for germination are: a **temperature between 17°C and 30°C** providing the warmth for enzymes in the seed to work efficiently; **water** to reactivate the dehydrated protoplasm and its enzymes and to enlarge the cell vacuoles; **oxygen** to allow aerobic respiration to release enough energy for the many reactions needed to make more protoplasm. With a suitable temperature, water and enough oxygen, all viable seeds can grow.

The stages of germination are shown in Figure 42.2. The dry seed absorbs water through its testa and its micropyle: the cells enlarge, the testa splits and enzymes begin to digest the stored food. Digestion of stored food is described in Unit 23. The first growing region to produce new cells and increase in size is the embryo root or radicle. Once the testa has split, the radicle grows downwards under the stimulus of gravity, controlled by its growth hormones. Water is now needed to transport the digested food from the cotyledons to all the growing regions. After the radicle has emerged, much more water can be absorbed through the root hairs that grow on its surface.

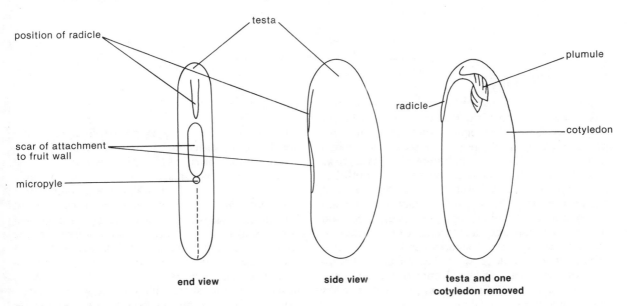

Fig. 42.1 French bean: dormant seeds

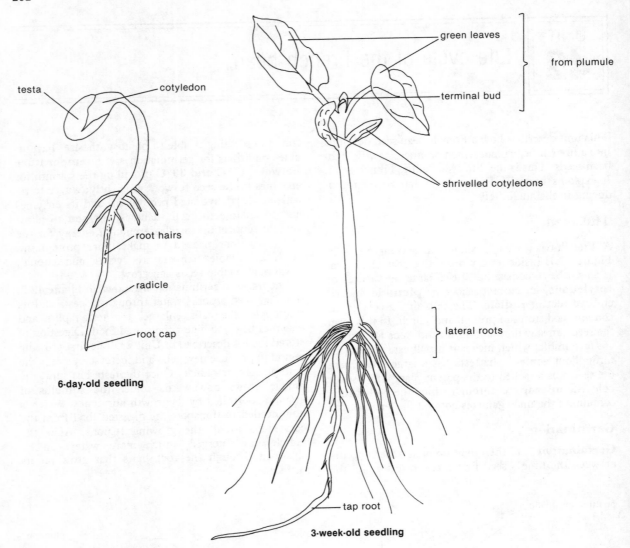

Fig. 42.2 French bean: stages of germination

The next growing region to develop is the part that lies between the top of the radicle and its attachment to the cotyledons. This region grows against the force of gravity, carrying the rest of the seed upwards in a hooked position as shown in Figure 39.1. As it grows through the soil, the testa is usually rubbed off, but the plumule remains protected between the two fleshy cotyledons. So long as the cotyledons remain attached to the seedling, they provide digested food for any part that needs it. Once above the surface of the soil the seedling straightens and the first two leaves of the plumule grow and become green. Photosynthesis begins and the cotyledons shrivel and drop off soon afterwards. In summer, germination takes about three weeks.

Growth

Once photosynthesis begins, growth is rapid. There are dozens of different varieties of French bean: a typical dwarf variety, *Masterpiece*, is described and shown in the diagrams in this unit. The plumule develops into a branched bushy shoot, about 30 cm high with about twenty large green **compound leaves** each consisting of three **leaflets**. A single leaf is shown in Figure 42.3. About seven weeks after the start of germination, i.e. about four weeks after the end of germination and the beginning of photosynthesis, the plant is mature and white flowers develop on the shoot. Flowers grow in clusters usually of between four and six; they open one after

Fig. 42.3 French bean: compound leaf

Fig. 42.4 French-bean plant

Flowers

The flowers contain the organs of sexual reproduction. French beans are **bisexual** or **hermaphrodite**: both male and female gametes develop on the same plant, male gametes from **pollen grains** and female gametes in **ovules**, as described in Unit 41. Flowers are small and white in the *Masterpiece* variety, though they are different colours in other varieties. Two small green scales, called **sepals**, protect the young flower bud. There are five white **petals**, some fused to one another and some free. The central pair of petals are twisted inside the flower in a corkscrew shape. Different stages and different views of French-bean flowers are shown in Figure 42.5. The large upright petal at the back is the **standard**; the single petals on each side are the two **wings**; the two twisted petals at the centre are the two **keel** petals. These names are descriptive, which makes them easier to remember.

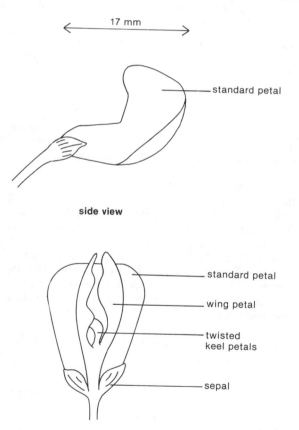

Fig. 42.5 French bean: flowers

the other, the bottom flowers on the stem opening first. There are usually four or five clusters of flowers on a plant. Figure 42.4 shows the external appearance of a dwarf French-bean plant.

Stamens

There are ten stamens in the French-bean flower. The stamen's function is to produce pollen from which male gametes develop for sexual reproduction. Each stamen consists of a swollen head, the **anther**, in which the thousands of pollen grains develop, and a stalk, the **filament**, which supports the anther. Nine of the ten filaments are fused together to form a canoe-shaped trough lying inside the twisted keel petals. The tenth filament lies freely along the top of the 'canoe'.

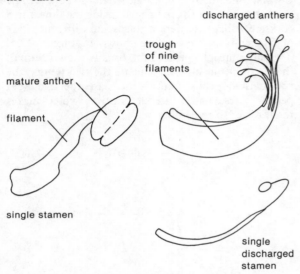

Fig. 42.6 French bean: mature and discharged stamens

Figure 42.6 shows the mature form of a stamen in the young flower before the petals have opened and several stamens in an older flower with opened petals. By the time the flower opens the anthers are empty, shrivelled and dry: the pollen grains have been discharged inside the twisted keel before the rest of the flower has matured. (Flowers whose stamens mature and release pollen before the female parts of the flower have fully matured are called **protandrous**.)

Carpel

There is only a single carpel in the French-bean flower. The carpel's functions are to produce eggs (female gametes) inside the ovules (reproductive organs), to trap pollen grains, to protect and feed the pollen tubes which grow from the pollen grains, to provide suitable conditions for fertilisation in the ovules, to provide food and protection for a growing embryo inside each ovule, and to protect the ovule as it grows into a seed.

The carpel lies at the centre of the flower. Its swollen base, called the **ovary**, is surrounded by the trough of filaments which in turn is surrounded by the twisted keel petals. The ovary matures after the stamens and is still small even when the petals have fully opened. Each ovary contains about five **ovules**. The other end of the carpel narrows to form a twisted **style**, which ends at a slightly swollen sticky **stigma**. These parts also lie in the twisted keel petals. It is because the stigma is sticky that pollen grains will become attached to it. The style is the part of the carpel through which the pollen tubes grow. Figure 41.6 in Unit 41 shows the shape of a young carpel in the fully opened flower; Figure 42.7 shows an older carpel which has developed into a **pod** containing four ovules which have developed into **seeds**. (It is much easier to see the arrangement of the former ovules in the pod.)

Figure 42.8 shows a group of pods, each of which has developed from a flower, together with a single opened flower. It shows the sequence of development along the stem, the oldest pod at the bottom, the youngest at the top.

Pollination

Before fertilisation can take place, pollen grains must be transferred from the anther where they are formed to the stigma where they will grow. This is the process of **pollination**. Pollen escapes from the anthers even before the flower has opened. Insects no bigger than 1 or 2 mm in length are able to get

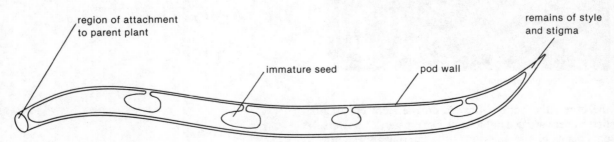

Fig. 42.7 French bean: L.S. young pod

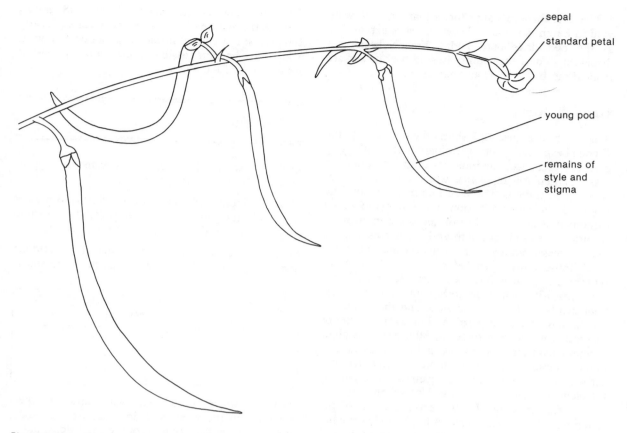

Fig. 42.8 *French bean: groups of pods and flower*

between the keel petals and crawl around inside. As they move about they unknowingly carry pollen on their bodies. When the stigma is mature, pollen grains get stuck to it from an insect's body.

If a flower's stigma ceases to be receptive to pollen before its own pollen is released, it can be pollinated only by another flower. In the French bean, however, the pollen is formed and released before the stigma becomes mature, so that a flower can be pollinated by itself: pollen may be discharged from its anthers directly on to its own stigma; or an insect may carry pollen to its own stigma. In the French bean less than 2 % of flowers are cross-pollinated.

Growth of the pollen tube down the style to the ovules, and fertilisation to form a zygote, are similar in all flowers and are described in Unit 41. In less than a week the carpel of the French bean increases in length from 1 cm to 15 cm. The ovary becomes the pod or **fruit** of the plant and the ovules inside develop into seeds. A fruit is simply a ripened ovary containing seeds. At first the seeds are soft and green; as they age, they grow bigger, the embryo inside them takes shape and grows to fill the seed, the food stores in the cotyledons increase and water is withdrawn from the seed, which becomes hard, brown and dehydrated. The maturing of the seed takes about six weeks.

As the seeds harden and mature, the green pods surrounding them become yellow and dry. During this period of **senescence**, or old age, the whole plant declines: the leaves change from green to yellow and wither as the plant begins to die.

Dispersal

During the winter the pods disintegrate and the seeds fall to the ground. French-bean seeds are larger than those of most flowering plants and are not usually carried far from the parent plant. Unlike a mature plant, a seed can withstand the conditions of winter. Because the testa is tough and thick, it protects the embryo inside from drying out, from being infected and from being eaten by small animals.

The French bean's ancestor, a plant still growing wild in South America, releases its seeds explosively as the pod dries out. Today French beans are dispersed all over the country in seed packets and are planted out by students and gardeners.

Growth curve

The graph in Figure 42.9 shows the dry mass of the French-bean plant from germination to death. Dry mass is a more accurate indicator of growth than wet mass, particularly in plants, which contain a lot of water. The curve is typical of flowering plants. The slight fall during germination is due to loss of stored carbohydrate as carbon dioxide and water in vigorous respiration. As soon as photosynthesis begins, the dry mass increases rapidly. At maturity the mass increases much more slowly than before: this is the stage of reproduction, when flowers, seeds and fruits form. After maturity (which in trees continues for years) photosynthesis declines and stops and there is again loss in mass due to respiration. This is the period of senescence which leads to death. After death the plant becomes food for saprophytes; it first forms humus in the soil and is eventually completely converted into carbon dioxide, water and inorganic ions as part of the element cycles described in Units 9 and 10.

Note that in Figure 42.9 the vertical axis of the graph has a logarithmic scale. A logarithmic scale has to be used if both very small and very large numbers are to be shown accurately. A single seed weighs only 0.4 grams and at germination loses 0.08 grams, whereas the mature plant reaches a dry mass of nearly 100 grams: if a normal linear scale had been used, it would have been impossible to show a fall in mass at germination of 0.08 grams.

Questions

Q 42.1 Explain why seeds need (a) water, (b) oxygen, (c) warmth in order to germinate.

Q 42.2 Explain why (a) inorganic fertilisers are *not* needed for the germination of seeds and (b) inorganic fertilisers *are* needed for the growth of seedlings.

Q 42.3 The table shows the dry mass of cotyledons and embryos from batches of ten germinating seeds.

Days after start of germination	Dry mass from ten germinating seeds (g)	
	Cotyledons	Embryos
2	4.0	0.1
4	3.2	0.7
6	1.7	1.6
8	1.1	2.7
10	0.9	3.3

(a) On a single set of axes, with days after germination on the x-axis and dry mass on the y-axis, draw graphs of the dry mass of the cotyledons, of the embryos and of the two combined.

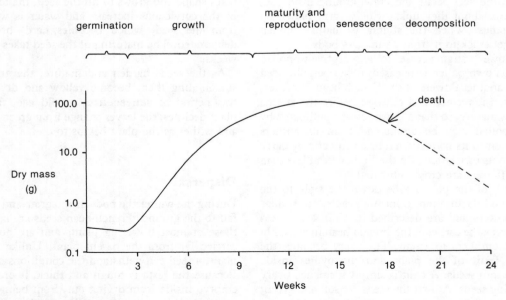

Fig. 42.9 French bean: growth curve

(b) Explain the reasons for (i) the loss in dry mass of the cotyledons, (ii) the increase in dry mass of the embryos, (iii) the loss in combined dry mass up to day 6, (iv) the increase in combined dry mass from day 6 onwards.

PRACTICAL WORK

Experiment 42.1 To determine the conditions needed for the germination of French-bean seeds

Note for teachers
French-bean seeds quickly lose their viability and should not be stored for longer than a year.

Materials required by each student
4 petri dishes
1 thermometer
1 spirit marker
absorbent cotton wool
40 French-bean seeds soaked in water for the previous 24 hours

Materials required by the class
1 dark cupboard in the laboratory
1 refrigerator

Method
1. Label the bottoms of the four petri dishes A, B, C and D and add your initials.
2. Line the bottom half of each petri dish with cotton wool.
3. Put ten soaked French-bean seeds on the cotton wool in dish A.
4. Thoroughly moisten the cotton wool in dishes B, C and D. Put ten soaked seeds in each dish.
5. Put the lids on the dishes and leave them for five days. Leave:
   Dish A in a warm well-lit part of the laboratory;
   Dish B in a warm well-lit part of the laboratory;
   Dish C in the closed cupboard in the laboratory;
   Dish D in the refrigerator.
6. After five days record in a table the number of seeds in each dish which have germinated.
7. Record in your table the temperatures of the room, cupboard and refrigerator.

Interpretation of results
1. What do the results in dishes A and B indicate about the importance of water in germination?
2. What do the results in dishes B and C indicate about the importance of light in germination?
3. What do the results in dishes C and D indicate about the importance of warmth in germination?
4. Use your answers to the above questions to list the conditions which are needed for successful germination in French beans.
5. Suggest one other external condition that may be necessary for seed germination. Devise an experiment to test whether it is necessary.

Experiment 42.2 To determine whether a flower is capable of self-pollination

Materials required by each student
1 pair of forceps
3 small clear polythene bags
adhesive tape
healthy potted French-bean plants to provide six unopened flower buds

Method
1. Find six unopened flower buds on a French-bean plant and use the forceps to remove any other flowers or flower buds.
2. Cover three of the unopened flower buds with bags. Stick adhesive tape on the bags so that gases can diffuse in and out of them but it is difficult or impossible for insects to get into them. Leave the other three unopened flower buds uncovered.
3. Leave the potted plants in a warm well-lit place for a few weeks, watering them from time to time with equal volumes of water.
4. Examine the plants at regular intervals and record the time at which fruits (bean pods) appear in each flower.

Interpretation of results
1. Explain why only unopened flower buds were used in this experiment.
2. Why were three of the flower buds enclosed within bags?
3. Why was it important that gases could diffuse into and out of the bags?
4. The results of your experiment do not prove that cross-pollination occurred in the flowers that were open to the air. Explain why not and suggest how the experiment could be improved to ensure that cross-pollination does occur in the flowers open to the air.
5. What can you conclude about the ability of the French bean to self-pollinate its flowers?

UNIT 43: Sexual reproduction in humans 1

In humans fertilisation is the fusion of a sperm and an egg to form a zygote. Half the zygote's inherited material (DNA) comes from the sperm and half from the egg, i.e. half from the father and half from the mother. A zygote develops into an individual who, when mature, can supply more sperms or eggs for fertilisation: the human life cycle is described in Unit 41. After reduction division, described in Unit 47, there are unlikely to be many sperms or eggs with the same combination of chromosomes. The mixing of the two parents' chromosomes at fertilisation makes it even more unlikely that there should be two zygotes with the same combination of chromosomes. It is possible that division after zygote formation will result in identical twins, triplets, etc. This, however, is asexual reproduction.

Reproductive system

The human reproductive system is made up of the organs that produce the gametes (the sperms and eggs), the organs that ensure fertilisation and, in the female, the organs that ensure development of the zygote into an embryo and finally into a baby.

Male

Figure 43.1 shows the male reproductive system, as well as the bladder from the urinary system, from one side. In reality there are a **testis**, an **epididymis**, a **vas deferens**, a **seminal vesicle** and a **Cowper's gland** on each side. The **scrotum**, **penis**, **urethra** and **prostate gland** are single and lie along the midline of the body. The two testes are carried in the sac

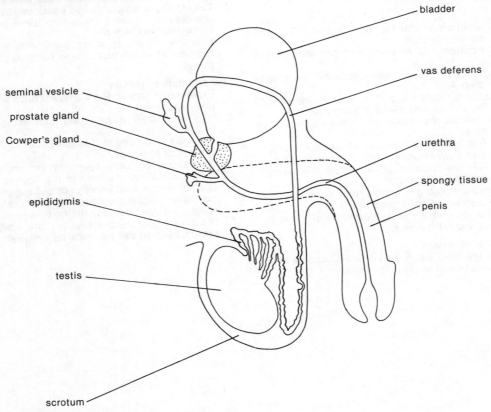

Fig. 43.1 Human male: reproductive system (side view)

called the scrotum. Each testis consists of a mass of extremely fine tubes which produce sperms and have an overall length of 100 metres. Sperms can be made at a rate of 100 million a day, but only if the temperature is about 35°C. The testes are kept cooler than most parts of the body, which have a normal temperature of about 37°C. As sperms mature, they are moved from a testis into an epididymis, where they are stored and further matured until, after about three weeks, they are capable of swimming. They do not swim until they are passed out of the penis (**ejaculated**) in a fluid called **semen** (or **seminal fluid**). If they are not passed out, they die and are reabsorbed in the epididymis.

Before the sperms can be passed out, spongy tissues of the penis become engorged with extra blood with the result that the penis becomes erect. The valve at the bladder opening remains closed. Various secretions are added to the sperms. The first secretions are from Cowper's glands and probably clean the urethra. Prostate-gland secretions pass through about 30 small openings into the urethra. Secretions from each seminal vesicle pass through its own duct into a vas deferens, which is shut off by a valve until just before the sperms are pushed along from an epididymis. Remember that there are a testis, an epididymis, a seminal vesicle and vas deferens on each side. The mixtures of sperms and seminal-vesicle secretions from the two sides join at the beginning of the urethra. Semen finally passed out is a mixture of secretions from the two Cowper's glands and the prostate gland (containing proteins, enzymes and salts), secretions from the seminal vesicles (containing sugar and enzymes) and 200 to 300 million sperms (which make up less than 1 % of the volume of semen). Muscular contractions of the walls of all the tubes move their contents through the system.

Female

Figure 43.2 shows the female reproductive system from the front. **Ovaries**, which produce the eggs, lie just above the groin; the **uterus** or womb, where fertilised eggs develop, lies along the mid-line of the body behind the pubic bones. The bladder lies in front of the uterus and **vagina**, with the opening from the urethra in front of the opening from the vagina. At the front of the urethra is the **clitoris**, a small protrusion which corresponds to the penis in the male. Since the clitoris does not perform the function of the penis, it is like male nipples, which do not perform the function of breastfeeding. But the clitoris is not functionless since it plays a part in sexual arousal. Note that the female has separate urinary and reproductive systems, whereas in the male the urethra is part of both. After puberty the ovaries release eggs at the rate of about one every four weeks until the **menopause** at about the age of 50.

Inside the ovary is connective tissue containing eggs at different stages of development. When an egg matures, it bursts out of the surface of the ovary. This is **ovulation**. (Reduction division, which began before birth, is not completed till after ovulation.) Where the egg used to be in the ovary there is now a fluid-filled sac (the **corpus luteum**, i.e. yellow body) whose walls begin to secrete progesterone as well as continuing to secrete oestradiol. (Secretion of

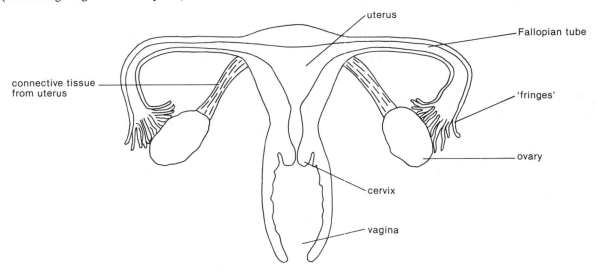

Fig. 43.2 *Human female: reproductive system (front view)*

oestradiol is continuous. Progesterone and oestradiol are the two female sex hormones.) Swaying movements of the blood-filled fringes at the enlarged end of the **Fallopian tube** waft the egg, still surrounded by some of the cells which helped to feed it in the ovary, through the fluid that surrounds the abdominal organs into the Fallopian tube. Muscular movements of the tube, together with beating movements of fine hairs that line the tube, move the egg towards the uterus. Eggs are usually released alternately from each ovary. Occasionally two or more are released either from one ovary or from the two ovaries: if two or more are fertilised, they grow into non-identical twins, triplets, etc.

The menstrual cycle is typically a four-week process in which an egg is released from an ovary (ovulation) and the uterus is prepared for a possible pregnancy. At the end of four weeks, if the egg has not been fertilised, the cycle starts again. Figures 43.3 and 43.4 show the events of a typical cycle.

On day 1 the uterus lining containing an unfertilised egg from the previous cycle begins to be shed. There is bleeding which lasts for a few days. This is **menstruation**, a **menstrual period** or a period for short. The start of a period is also the time when a new egg begins to mature in an ovary and when its surrounding cells prepare to secrete oestradiol. On day 5, when a period is just ending, the secretion of oestradiol increases and stimulates the formation of a new lining to the uterus. By day 13 the egg is nearly mature and the cells surrounding it reduce the oestradiol secretion. On day 14 the egg is released (ovulation) and the cells lining the eggless sac in the ovary soon increase their secretion of oestradiol and begin to secrete progesterone. Both these hormones continue the thickening of the uterus wall by giving it an increased blood supply. The egg moves down the Fallopian tube and reaches the uterus on day 19. About day 22 the secretion of the sex hormones begins to decline and on day 28 the unfertilised egg and the uterus lining start to be shed. You can follow these events in Figures 43.3 and 43.4. Menstrual cycles do not necessarily last four weeks: they vary from three to eight weeks and even longer.

Insemination and fertilisation

The mature egg is capable of being fertilised only during the first day or two after it is released from the ovary. For fertilisation to take place sperms must reach it when it is still high up in the Fallopian tube.

At sexual arousal the penis enlarges and firms as the volume of blood increases within it, while the vagina and clitoris both enlarge and the vagina secretes a lubricating fluid. Sexual intercourse begins when the penis is put into the vagina: regular thrusting movements of the penis result in the release of semen at male **orgasm**. This is **insemination** of the female. Female orgasms also occur but, like the clitoris (and unlike male orgasms), they are not essential for fertilisation.

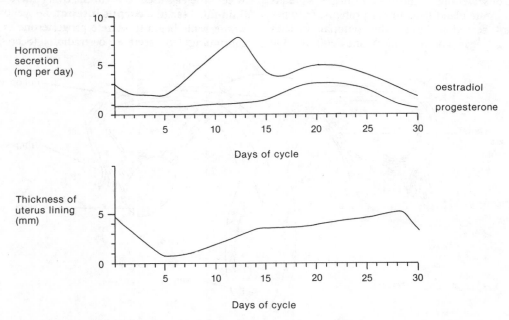

Fig. 43.3 Hormone secretion and uterus lining in a menstrual cycle

For fertilisation to take place sperms must now swim all the way to the egg. Proteins, enzymes, sugars and salts in semen provide the sperms with food and stimulate swimming. Acid secretions from the vagina are unsuitable for sperms: a quarter of them die within an hour in the vagina. By swimming away from the acid conditions and against a fluid current the sperms go towards and through the **cervix**, the narrow neck of the uterus, a journey which takes them several hours. Mucus in the cervix screens the sperms and ensures that only healthy ones arrive in the uterus. Not many of the 200–300 million sperms get as far as this.

Sperms can stay at least two days in the cervix but at the end of the first 24 hours some are already half way up the uterus. They swim into both Fallopian tubes, against the fluid current created by the fine hairs that line the tubes. Only a few hundred reach an egg: to have completed the journey those few hundred must be strong, vigorous and able to withstand acid.

On reaching an egg high up in the Fallopian tube the sperms try to push through its membrane: they butt at it, brush their tails on it and move it about until one finally gets in. The membrane immediately changes texture, preventing other sperms from entering. Inside the egg the sperm tail is lost while the sperm head swells and bursts, shooting the DNA from its haploid nucleus towards the haploid nucleus of the egg. The two nuclei fuse to form a diploid zygote: this is fertilisation.

Implantation

Twenty-four hours after fertilisation the fused nucleus has divided into two. The zygote then divides into two cells, the two cells divide into four, the four into eight, and so on. Divisions continue while the embryo travels down the Fallopian tube to the uterus. The embryo embeds itself in the thickened uterus wall. This is **implantation**. The embryo begins to secrete a hormone which travels in the mother's blood to the fluid-filled sac in the ovary to stimulate higher levels of oestradiol and progesterone. Look at Figure 43.3 and you will see that in a normal menstrual cycle oestradiol and progesterone begin to decline on about day 22. If this happens during pregnancy, the uterus lining is shed and the embryo is lost. For various reasons about 60% of fertilised eggs do die.

Figure 43.4 shows a typical menstrual cycle and an incomplete cycle in which an egg is fertilised.

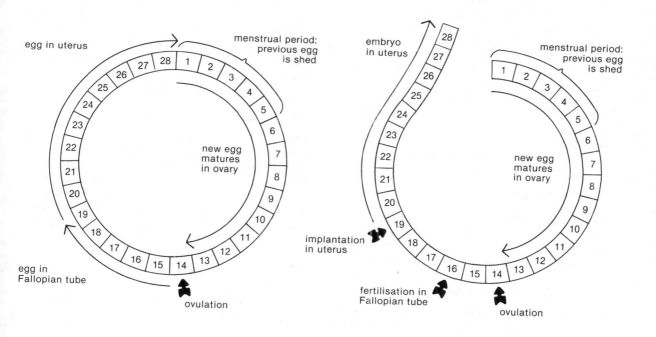

Fig. 43.4 Menstrual cycle and events following fertilisation

Test-tube babies

By testing the amount of hormones in urine, surgeons can work out when an egg is ready to be released from an ovary. With a microscope and a fine pipette passed through the abdominal wall, they can suck up an egg from the ovary just before ovulation. They keep the egg in a series of culture solutions and add sperms to it. About three days after fertilisation, when the embryo is just a few cells, it is transferred via the cervix into the uterus where it embeds and develops normally into a baby, a so-called test-tube baby. This process may enable an otherwise infertile woman to have a baby.

Artificial insemination

Artificial insemination occurs when sperms are placed in the vagina not by the penis but by some instrument. It can be used when a woman wants a baby but her husband is infertile. The semen usually comes from an unknown donor. Sometimes, however, a woman can be artificially inseminated by her husband's semen even though she has not been able to conceive naturally (perhaps because her husband's semen contains too few active sperms).

Artificial insemination was first used in farm animals to improve their genetic stock and is most common in cows. Prize bulls pass on DNA coding for many desirable features to their offspring. Artificial insemination enables a prize bull to have far more offspring: one fathered 80 000 calves in a single year. Semen is obtained by stimulating a bull to ejaculate. It can then be transported to farms and artificially inserted in cows' vaginas. Refrigerated, it can be stored for many years and be used to inseminate cows even after the bull who provided it has died.

Contraception

Contraception is the prevention of pregnancy without giving up intercourse. Without contraception all countries would now have a rapidly rising population. Figure 43.5 shows different contraceptive devices.

Oral contraceptives (the pill)

Many different kinds of oral contraceptive are on the market, but all contain synthetic versions of female sex hormones and are taken by women. The hormones either prevent an egg from being released from the ovary (ovulation) or prevent an egg from embedding in the uterus wall (implantation). Used according to the manufacturer's instructions, the pill is very reliable in preventing pregnancy, but some women suffer unpleasant side-effects. Some versions of the pill have been linked with an increased risk of thrombosis and cancer. Regular monitoring of women on the pill is essential.

Spermicides

Spermicides are chemicals, sold in the form of jellies, pastes and pessaries, which kill sperms. Used alone, they are unreliable. Spermicides are unlikely to have serious side-effects.

Sheath (condom or French letter)

The sheath, usually made of thin rubber, is unrolled over the man's erect penis just before intercourse. It prevents semen containing sperms from entering the vagina at ejaculation. It is not very reliable and should be used with a spermicide: some sheaths have a coating of spermicide. A sheath reduces the pleasure of intercourse, but it has no side-effects. It reduces infection by sexually transmitted disease (STD).

Intra-uterine device (IUD or coil)

The coil is a small curved piece of plastic in one of a variety of shapes, such as a spiral, loop or ring. Having been placed in the uterus by a specially trained doctor, it prevents an egg from embedding in the uterus wall. In some women it produces excessive bleeding at menstruation, and it increases the risk of infections of the cervix. If it remains in place, it is very reliable for two years or more.

Cap (diaphragm)

The cap fits over the woman's cervix at the top of the vagina. The right size and shape have to be prescribed by a specially trained doctor. The cap has to be put in place by the woman some time before intercourse and should remain in place for several hours afterwards. It prevents sperms from swimming through the cervix, but is not very reliable and should be used with a spermicide. It has no side-effects.

Rhythm method (safe period)

The rhythm method is avoidance of intercourse at the time of ovulation. It is reliable, if at all, only when a woman has a regular menstrual cycle of a known number of days.

Sterilisation

The usual method of sterilising women is to block or cut the Fallopian tubes so that no sperms can reach an egg and no egg can reach the uterus. In men the two vasa deferentia are cut and tied so that sperms cannot be passed into the semen. Both these methods are very reliable but they are drastic and are rarely reversible.

Withdrawal (coitus interruptus)

Withdrawal of the penis from the vagina just before ejaculation is the oldest method of contraception and is recorded in the Bible: Genesis 38:9 describes how Onan, not wanting to have a child by his wife, spilled his 'seed' on the ground. Withdrawal is an unreliable contraceptive method and reduces the pleasure of intercourse.

Use of contraception

About three million women in the UK are on the pill and about half a million have an IUD. About 2.8 million men use sheaths. Almost one million men and women have been sterilised. These four methods are used by about two-thirds of couples where the women are in the fertile age range.

Further improvements are constantly made to contraceptive techniques: injections of slow-release hormones in women have been tried; so have a pill for men and a morning-after pill for women. Better spermicides are discovered all the time.

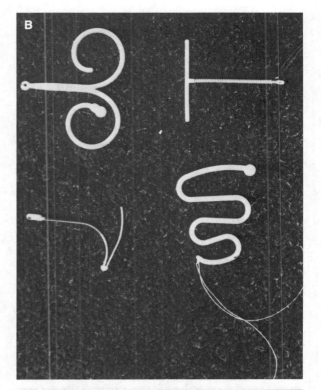

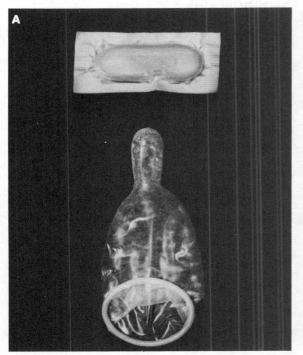

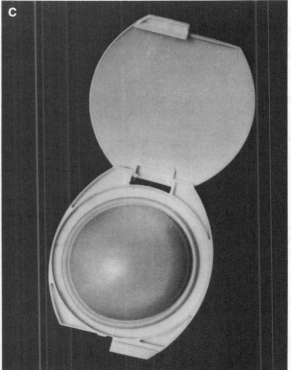

Fig. 43.5 Contraceptive devices: A. Sheath B. Intra-uterine devices C. Cap in its opened case

Sexually transmitted disease (STD)

Syphilis and **gonorrhea** have been known for centuries. They are rarely passed from one person to another except by sexual intercourse. Such diseases used to be known as **venereal diseases** or VD. Now the term **sexually transmitted diseases** or STD is used for diseases passed from one person to another by any kind of sexual contact, not only by sexual intercourse. Sexually transmitted diseases are caused by a variety of organisms: viruses, bacteria, fungi, single-celled animals, mites and insects such as lice. Syphilis and gonorrhea are both caused by bacteria and have been much reduced by antibiotics, notably penicillin. In Britain, notified cases of syphilis and gonorrhea have remained steady during the last ten years at a mere 15 % of all sexually transmitted diseases. Other genital and urinary infections, such as **herpes**, continue to increase. New sexually transmitted diseases, such as **AIDS** (acquired immune-deficiency syndrome), occasionally appear. Nearly all sexually transmitted diseases are curable.

Symptoms of various sexually transmitted diseases include: sores, ulcers and rashes in the genital region; discharges from the penis or vagina; pains and burning feelings on passing urine; frequent passing of urine; swollen glands in the groin; itching and soreness of the penis, vagina and anus. There are of course many causes of these symptoms apart from sexually transmitted diseases. But anyone who has such symptoms should see a doctor or go to an STD or Family Planning Clinic.

The treatment of sexually transmitted diseases varies with the cause. Antibiotics, especially penicillin, are used extensively; sulphur-based drugs are also used. Mites and lice are treated by painting the body with chemicals. Tracing the sexual contacts of infected people and treating them have helped to reduce the spread of these diseases.

The only sure method of avoiding a sexually transmitted disease is not to have sexual contact of any kind. Correct use of the sheath reduces the chances of getting or passing on these diseases. Washing the genitals before and after intercourse also helps to prevent infection.

Questions

Q 43.1 Each of the following processes takes place in the female before pregnancy: reduction division; ovulation; insemination; fertilisation; implantation. Explain what each term means and say where in the female body each process occurs.

Q 43.2 Assume that: a menstrual cycle lasts 28 days; ovulation takes place on day 14; eggs can be fertilised up to three days after ovulation; sperms can live in the female reproductive organs for up to seven days. What days are *not* 'safe', i.e. when must intercourse *not* take place if pregnancy is to be avoided.

Q 43.3 Explain why the 'safe period', the time when fertilisation is impossible, cannot be known for certain.

Q 43.4 Reread the section in this unit on 'test-tube babies' and answer the following questions.
(a) How do hormones get into urine?
(b) Why should it be easier for a surgeon to get an egg just before ovulation than after?
(c) Suggest what features of the culture solution are essential for success.
(d) Why wait three days before transferring the embryo to the uterus?

PRACTICAL WORK

Note for teachers
There is no practical work in this unit. Its subject-matter is likely to stimulate discussion and it contains plenty of questions.

UNIT 44
Sexual reproduction in humans 2; sexual reproduction in trout

Pregnancy

Pregnancy lasts from fertilisation to birth. From fertilisation to the birth of a fully developed baby takes about 38 weeks, or a little under nine calendar months. When the embryo arrives in the uterus, it is a small ball of cells less than 1 mm across. It embeds itself in the lining of the uterus, which is now richly supplied with blood vessels. The **placenta** develops immediately. The placenta is a region of the developing embryo in contact with the uterus wall: it develops **villi** (finger-like processes) which increase the surface area in contact with the wall. Exchange of material between the embryo and the mother's blood takes place across the placenta. At two weeks the embryo is 2 mm long, at six weeks 1 cm long, at eight weeks 2 cm long and at four months 10 cm long. Rudimentary arms and eyes are visible at four weeks and slight movements of the hands and fingers at six weeks. By eight weeks the embryo has most of its organs well-formed and henceforth can be called a **foetus**. The remaining 30 weeks of pregnancy are devoted to growth. Different regions grow at different rates: the brain and head grow quickly while the limbs grow slowly. At the end of pregnancy the foetus consists of six million million cells.

Figure 44.1 shows a twelve-week foetus in the uterus. The placenta is now a flat disc attached to the uterus wall by villi and to the foetus by the **umbilical cord** at its **navel**. The cord consists of connective tissue containing a loose jelly which surrounds two arteries and one large vein. By this time the foetus has a well-developed heart, which pumps blood to the placenta. Figure 44.1 also shows the foetus surrounded by fluid in a sac called the **amnion**. The **amniotic fluid** ('waters') cushions and supports the foetus and allows it unrestricted movement. Because the fluid is incompressible, the foetus is protected from physical damage.

As soon as the placenta has formed, it starts to produce oestradiol and progesterone at a far higher level than they are produced in the menstrual cycle. These hormones prevent more eggs from maturing in the ovaries and increase the size of the milk-producing glands in the breasts.

At the placenta the arteries from the foetus divide into smaller and smaller branches and finally into capillaries in the normal way. The capillaries in the villi are the only means of exchanging materials between foetus and mother: everything the foetus needs and everything it has to get rid of diffuses between the capillaries in the villi and the blood of the mother. During pregnancy the mother's blood flows into large spaces in the wall of the uterus. The villi lie bathed in these blood 'lakes'. Diffusion takes place because there are concentration gradients: substances needed by the foetus will be in lower concentration in its blood; waste products will be in higher concentration in its blood. There is no mixing of the blood of the foetus and the blood of the mother (except at the end of pregnancy when the placenta tissues begin to break down). The foetus's blood circulation is entirely separate from the mother's.

Figure 44.2 shows some of the products that pass across the placenta. There is danger in the fact that any substance in solution, provided that its molecules are small enough, can diffuse across the cells to the foetus's blood if there is a concentration gradient. A pregnant woman must take care what she eats and drinks and what gets into her blood. Something that

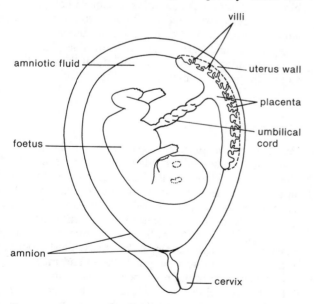

Fig. 44.1 Twelve-week foetus in uterus

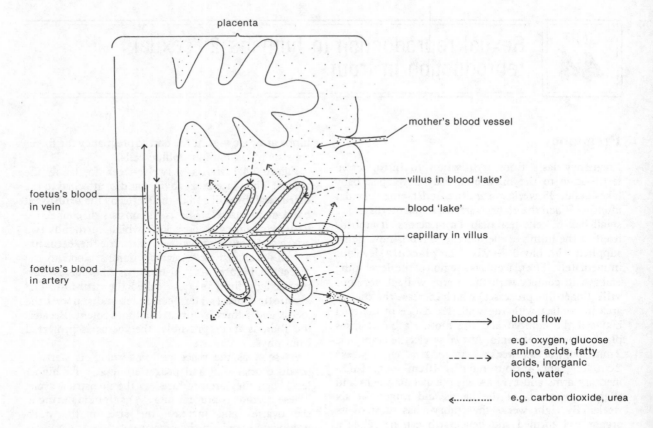

Fig. 44.2 Diffusion between foetus's and mother's blood

may not harm her may harm the foetus, though the placenta does prevent bacteria, most viruses and many harmful substances from passing to the foetus.

Birth

After about 38 weeks, in a process called **labour**, the muscles in the uterus wall contract every few minutes. The cervix, the neck of the uterus, **dilates**, i.e. opens. The width of the opening of the cervix gives an idea of how soon the baby will be born: it is not born until the cervix is fully dilated. The amniotic fluid passes out when the amnion breaks either before or during labour (breaking of the 'waters'). Muscles of the uterus wall contract more frequently until, together with contractions of muscles of the abdomen wall, they push the baby out, usually head first, through the vagina. Even after birth the umbilical cord remains attached from the baby's navel to the placenta and the baby continues to get oxygen through the blood vessels in the umbilical cord. At birth the baby takes its first breath of air, though it has been practising 'breathing' the amniotic fluid for months while in the uterus. The baby's blood circulation changes at birth: more blood passes to the lungs for oxygen; the circulation to the placenta along the umbilical cord stops. The umbilical cord is tied and cut and hardly bleeds at all. The placenta, called the after-birth, is later pushed out through the vagina by further contractions of the uterus.

Parental care

As soon as the placenta separates from the uterus wall, the secretion of oestradiol and progesterone into the mother's blood stops. This signals the start of other hormone secretions in the mother which encourage the production of milk in the enlarged breasts. The milk in the breast is 'let down' when the baby begins to suck: milk is then forced in fine streams from the nipple. Like many newly born mammals, the human baby is helpless and must not only be given milk but must be kept warm and clean. The human child has a far longer period of dependence on its parents than any other mammal.

Sexual reproduction in trout

Humans and trout are both vertebrates, i.e. both belong to the group of animals with backbones or vertebral columns. Both reproduce sexually by producing sperms and eggs which fuse to form zygotes, but the ways in which they do so could not be more different.

Humans usually produce only one egg at four-weekly intervals throughout the year, have fertilisation inside the body of the female, have internal development of the embryo for nearly nine months and have years of parental care of the child before it is independent. Trout **spawn**, i.e. shed eggs, every few days and produce getting on for a hundred at a time, do so only during a winter breeding season lasting a few months, have external fertilisation and, after covering the fertilised eggs, desert them. Whereas a woman can have only about one offspring a year, a female trout can have thousands.

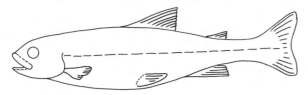

Fig. 44.3 Rainbow trout

Figure 44.3 shows an adult rainbow trout. Before spawning, the female chooses a place in the gravel bed of a river where she takes about two days to scrape out a hollow nest, called a **redd**, which is slightly longer and deeper than her own body. Her nest-building activity attracts male trout. The strongest or most dominant male drives other male trout away and both he and the female drive other fish away. (Fish eat one another's eggs.) In courtship the male nudges the female with his snout and quivers and swims from side to side over her as she rests in the hollow of the redd. After several hours, lying side by side with their genital openings close to each other, she sheds her eggs and he his sperms, called **milt**, at exactly the same moment. The eggs, which are orange and 4 mm across, fall to the bottom of the nest surrounded by white clouds of milt: sperms swim through the water in the nest to reach the eggs and fertilise them. Zygotes stay together in a compact group and the female sweeps gravel over them with fanning movements of her tail. The male leaves the redd after the burial of the eggs; the female may stay a few hours before she leaves too.

A trout zygote grows into an embryo on the food provided by the large yolk of the egg. After about four months the egg hatches and a free-swimming trout emerges. For about a month the small trout feeds on the remainder of the egg yolk inside itself. After that it finds its own food in the form of small animals in the water. Trout grow to maturity in two to three years and can live through more than one breeding season. But a trout's chances of surviving long enough to breed are obviously small: a female trout has thousands of offspring, yet only two of them need survive to breed and maintain the trout population.

Trout are carnivores: the adults eat small fish and insects. Trout are eaten by other carnivores and by omnivores, including humans, or they die and form the food of decomposers.

Questions

Q 44.1 Describe the differences in humans between the blood of the mother arriving in the blood 'lakes' of the placenta and the blood of the foetus arriving in the capillaries of the placenta.

Q 44.2 Why would it be dangerous for the blood of the human mother to pass into the foetus?

Q 44.3 Make a table like the one illustrated, but larger, and complete it to show the differences between human and trout reproduction.

	Reproduction	
	Human	Trout
(a) Number of eggs per year		
(b) Frequency of ovulation		
(c) Where the egg is fertilised		
(d) How food is obtained by		
(i) a zygote		
(ii) an embryo		
(iii) a young free-living offspring		
(e) Protection given to		
(i) a zygote		
(ii) an embryo		
(iii) a young free-living offspring		

PRACTICAL WORK

Note for teachers
There is no practical work in this unit. Teachers may like to complete the table in Question 44.3 in class.

UNIT 45: Inheritance and genetics

Only during the last hundred years have scientists begun to understand how it is that children resemble their parents. The study of inheritance and the variation in organisms owing to inheritance is called **genetics**. The nucleus of every cell of every organism contains very long, very narrow molecules of a substance called deoxyribonucleic acid or DNA. A DNA molecule may consist of hundreds and thousands of coded messages which instruct the cell to make hundreds and thousands of different enzymes. Each enzyme controls a chemical reaction which may influence the making of a structure, such as a horn in cattle, the making of a blood group, such as the ABO system, or a chemical process, such as the breakdown of a drug.

Genes and alleles

Each coded message, which is only a short piece of the DNA molecule, controls a particular inherited feature. This short piece of DNA is a **gene**. There are variations of a gene, called **alleles**, which are chemically slightly different from one another. For example, there is a gene which controls horn development in cattle: one allele allows horns to develop, the other does not. The DNA of these two alleles is slightly different chemically.

It is vital that different cells of one organism should contain the same alleles, i.e. the same coded messages in the form of the same DNA. Otherwise you could have, for example, a cow with a horn growing on one side of its head but not on the other. The nucleus divides by mitosis, a process described in Unit 40, which ensures that an exactly similar set of chromosomes containing exactly similar DNA is eventually formed in each resting nucleus after cell division.

Large numbers of coded messages are contained in every cell nucleus, but only a fraction of these will ever be needed in a particular cell. For example, only the cells on the head of the cow will need the coded message for making or not making a horn; only the cells in bone marrow which make the red blood cells will need the coded message for making or not making the A and B antigens; only cells concerned with drug breakdown will need the coded message for breaking down or not breaking down a drug.

When cell division by mitosis leads to asexual reproduction, the offspring contain the same DNA, genes and alleles as their single parent: they therefore inherit exactly the same characteristics as their parent.

Diploid organisms

Unit 41, on life cycles involving sexual reproduction, describes how the chromosomes of a diploid organism such as a human are inherited half from one parent and half from the other. The chromosomes are in pairs which (with rare exceptions) are the same size and shape: one of each pair comes from one parent, the other from the other parent. Since the chromosomes contain the DNA molecules which are made up of the alleles that control the different characteristics, half the alleles that are inherited by sexual reproduction come from one parent and half from the other parent. Figure 45.1 shows a diploid nucleus of a fruit fly: it contains four paired chromosomes, a total of eight chromosomes of which one of each pair has been inherited from one parent and the other of each pair from the other parent. Alleles are arranged lengthwise along a chromosome.

The two alleles of a gene are in exactly the same position on the two chromosomes of a pair. Whether or not the two alleles are the same or different, they control the same inherited characteristic of the organism.

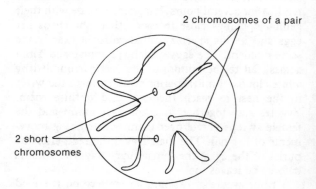

Fig. 45.1   Fruit fly: diploid nucleus n = 4

Homozygous and heterozygous

If the two alleles of a gene in a diploid nucleus are the same, they are **homozygous** and the organism is homozygous for that allele (or is a **homozygote** for that allele). If the two alleles are different, they are **heterozygous** and the organism is heterozygous for those alleles (or is a **heterozygote** for those alleles). Take the example of horned and hornless cattle. If both alleles of the pair are for horns or if both are for hornlessness, they are homozygous. If one allele of the pair is for horns and the other for hornlessness, they are heterozygous. But what happens to the heterozygous animal? Does it develop a horn on one side and not the other? Does it develop only a small horn on both sides? Does it develop two normal horns? Does it develop no horn? The only way to find out is to cross, i.e. mate, a homozygous horned animal with a homozygous hornless animal. An allele for horns will pass into the haploid gamete on the single chromosome that results from reduction division in the horned animal; an allele for hornlessness will pass into the haploid gamete that results from reduction division in the hornless animal. After fertilisation the diploid zygote will contain a pair of chromosomes of which one will have an allele for horns and the other an allele for hornlessness. The calf that grows from this zygote in fact becomes a hornless adult. Despite having an allele for horn development, the animal will show no trace of horns.

The allele for hornlessness completely masks the allele for horns: the allele for hornlessness is **dominant** while the allele for horn development is **recessive**.

Alleles do not always work in this way. Sometimes an organism is influenced by both alleles of a pair, when they are said to be **co-dominant**. For example, antirrhinums with one allele of a pair for white flowers and the other for red flowers will have pink flowers.

To find out whether one allele of a pair is dominant and if so which, or whether the two are co-dominant, we may have to carry out a number of crossings, i.e. matings, and see what kinds of offspring they result in.

Conventional terms and symbols

The alleles for hornlessness and horns can be represented as H and h. A gene is represented by a single letter of the alphabet. A capital letter stands for a dominant allele and a small letter for a recessive allele. You can choose any letter you like to represent any gene, but you must always give a key to show what your letter represents. It is sensible to avoid letters that look alike in their capital and small forms. A and a and B and b are good letters to choose, C and c are not. Do not choose X and x for a pair of alleles because × is used to indicate a mating or cross between two organisms. The symbol for male is ♂ and the symbol for female is ♀. Two alleles of a pair are shown one above the other with a line between them, e.g. $\frac{a}{a}$ or $\frac{a}{A}$. In organisms being crossed it is not usually known which allele of a pair came from the male gamete and which from the female gamete. If it is known, the allele from the male is written above the line and the allele from the female below the line.

There are two other terms you must know: **genotype** refers to an organism's alleles and **phenotype** to its characteristics. Two organisms which are the same phenotype may be different genotypes: of two hornless cows, one may be $\frac{H}{H}$ and the other $\frac{h}{H}$ or $\frac{H}{h}$. Two organisms which are the same genotype may be different phenotypes: of two $\frac{h}{h}$ (horned) cows, one may have horns while the other has lost them in an accident. When we are discussing genetics, we are not interested in changes in the phenotype resulting from the life it leads. We ignore them. We assume that the phenotype of an $\frac{h}{h}$ genotype is horned.

Figure 45.2 represents the crossing of a homozygous horned bull $\left(\frac{h}{h}\right)$ and a homozygous hornless cow $\left(\frac{H}{H}\right)$. Their offspring must all be the same genotype $\left(\frac{h}{H}\right)$ and the same phenotype (hornless): because hornlessness is dominant, whenever a genotype includes an H the phenotype must be hornless. In Figure 45.3 the genotype of an offspring is shown in the square, called a **Punnett square** after its inventor. Outside the square are the only possible alleles from the two parents, with the

phenotypes of parents   horned ♂   hornless ♀
genotypes of parents $\frac{h}{h}$ × $\frac{H}{H}$

H represents the allele for hornlessness
h represents the allele for horns

Fig. 45.2 Crossing of a homozygous horned bull with a homozygous hornless cow

allele in the male gamete on the left and the allele in the female gamete on the right. (The alleles are written in either order when it is not known which comes from the male gamete and which from the female.)

phenotypes of parents   horned ♂   hornless ♀

genotypes of parents $\dfrac{h}{h}$ × $\dfrac{H}{H}$

allele in male gametes   h   H   allele in female gametes

$\dfrac{h}{H}$

genotype of offspring $\dfrac{h}{H}$

phenotype of offspring   hornless

Fig. 45.3 Crossing of a homozygous horned bull with a homozygous hornless cow

Figure 45.4 shows how the alleles on the chromosomes in the nuclei of the parents in Figure 45.3 behave at reduction division and fertilisation.

Figure 45.5 represents the crossing of a homozygous horned bull $\left(\dfrac{h}{h}\right)$ and a heterozygous hornless cow $\left(\dfrac{H}{h}\right)$. Again the offspring must inherit an h allele from the bull, but they may inherit either an H or an h allele from the cow. There are two possible genotypes of the offspring $\left(\dfrac{h}{H} \text{ and } \dfrac{h}{h}\right)$ and two possible phenotypes (hornless and horned). Note that the cow's allele from its father is written on the left and its allele from its mother on the right.

phenotypes of parents   horned ♂   hornless ♀

genotypes of parents $\dfrac{h}{h}$ × $\dfrac{H}{h}$

allele in male gametes   h   H   alleles in female gametes

$\dfrac{h}{H}$ h

$\dfrac{h}{h}$

genotypes of offspring $\dfrac{h}{H}$ $\dfrac{h}{h}$

phenotypes of offspring   1 hornless : 1 horned

Fig. 45.5 Crossing of a homozygous horned bull with a heterozygous hornless cow

Figure 45.6 represents the crossing of a heterozygous hornless bull $\left(\dfrac{H}{h}\right)$ and a homozygous hornless cow $\left(\dfrac{H}{H}\right)$. Again there are two possible genotypes of the offspring $\left(\dfrac{H}{H} \text{ and } \dfrac{h}{H}\right)$, but there is only one possible phenotype (hornless). Note that the bull's allele from its father is written on the left and its allele from its mother on the right.

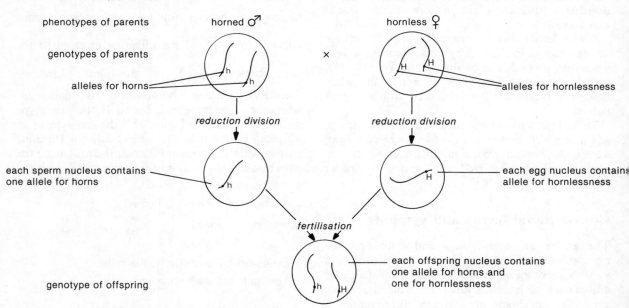

Fig. 45.4 Alleles at reduction division and fertilisation

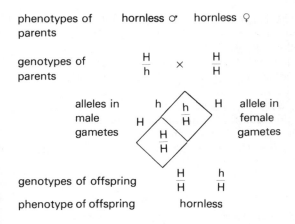

Fig. 45.6 Crossing of a heterozygous hornless bull with a homozygous hornless cow

Lastly, Figure 45.7 represents the crossing of a heterozygous hornless bull $\left(\frac{h}{H}\right)$ and a heterozygous hornless cow $\left(\frac{H}{h}\right)$. You can see that there are four possible combinations of alleles, yet only three possible genotypes and two possible phenotypes. It is important to realise that the genotypes $\frac{H}{h}$ and $\frac{h}{H}$ are the same: it makes no difference inside the animal which allele has come from the male and which from the female. As crosses get more complicated, you can see the advantage of using Punnett squares.

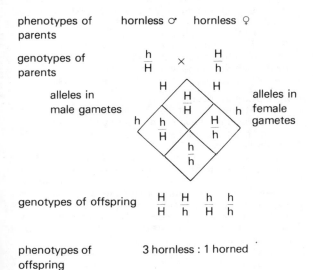

Fig. 45.7 Crossing of a heterozygous hornless bull with a heterozygous hornless cow

Chance or probability

The crossings in Figures 45.5 and 45.7 both result in two possible phenotypes but in different proportions. In Figure 45.5 there is an equal chance that an offspring will be hornless or horned, i.e. the chance that it will be either hornless or horned is one in two (50 % or 0.5). In Figure 45.7 there is only one chance in four (¼, 25 % or 0.25) that an offspring will be horned: $\frac{h}{H}$, $\frac{H}{H}$ and $\frac{H}{h}$ are all hornless; only $\frac{h}{h}$ is horned. Wherever there is random fertilisation between heterozygotes, the chance that an offspring will have a recessive characteristic is 0.25 (0.25 is the best way of writing it) and the chance that it will have a dominant characteristic is 0.75. This does not mean that, if a heterozygous hornless cow has four calves by a heterozygous hornless bull, three will be hornless and one will develop horns. That is only the most likely result. Any combination of hornless and horned offspring is possible. Even four horned calves are possible, though that is the least likely result.

If you toss a coin twice, you cannot rely on it to come down heads once and tails once. If, however, you toss a coin a thousand times, you are likely to get something near to 500 heads and 500 tails. Similarly, if you study a thousand offspring with both parents heterozygous for hornlessness, you are likely to find something near 750 hornless and 250 horned. The greater the number you study, the closer the ratio should come to 3:1. Also, if you are studying offspring of unknown heredity, the closer that two alternatives come to a ratio of 3:1, the more likely is it that the chances of their occurring are in fact 3:1.

How, when we want to conduct experiments about horn development, do we know we have, say, a heterozygous bull to start with? A hornless bull can be either heterozygous $\left(\frac{h}{H} \text{ or } \frac{H}{h}\right)$ or homozygous $\left(\frac{H}{H}\right)$.

We may be able to tell which it is from its pedigree, which is a record of several generations of the animals it has been bred from. If not, we must experiment. We might mate it with twenty homozygous horned cows. Finding the cows is no problem. Because horns are recessive, any horned cow must be homozygous $\left(\frac{h}{h}\right)$. If any of the calves is horned, we know for certain that our bull is heterozygous: if he had only alleles for hornlessness, which is dominant, all his calves would be hornless. If in fact all their calves are hornless, it is virtually certain that the bull is homozygous.

Inheritance from parents of different genotypes for one characteristic is **monohybrid** (*mono* means *one* and *hybrid* means *of different inheritance*). The ratios 1:1 and 1:3 occur so often in monohybrid inheritance only because each allele of a heterozygous pair has an equal chance of appearing in a gamete and because there are equal chances of all possible combinations of alleles, i.e. there is **random fertilisation**.

Questions

Q 45.1 A horned bull is mated with a horned cow. The allele for horn development is recessive. Draw a diagram to show the cross and the genotype and phenotype of their calf. Give your key.

Q 45.2 In cattle the allele for red coat colour is recessive. A heterozygous black-coated bull is mated with a red-coated cow. Draw a diagram to show the genotype and phenotype of the calves that can result from this cross. Give your key. What is the probability that their first calf will be red-coated?

Q 45.3 A herd of thirty ewes (female sheep) were mated with a black ram (male sheep). The allele for black wool is recessive to the allele for white wool. Some ewes had white lambs and some had black lambs.
(a) Give your key for the two alleles.
(b) What are the genotypes of (i) the black ram and (ii) the white ewes that had black lambs?
(c) What are the two possible genotypes of the ewes that had white lambs? Explain why there are two possible genotypes.

Q 45.4 Labrador retrievers are black or yellow. Their coat colour is controlled by a single pair of alleles. A black dog (male) is mated with a black bitch (female). Eight puppies are born in the litter: four are yellow and four are black.
(a) Which coat colour is dominant?
(b) Which of the ten dogs (two parents and eight puppies) are definitely homozygous for coat colour?
(c) Which of the ten dogs (two parents and eight puppies) are definitely heterozygous for coat colour?

Q 45.5 Mohan, a male 'white' tiger, was captured in India. ('White' tigers are off-white with light grey-brown stripes.)
(a) Assuming that the normal yellow colouring is controlled by a dominant allele Y and the white colouring by a recessive allele y, show the probable genetic composition of the parents and cubs in three of Mohan's matings. (i) When Mohan was mated with an unrelated yellow tigress, all the cubs were yellow. (ii) When Mohan was mated with one of his yellow daughters, some cubs were yellow and some were white. (iii) When Mohan was mated with a white tigress, all the cubs were white.
(b) Show the genetic composition of the cubs that could result from a mating of two of Mohan's yellow offspring with one another.
(c) After Mohan's death, how could you be sure of getting white tiger cubs?

PRACTICAL WORK

Experiment 45.1 To investigate inheritance in maize cobs

Note for teachers
Maize cobs showing ratios of starchy:waxy, coloured:colourless and smooth:wrinkled seeds are obtainable from most biological suppliers. Of these, the coloured:colourless phenotypes are most easily distinguished by young students. Cobs with approximate 3:1 and 1:1 ratios should be used.

Materials required by the class
maize cobs with approximate 3:1 and 1:1 ratios in the phenotypes of the seeds

Method
1. Examine one of the maize cobs and look for any obvious difference in the appearance of the seeds. Some may appear waxy while others do not, or there may be a difference in the colour or texture of the seeds. Record the two types of seed with the most obvious difference between them.
2. Record the number of each of the two different types of seed on the maize cob. Count row by row until you have counted all the seeds on the cob.
3. Repeat instructions 1 and 2 with a second maize cob.

Interpretation of results
1. Calculate the ratio of the two types of seed on each cob by dividing the larger number by the smaller. For example, 78 coloured seeds and 27 colourless seeds would give a ratio of 78:27 or 2.9:1.
2. The seeds on each cob are the offspring formed by the fertilisation of the ovules of one parent plant (which then formed the cob) by pollen from a second parent plant. Use your knowledge of genetics to explain the probable genotypes of the parent plants of the seeds on each cob. Only one pair of alleles were involved in each difference in the appearance of the seeds.
3. Suggest why the ratios you found do not exactly correspond to those you might expect from the crosses you have deduced in 2.

UNIT 46

Human inheritance

Even in cattle most characteristics are controlled by more than one gene. Human genetics is more complicated still. Human eye colour, for example, is controlled by at least three genes. Often we are not sure how many genes control a characteristic. One characteristic that does seem to be controlled almost entirely by a single gene is 'tongue-rolling'.

Tongue-rolling

Some people can roll their tongues upwards at the sides in the shape of a U, as in the photograph in Figure 46.1, though it may take them some time to learn to do so. Others cannot do it even after hard practice. Most people fall readily into one of these two categories.

The allele for ability to roll the tongue (R) is dominant over the allele for the inability to do it (r). There are therefore two phenotypes, tongue-rolling and non-tongue-rolling, and three genotypes, homozygous and heterozygous tongue-rolling and homozygous non-tongue-rolling. Figure 46.2 shows some possible crosses involving these alleles. Remember that $\frac{R}{r}$ and $\frac{r}{R}$ represent the same genotype.

R represents the allele for tongue-rolling
r represents the allele for non-tongue-rolling

I Both parents are homozygous tongue-rollers:

genotypes of parents $\quad \frac{\male\ R}{R} \times \frac{\female\ R}{R}$

allele in male gametes $\quad$ R $\quad$ R $\quad$ allele in female gametes

genotype of offspring $\quad \frac{R}{R}$

phenotype of offspring $\quad$ tongue-rolling

All children would be the genotype $\frac{R}{R}$ and the phenotype tongue-rolling

II Both parents are homozygous non-tongue-rollers:

genotypes of parents $\quad \frac{\male\ r}{r} \times \frac{\female\ r}{r}$

allele in male gametes $\quad$ r $\quad$ r $\quad$ allele in female gametes

genotype of offspring $\quad \frac{r}{r}$

phenotype of offspring $\quad$ non-tongue-rolling

All children would be the genotype $\frac{r}{r}$ and the phenotype non-tongue-rolling

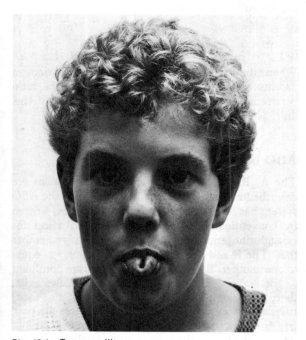

Fig. 46.1 Tongue-rolling

Fig. 46.2 Some tongue-rolling and non-tongue-rolling crosses

III The father is a homozygous non-tongue-roller and the mother a homozygous tongue-roller:

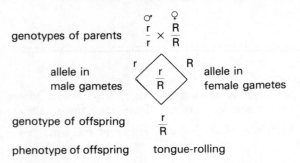

All children would be the genotype $\frac{r}{R}$ and the phenotype tongue-rolling.

IV Both parents are heterozygous tongue-rollers:

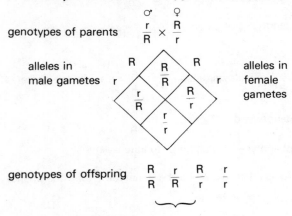

phenotypes of offspring 3 tongue-rolling : 1 non-tongue-rolling

There are three possible genotypes, $\frac{R}{R}$, $\frac{r}{R}$ or $\frac{R}{r}$, and $\frac{r}{r}$ and two possible phenotypes, tongue-rolling and non-tongue-rolling

Fig. 46.2 (cont.) Some tongue-rolling and non-tongue-rolling crosses

A non-tongue-roller must be homozygous $\frac{r}{r}$. A tongue-roller may be homozygous $\frac{R}{R}$ or heterozygous $\frac{r}{R}$ or $\frac{R}{r}$. As with horned and hornless cattle, we may be able to tell the genotype of tongue-rollers when they have offspring of their own. A tongue-roller with a non-tongue-rolling child must have supplied a non-tongue-rolling allele and must therefore be heterozygous $\frac{r}{R}$ or $\frac{R}{r}$.

Ear-lobe inheritance

In some people the lowest point of the ear lobe is attached to the side of the head; in others the ear lobe hangs freely below its attachment to the head. Some research workers believe that ear-lobe attachment is controlled by one gene with two alleles, the allele for free-hanging lobes being dominant to the allele for attached lobes. It is true that many ear-lobe attachments are not obviously either free-hanging or attached but are somewhere between the two. Nonetheless many cases of ear-lobe inheritance obey the simple genetic rules described for tongue-rolling.

Let F represent the allele for free ear lobes and f the allele for attached ear lobes. Work out all the possible combinations of alleles in the offspring of each of the combinations of parents:

(i) $\frac{F}{F} \times \frac{F}{F}$,   (ii) $\frac{f}{f} \times \frac{f}{f}$,   (iii) $\frac{F}{F} \times \frac{f}{f}$,
(iv) $\frac{F}{f} \times \frac{f}{F}$,   (v) $\frac{F}{f} \times \frac{F}{F}$,   (vi) $\frac{f}{f} \times \frac{f}{F}$.

When the possible children of a combination of parents are different phenotypes, express the ratio between them. The answers are at the end of this unit.

ABO blood system

The A, B, AB and O blood groups in humans are described in Unit 26. The gene that controls the ABO system has not two but three different alleles, written by convention as I^A, I^B and i. I and i stand for isoagglutinogen, but you do not need to remember that. The I^A and I^B alleles are co-dominant, i.e. equal to one another. Both the I^A and I^B alleles are dominant to the i allele. Though three different alleles exist, there can be only two in any one genotype. The usual rules apply: a gamete has one allele; a zygote has two, one from each parent, and develops into a diploid human being. Consider some pairs of alleles:

Genotype	Phenotype	
$\frac{I^A}{I^A}$, $\frac{I^A}{i}$ or $\frac{i}{I^A}$	Blood-group A (A antigen on red blood cells)	I^A is dominant
$\frac{I^B}{I^B}$, $\frac{I^B}{i}$ or $\frac{i}{I^B}$	Blood-group B (B antigen on red blood cells)	I^B is dominant
$\frac{I^A}{I^B}$ or $\frac{I^B}{I^A}$	Blood-group AB (both A and B antigens on red blood cells)	I^A and I^B are co-dominant
$\frac{i}{i}$	Blood-group O (neither A nor B antigen on red blood cells)	homozygous recessive alleles determine the phenotype

If you are blood-group O, your genotype must be $\frac{i}{i}$. If you are blood-group AB, your genotype must be $\frac{I^A}{I^B}$ or $\frac{I^B}{I^A}$. If you are blood-group A, you do not know if your genotype is $\frac{I^A}{I^A}$ (homozygous) or $\frac{I^A}{i}$ or $\frac{i}{I^A}$ (heterozygous). If you are blood-group B, you do not know if your genotype is $\frac{I^B}{I^B}$ (homozygous) or $\frac{I^B}{i}$ or $\frac{i}{I^B}$ (heterozygous).

Mrs Smith and Mrs Brown had babies at about the same time at the same hospital. When they left hospital with their babies, Mrs Smith found that hers was labelled Brown and Mrs Brown that hers was labelled Smith. Were they given the wrong babies or were the babies given the wrong labels? Both babies and all four parents were given ABO blood tests. The baby labelled Brown was group O $\left(\frac{i}{i}\right)$ and the baby labelled Smith was group A $\left(\frac{I^A}{I^A} \text{ or } \frac{i}{I^A} \text{ or } \frac{I^A}{i}\right)$. Since both Mrs Smith and Mrs Brown were group O $\left(\frac{i}{i}\right)$, both could have had either the baby labelled Brown $\left(\frac{i}{i}\right)$ or the baby labelled Smith $\left(\frac{I^A}{i}\right)$. But Mr Smith was group AB, which meant the Smith's baby could be only group A or group B, and Mr Brown was group O, which meant the Brown's baby had to be group O. The labels on the babies were correct. The mothers were given the wrong babies when they left hospital. The genetics of the Brown and Smith crosses are shown in Figure 46.3.

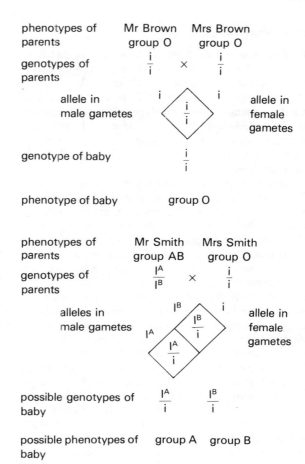

Fig. 46.3 Brown and Smith crosses

Other human examples of single-gene inheritance

Sometimes blue and brown eyes are given as examples of inheritance by a pair of alleles but this is an over-simplification: there are many different shades of eye colour between blue and brown.

Drug clearances are good examples but they are not easy to demonstrate. There is a drug called isoniazid which kills the bacteria that cause tuberculosis. It is broken down in the body by a particular enzyme. The gene that controls the production of this enzyme has two alleles. The recessive allele cannot make the enzyme which breaks the drug down: the drug accumulates in homozygotes with two recessive alleles and can poison them. The dominant allele can make this enzyme: both homozygotes with two dominant alleles and heterozygotes can break down the drug.

Sex inheritance

Like other inherited characteristics, sex is genetically determined, but in humans a whole chromosome with many genes is involved. Of the 23 pairs of chromosomes in humans, 22 always match one another in length, while the 23rd pair may or may not be the same length. If they are different lengths, the longer is called an X chromosome and the shorter a Y chromosome. If they are the same length, both are X chromosomes.

Females have two X chromosomes, written as XX. Look at the cut-outs of the chromosomes in Figure 41.3: there is a single long X chromosome and a single short Y chromosome. These cut-outs were therefore taken from the cell of a male. That approximately the same numbers of males and females are born is due to the XX and XY chromosomes. All the diploid cells of a woman contain two X chromosomes; all the diploid cells of a man contain both an X and a Y chromosome. All the eggs produced by a woman will contain one X chromosome after reduction division. Half the sperms produced by a man will contain an X chromosome and half a Y chromosome. Whether an X-carrying sperm or a Y-carrying sperm fertilises the egg determines the sex of the zygote and hence of the child. This is shown diagrammatically in Figure 46.4.

The convention of putting the male above or to the left of the female, which is used for alleles, is not followed for the sex chromosomes. Remember that X and Y are always written from left to right in their order in the alphabet.

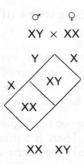

Fig. 46.4 *Chromosomes in sex inheritance*

Questions

Q 46.1 Of three children in a family, two were tongue-rollers and one was not. The mother was a tongue-roller, the father not. Let R represent the tongue-rolling allele and r the non-tongue-rolling allele.
(a) What are the genotypes of the three children and their parents? Explain your answers.
(b) Three of the grandparents of these children were not tongue-rollers. What is the phenotype of the fourth grandparent? Explain your answer.

Q 46.2 A woman whose blood group is A has three children by the same husband. The children's blood groups are O, A and B respectively. What are the genotypes of this woman, her husband and each of the three children? Explain your answers.

Q 46.3 A woman belonging to blood-group A had a group O child. The man she thought was the father of the child belonged to blood-group B. Could he have been the father of this child? Explain your answer.

Q 46.4 Four babies were born in the same hospital. Owing to a power cut their identities were confused before they could be given to their parents. The babies' blood groups were A, B, O and AB. The blood groups of the parents were:
   Mr and Mrs Farmer   A × B
   Mr and Mrs Miller   B × O
   Mr and Mrs Draper   O × O
   Mr and Mrs Carter   AB × O
Name the parents of each of the babies. Explain your answers.

PRACTICAL WORK

Experiment 46.1 To demonstrate reduction division with the use of bead models

Materials required by each student
46 poppit beads (23 of each of two colours)
a small piece of modelling clay

Method
1. Make two fifteen-bead chains, one of each colour, and stick a piece of the modelling clay on each. The chains represent a pair of chromosomes in a diploid nucleus. The pieces of modelling clay represent the centromeres of the chromosomes.

2 Make two eight-bead chains of the same two colours and stick a piece of modelling clay on each. These represent another pair of chromosomes, each with a centromere, inside the same diploid nucleus.

3 Make an exact copy of each of the four chains, lay each copy alongside its original and stick it to the modelling clay on the original. This represents the chromosome replication that occurs inside the nucleus before it divides: each chromosome becomes two chromatids joined only by a centromere.

4 Lay the two pairs of long chains side by side and the two pairs of short chains side by side so that the two groups of four chains are in line with one another in front of you. Move one pair of long chains to your left and the other to your right; move one pair of short chains to your left and the other to your right. These movements represent the first stage in reduction division.

5 You now have a pair of long chains and a pair of short chains on each side of you. Are the chains on your left made of beads of the same colour or of different colours? If they are made of beads of the same colour, need they have been? If they are made of beads of different colours, need they have been?

6 Split the modelling clay on the four pairs of chains. Move one of the long chains on your left away from you and the other towards you; move one of the short chains on your left away from you and the other towards you; move one of the long chains on your right away from you and the other towards you; move one of the short chains on your right away from you and the other towards you. These movements represent the second stage of reduction division: each chromatid has now become a new chromosome; the four couples of chains you have formed, each with one long and one short, represent haploid nuclei at the end of reduction division.

7 You have mimicked the events that occur when one diploid nucleus forms four haploid nuclei (instead of, as in mitosis, forming two new identical nuclei). These events are necessary for the formation of gametes: gametes have haploid nuclei.

Interpretation of results

1 What do the beads in each chain represent? Are they a good representation? Explain your answer.
2 What was the diploid chromosome number in your model cell?
3 Do your model gamete nuclei have the same number of chromosomes as the original model nucleus? If not, explain the advantage of the difference in chromosome number.
4 Are your model gamete nuclei identical to one another? If not, how many different types have you formed? Use your model to work out the highest number of different gamete nuclei that could be formed from a diploid nucleus with two different pairs of chromosomes.
5 How many different genotypes would occur if all your model gametes were randomly fertilised by the model gametes made by another member of the class using beads of two different colours from yours?
6 Do your answers to 4 and 5 suggest that reduction division leads to constant form or to variation in offspring?

Problem answers

The possible combinations of alleles in the offspring of the parents with free ear lobes and attached ear lobes are:

(i)

genotypes of parents $\dfrac{F}{F} \times \dfrac{F}{F}$ (♂ × ♀)

allele in male gametes   F $\dfrac{F}{F}$ F   allele in female gametes

genotype of offspring $\dfrac{F}{F}$

phenotype of offspring   free ear lobes

(ii)

genotypes of parents $\dfrac{f}{f} \times \dfrac{f}{f}$ (♂ × ♀)

allele in male gametes   f $\dfrac{f}{f}$ f   allele in female gametes

genotype of offspring $\dfrac{f}{f}$

phenotype of offspring   attached ear lobes

(iii)

genotypes of parents $\dfrac{F}{F} \times \dfrac{f}{f}$ (♂ × ♀)

allele in male gametes   F $\dfrac{F}{f}$ f   allele in female gametes

genotype of offspring $\dfrac{F}{f}$

phenotype of offspring   free ear lobes

(iv)

genotypes of parents $\frac{F}{f} \stackrel{\male}{\times} \stackrel{\female}{\frac{f}{F}}$

alleles in male gametes   F, F   alleles in female gametes   f, f

Punnett square offspring: $\frac{F}{f}, \frac{f}{f}, \frac{F}{F}, \frac{F}{f}$

genotypes of offspring $\frac{f}{f} \quad \frac{F}{f} \quad \frac{f}{F} \quad \frac{F}{F}$

phenotypes of offspring   1 attached ear lobes : 3 free ear lobes

(vi)

genotypes of parents $\frac{f}{f} \stackrel{\male}{\times} \stackrel{\female}{\frac{f}{F}}$

allele in male gametes   f   alleles in female gametes   f, F

genotypes of offspring $\frac{f}{f} \quad \frac{f}{F}$

phenotypes of offspring   1 attached ear lobes : 1 free ear lobes

(v)

genotypes of parents $\frac{F}{f} \stackrel{\male}{\times} \stackrel{\female}{\frac{F}{F}}$

alleles in male gametes   F, f   allele in female gametes   F

genotypes of offspring $\frac{f}{F} \quad \frac{F}{F}$

phenotype of offspring   free ear lobes

UNIT 47

Genetic variation

In any population of sexually reproducing organisms very few are exactly alike. No two human beings, not even 'identical twins', are exactly alike. There are two reasons for variation: one is **genetic**, that organisms inherit different alleles; the other is **environmental**, that organisms are affected by differences in their surroundings, i.e. by differences in the lives they lead. This unit deals with genetic variation.

Except for identical twins, triplets, etc., even children born to the same mother and father inherit different characteristics. The explanation lies in reduction division and sexual reproduction.

Reduction division

At reduction division only one of each pair of chromosomes passes into a gamete. Look at the diploid nucleus in Figure 47.1 showing one pair of chromosomes. One of these chromosomes (shown as a continuous line) was inherited by that organism from one of its parents while the other (shown as a dotted line) was inherited from the other parent. Figure 47.1 shows that at reduction division, if there is only one pair of chromosomes, it is possible for only two different haploid nuclei to form.

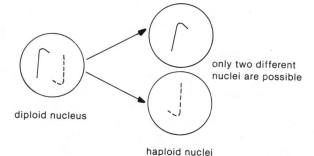

Fig. 47.1   Reduction division where n = 1

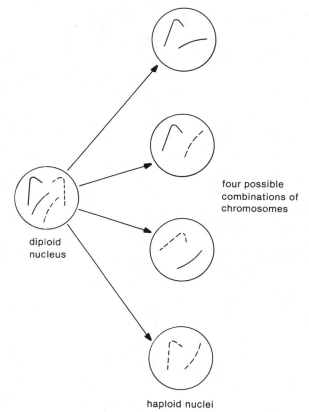

Fig. 47.2   Reduction division where n = 2

Look at Figure 47.2. Here the diploid nucleus has two pairs of chromosomes, one of each pair (shown as a continuous line) inherited from one of its parents, the other of each pair (shown as a dotted line) inherited from the other parent. At reduction division four different combinations of chromosomes are possible in the haploid nuclei.

Look at Figure 47.3. Here the diploid nucleus has three pairs of chromosomes. Again the chromosomes from one parent are shown as continuous lines and those from the other parent as dotted lines.

Try drawing the possible combinations where $n = 4$. Can you see a pattern emerging? With only one pair of chromosomes, the different possible haploid nuclei are two; with two pairs of chromosomes, the different possible haploid nuclei are four; with three pairs, the number is eight, with four pairs sixteen, with five pairs 32. You should recognise the series 2, 4, 8, 16, 32 as 2^1, 2^2, 2^3, 2^4, 2^5, or 2^n where n equals the haploid number of chromosomes, i.e. the number of different chromosomes.

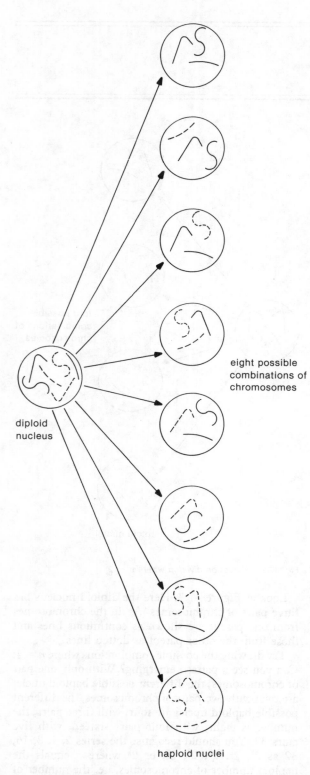

Fig. 47.3 Reduction division where n = 3

In humans there are 23 different chromosomes: the number of possible combinations of these 23 chromosomes is therefore 2^{23}. This works out at 8 388 608 different possible haploid nuclei.

Each chromosome of a pair has the same chance of going into a haploid nucleus with each of the chromosomes of any other pair. We say that chromosomes are distributed by chance or that they experience **independent assortment**.

Sexual reproduction

Look at Figure 47.2 again. When there are two pairs of chromosomes ($n=2$), four different combinations of chromosomes are possible in the haploid gamete nuclei. Suppose we call these four gametes A, B, C and D. At fertilisation during sexual reproduction these four gametes could fuse with any one of four other gametes, E, F, G and H, produced by another organism. This is shown in a Punnett square in Figure 47.4.

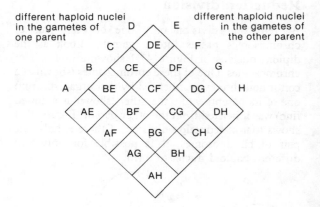

Fig. 47.4 Fertilisation where n = 2

Each pair of letters represents the fusion of two different gametes to form a zygote. When therefore there is random fertilisation and $n=2$, four different gametes from one parent can fuse with four different gametes from the other parent and there are sixteen possible zygotes. The mathematical formula for the possible zygotes when there is random fertilisation is 2^{2n}.

In humans, where $n=23$, the number of possible zygotes is $2^{2 \times 23}$ or 2^{46}, a figure of more than 70 million million, i.e. more than 70 000 000 000 000. This figure represents the possible combinations of chromosomes of any two gametes from a man and a woman. No wonder members of the same family are different from one another. Non-identical twins are as different from one another as any other brothers and sisters except that they are the same age. Non-identical twins are formed from different eggs and different sperms.

Identical twins are formed from the same egg and the same sperm. After fertilisation the zygote divides and redivides by mitosis and at each division the DNA contained in the chromosomes of the zygote is exactly duplicated. Each individual twin develops from one or more similar cells of the divided zygote: they inherit exactly the same alleles and are genetically identical. Though the formation of the zygote is sexual reproduction, its separation after mitosis is asexual reproduction.

Mutation

Though independent assortment of chromosomes (at reduction division) and random fertilisation explain an immense amount of genetic variation, they do not explain how entirely new characteristics appear in organisms. They do not explain why among moths that have been white for generations a black moth may appear. They do not explain how humans and apes have evolved from common ancestors yet are now so different that they cannot interbreed.

Entirely new characteristics are possible in organisms only because genes and chromosomes do occasionally change. A change in the inherited material of either a gene or a chromosome is called a **mutation**. *Mutation* is another word for *change*. Though mutations can occur in any of an organism's cells, they will affect offspring of sexual reproduction only if they occur in the cells forming the gametes or in the gametes themselves.

Gene mutation

If part of a DNA strand changes its chemical structure, a gene is changed and a new allele is formed: this is a **gene mutation**. A gene mutation is easiest to understand in a virus. Different viruses cause many human illnesses from the common cold to polio. It is difficult to say whether or not viruses are organisms. They are small collections of complex molecules on the verge of non-living and living matter. A virus may consist of only a few genes, just a strand of DNA in a protein coat. Mutations are frequent in the genes of viruses that cause the common cold and influenza. This explains why there are no readily available vaccines with which we can be inoculated against colds and flu: a vaccination effective against one strain of a virus may not be effective against a new strain formed after a gene mutation.

All the different alleles of genes that exist today have been formed by mutations of other alleles at some time in the past. Most, but not all, mutations are harmful to the organisms affected by them. A mutation that allows a virus to overcome a vaccine is clearly helpful to it: it increases the likelihood that the virus and its offspring will survive while other viruses are overcome by the effects of the vaccine.

A famous gene mutation probably occurred in Queen Victoria or in one of her parents. Queen Victoria handed down the allele for haemophilia (a disease in which the blood does not clot readily) to one of her sons and to two of her daughters. Her daughters handed it down to many of the royal families of Europe.

Chromosome mutation

Chromosome mutations take various forms: the chromosome content of a cell may be changed by twisting parts of its DNA, by losing or gaining DNA, even by losing or gaining a whole chromosome. In humans, because each chromosome contains many genes, most chromosome mutations are so drastic as to result in death. But one well-known human chromosome mutation can result in a child with Down's syndrome. This mutation gives the zygote an extra chromosome. Thus somebody with Down's syndrome has 47 chromosomes instead of the usual 46 (23 pairs). Figure 41.3 shows the 23 pairs of chromosomes in a normal human. A person with Down's syndrome has three chromosomes in place of the third smallest pair.

Genetic engineering

It is now possible to change genes in ways we want. The process is called **genetic engineering**. By using appropriate enzymes biologists break a molecule of DNA, insert in it a piece of DNA (such as a single allele) from another organism, and join up the ends.

For the technique to work, the original DNA must be isolated undamaged. Biologists have found it easiest to work with the circular DNA molecules found in some bacteria: these are much shorter molecules than the usual long narrow molecules of DNA. Figure 47.5 shows how an allele from another organism, in the form of a short piece of DNA, can be inserted into circular DNA. The new DNA formed from the original DNA and the DNA of the allele is called **recombinant DNA**.

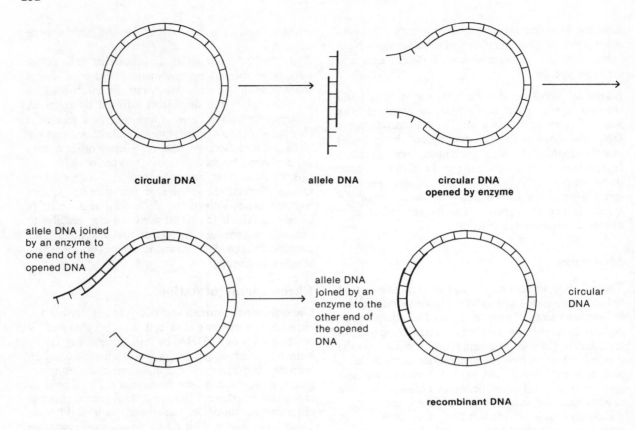

Fig. 47.5 Formation of recombinant DNA

The single human alleles that code for the making of the hormone insulin have been successfully inserted into circular DNA. The recombinant DNA has been introduced into bacteria which can be grown successfully in culture solution. Inside the bacteria the human alleles continue to code for the making of human insulin. The bacteria now make human insulin. When each of the bacteria divides to form a clone, the recombinant DNA is exactly reproduced. Bacteria now provide a good supply of human insulin.

Genetic engineering has exciting possibilities. Other human alleles code for antibodies. Can we get bacteria, by introducing these alleles into them, to produce human antibodies? Some bacteria have alleles that code for making ammonium ions from nitrogen gas in the atmosphere. Can we, by introducing these bacterial alleles into crop plants, do away with the need for nitrogen fertilisers? Can we reduce the number of handicapped children by substituting normal alleles for harmful alleles in human embryos?

Questions

Q 47.1 Explain the difference between identical and non-identical twins.

Q 47.2 Why are studies of identical twins of value to scientists?

PRACTICAL WORK

Experiment 47.1 To investigate the influence of light on the growth of genetically identical plants

Note for teachers
Plants produced by any propagation method in Experiment 40.1 will be genetically identical and the students' own plants may be used for this experiment.

The use of *Bryophyllum* plantlets is recommended because large numbers can be obtained at any time from a single parent.

Materials required by each student
1 250 cm³ beaker
2 small plant pots
2 plastic labels
1 small dibber (or pencil)
1 bowl
1 spirit marker

Materials required by the class
moss peat
sharp sand
garden loam
1 dark cupboard in the laboratory
6 plantlets from the leaf margin of a single *Bryophyllum* (air plant)

Method
1. Use the beaker to put two parts of sand, one part of garden loam and one part of moss peat in the bowl and mix them thoroughly.
2. Add water to the soil mixture until it is damp and forms a lump if you squeeze a handful of it together.
3. Fill the two plant pots with the soil mixture to within 1 cm of their rims.
4. Write the date and your initials on the plastic labels.
5. Use the dibber to make three small holes in the soil in each pot and plant one young plantlet taken from a *Bryophyllum* leaf in each hole. Gently press the soil around each plantlet to support it. Make sure that all the plantlets come from the same parent plant.
6. Stick the labels in the soil in the two pots.
7. Leave one of the pots in the laboratory in medium light (i.e. not near a window) and the other in the dark cupboard in the laboratory.
8. When the top 1 cm of soil in the pot in the light is dry, water both pots with the same volume of water.
9. Measure the length of the stems of the plants each week for several weeks.

Interpretation of results
1. Find the mean stem height of (a) the plants grown in the light and (b) the plants grown in the dark.
2. Which plants had grown more? Explain your result.
3. Explain why it is not possible that the difference in height was caused by genetic differences in the plants.
4. Suggest one way in which you could have measured the growth of the plants other than by measuring the lengths of their stems.

UNIT 48 — Variation due to the environment; natural and artificial selection

Environmental causes of variation

If you take leaf or stem cuttings from a single parent plant, they will contain exactly the same DNA in their nuclei. If you grow these genetically identical cuttings at different temperatures, in different light intensities, in different soils, with different amounts of water and fertiliser, with different amounts of carbon dioxide in the air surrounding them, you will get plants very different from one another. All these factors will influence the shape, size and mass of the plants that grow from the cuttings, i.e. they will affect the phenotypes but not the genotype.

Suppose you want to determine the effect of one environmental factor. You must use organisms that are genetically identical and are as nearly as possible identical in shape, size and mass. You must keep all the other environmental factors that might influence the organisms constant throughout your experiment (as in the experiment at the end of Unit 47).

Plants are normally used for such experiments because it is easy to get genetically identical material in the form of clones from a single parent, either by natural asexual reproduction or by vegetative propagation. Animal clones can be obtained by artifically dividing embryos, but this is much more difficult.

Biologists are trying to breed **pure-line** animals, i.e. animals that are not only genetically the same but are homozygous for all their alleles: strains of mice have been developed which are homozygous for many, though not all, alleles. In humans, only clones, i.e. identical twins, triplets, etc. are genetically identical. If you know any identical twins, you will know that they are in fact different. Their differences are produced by their environments, which include not only such things as the foods they eat and their relations with other people but their positions in the uterus before they were born.

Natural selection

All species of organisms reproduce offspring in far greater numbers than can survive and reproduce themselves. Animals such as trout produce thousands of offspring of which, under natural conditions when they are preyed on by bigger animals and may die from disease or starvation, very few survive to reproduce.

Even rabbits, which are mammals like ourselves, produce many more offspring than survive. A doe (female rabbit) in the wild has a breeding season lasting about six months. During that time she may have four litters each with five or six young: she produces between twenty and twenty-five rabbits a year. Most of these young are dead within a few months: only about ten per cent survive to the first winter and fewer than that survive to breed the following summer. As herbivores, rabbits are low in the food chain and have numerous predators such as foxes, stoats, and feral (wild) and domestic cats. They succumb to harsh weather; to diseases such as myxomatosis, tuberculosis and stomach and throat infections similar to humans; and to parasites such as roundworms, tapeworms and liverflukes which, even if they do not kill them, reduce their vigour and resistance to disease.

Only rabbits that are alert, fast, warm and resistant to disease are likely to survive long enough to have offspring. Thus the alleles responsible for alertness, speed, warmth and resistance to disease are more likely than others to be passed on to future generations. The species will tend to become better adapted or better fitted to its surroundings. This is the theory of **natural selection** which Charles Darwin (1809–82) explained in his famous book *The Origin of Species* published in 1859. Darwin also explained that today's plants and animals have **evolved** (developed) from primitive animals and plants and before that from single-celled organisms. This is why a bird's wing and a human's arm, though they have very different functions, have a similar bone arrangement. It is also why the human embryo at one time in its development has structures similar to the gill slits of a fish. But Darwin was unable to explain how organisms developed entirely new characteristics. We now understand that new characteristics are the result of mutations, i.e. of spontaneous changes in the genes that control inheritance.

Usually the greatest threat to rabbits comes from

their predators, when alertness and speed are the most important factors in ensuring their survival. In Britain between 1953 and 1955 the most important factor was resistance to disease. Myxomatosis arrived in 1953 and within two years killed 99 % of the rabbit population of England, Wales and Scotland. By the mid-70s many rabbits were found which had the genetic capacity to survive the disease. Even in 1953, alleles must have been present in some rabbits giving them resistance to myxomatosis. More of these rabbits and of their myxomatosis-resistant offspring will have survived than other rabbits. Although myxomatosis-resistant rabbits are not a new species, they show how genetic proportions change in a population as a result of natural selection. They are a small example of evolution.

Natural selection goes on around us all the time. In forests where there is dense ground cover by small plants, most seeds falling from trees are wasted: only an exceptionally vigorous seedling can survive in competition with the existing vegetation. But remember that the 'survival of the fittest', which is the principle of natural selection, does not mean survival of the strongest. It means the survival of the organisms best fitted to the conditions under which they live. A study near Oxford showed that land snails (*Cepaea nemoralis*) preyed on by song thrushes (*Turdus ericetorum*) were most likely to survive if they were difficult to see because their colouring merged with their surroundings, i.e. they were camouflaged.

Animals at the top of a food chain do not have predators. But they compete with other members of their species for food. Like all organisms they produce more offspring than can survive: only the best hunters or those best at finding or seizing food will in fact survive; the rest will die of starvation. Nestlings rely on food brought to them by the parent birds: if the parent birds cannot find enough food, only the strongest and most determined nestlings seize enough food to survive.

Natural selection also works through mating. Competition for mates is most evident in herd animals such as sheep, deer and seals: a strong healthy male which is a good fighter can keep a harem of up to 30 females, all of which will bear offspring containing his DNA. The weak males fail to reproduce and their DNA is not passed on to the next generation.

Artificial selection

For thousands of years humans have deliberately grown crops from seeds of high-yielding crops: not only are the alleles producing the best crops selected but, if a mutation of an allele produces an even better crop, it too is selected and remains in the population. When humans decide which plants or animals shall reproduce, selection is not natural but **artificial**.

The ancestor of the French bean is a dry bean with a twining stem and a dry pod wall which splits and releases its hard dry dormant seeds explosively. This bean, which grows wild in South America, is the original *Phaseolus vulgaris*. Merely by selecting seeds from plants with desirable characteristics, including those resulting from mutations, growers have developed the present dwarf French bean which is bushy, has a fleshy pod that does not split and contains fleshy seeds that will germinate immediately. In fact growers have developed several different varieties of French bean: some still climb like the wild bean; some have red or purple flowers instead of the usual white; some have purple pods instead of the usual green; some are more resistant to mildew than others; some grow better at low temperatures; some grow more quickly. You can buy seeds of the variety of French bean best fitted to the conditions in which you will grow them. All these different varieties belong to the same species as the original wild bean, *Phaseolus vulgaris*.

Dogs are an extreme example of artificial selection. Suppose that you had never before seen or heard of dogs and that you were looking at a Great Dane and a dachshund. You would surely take it for granted that they were different species. In fact all dogs belong to the same species (*Canis familiaris*), which means they can all interbreed. By selecting dogs with the characteristics that different people want, breeders have produced all the different varieties (i.e. breeds) of dog and have done so mainly within the last few hundred years.

The breeding of guide dogs for the blind is a recent example of artificial selection which has taken only a few years. When dogs were selected for training because they were alert and independent, regardless of their pedigrees, only two in ten were successful as guide dogs. Good guide dogs must also be responsive to subtle changes in the human voice, be hard-working and willing, realise when they are in danger yet not panic, and have confidence in themselves. These more subtle qualities cannot be picked out in a dog before training. Since eight out of ten dogs proved unsuccessful after training, a great deal of money had been wasted. A new programme was begun in which only dogs from successful families of guide dogs were selected for breeding. No less than nine out of ten of the offspring from these selected 'blood lines' make successful guide dogs.

Questions

Q 48.1 Gene changes in some bacteria produce strains resistant to penicillin. Whenever penicillin is used for some time, resistant bacteria flourish. How does this illustrate mutation, natural selection and evolution?

Q 48.2 Females of the fruit fly (*Drosophilia melanogaster*) will not mate with a male unless he performs the correct courtship dance. Some fruit flies inherit small wings which prevent them from performing a perfect courtship dance. Suggest why there are few small-winged fruit flies in natural populations.

Q 48.3 The common land snail, *Cepaea nemoralis*, is preyed on by song thrushes (*Turdus ericetorum*). The thrushes break open the snails' shells on flat stones, called anvils. The table shows the percentage of snails of two different colours found in a beech wood and the percentage of snail shells found near the anvils of thrushes preying in the same wood.

Colour of snail shells	Percentages	
	Snails	Snail shells near thrush anvils
Yellow	23	42
Brown-pink	77	58

(a) Which snails were less likely to be eaten by song thrushes?
(b) Suggest why these snails were better adapted to escape predation by song thrushes.
(c) Would you expect similar results in grassland surrounding this wood? Explain your answer.

PRACTICAL WORK

Experiment 48.1 To investigate the effect of competition for space on the germination of cress seeds

Note for teachers
Prepare the agar for use in this experiment by dissolving the following ingredients in 1 dm³ of distilled water and autoclaving it:

agar	15 g
K_2HPO_4	1 g
$Ca(H_2PO_4)_2 \cdot H_2O$	0.5 g
$MgSO_4 \cdot 7H_2O$	0.2 g
NaCl	0.1 g
$FeCl_3 \cdot 6H_2O$	0.01 g
$(NH_4)_2SO_4$	0.5 g

Pour this medium into sterile flat-bottomed specimen tubes to a depth of about 3 cm and fit each tube with a cotton-wool plug. Allow five tubes for each student.

Materials required by each student
5 flat-bottomed specimen tubes containing nutrient agar and cotton-wool plugs
1 weighing bottle
1 pair of forceps
1 fine paint brush
1 spirit marker
3 g cress seeds

Materials required by the class
balances to weigh up to 1.0 g (1 per 4 students)

Method
1 Copy the table.

Mass of seeds (g)	Number of seeds per tube	Number of seeds germinated per tube	Percentage success rate
0.10			
0.25			
0.50			
0.75			
1.00			

2 Label the tubes of agar 0.1 g, 0.25 g, 0.5 g, 0.75 g and 1.0 g and add your initials.
3 Weigh 0.1 g of cress seeds very accurately. Record the number of seeds in this sample. Sprinkle the seeds on the agar in the tube labelled 0.1 g. Replace the cotton-wool plug in the tube.
4 Weigh further samples of cress seeds of 0.25 g, 0.5 g, 0.75 g and 1.0 g and sprinkle them on the agar in the appropriately labelled tubes. Do not count the seeds in these samples but use the number of seeds in the 0.1 g sample to estimate the number of seeds in them. Record these estimates in your table.
5 Leave the tubes in a warm place away from direct sunlight.
6 After three to five days use the forceps to remove the seedlings that have grown in each tube. Record the number of seedlings as the number of seeds germinated in each tube.

Interpretation of results
1 Calculate the percentage success rates using the formula:
$$\frac{\text{number of seeds germinated per tube}}{\text{number of seeds per tube}} \times 100$$
Record them in your table.
2 Draw a graph with number of seeds per tube on the x-axis and percentage success rate on the y-axis.
3 Was the percentage success rate constant in the five tubes? If not, explain any trend your graph shows.

Appendix A Mathematical, physical and chemical background

Mathematical background

To interpret the mathematical data that you collect in biological investigations you need only to add, subtract, multiply and divide. With these skills you can calculate arithmetic means, percentages, rates, ratios and probabilities.

Arithmetic means

To find the arithmetic mean or average of a set of figures you add up the figures and divide by the number of figures. The mean of 11, 3, 9, 7 and 10 is:

$$\frac{11 + 3 + 9 + 7 + 10}{5} = \frac{40}{5} = 8$$

The mean of 7, 12, 10, 9, 3 and 6 is:

$$\frac{7 + 12 + 10 + 9 + 3 + 6}{6} = \frac{47}{6} = 7.83$$

To find the mean mass of new-born kittens you would weigh several litters and divide by the total number of kittens. It is not necessary to weigh each kitten individually: you can weigh all the kittens in a litter together if you have a large enough weighing machine. The mean mass gives you a useful guide to the mass of new-born kittens: it does not matter if none of the kittens weighs that amount.

Percentages

A percentage is a proportion of 100. To convert a fraction, which is a proportion of 1, to a percentage, multiply by 100.

$$\frac{1}{4} \times 100 = 25\%$$

$$\frac{1}{2} \times 100 = 50\%$$

$$\frac{48}{80} \times 100 = 60\%$$

$$\frac{2}{1} \times 100 = 200\%$$

Suppose that baby A weighs 2.5 kg and baby B weighs 5.0 kg. Baby A's weight is half baby B's while baby B's is twice baby A's; or baby A's weight is 50% of baby B's while baby B's is 200% of baby A's.

If 23 of 31 seeds have germinated, what is the percentage germination?

$$\frac{23}{31} \times 100 = 74.2\%$$

Percentages are useful when we want to compare different fractions. If in the same area plant species A is touched 52 times in 82 nail lowerings and plant species B is touched 39 times in 63 nail lowerings, which plant covers the greater area? (See *Point frame* in Unit 3 for an explanation of nail lowerings.)

$$\text{Coverage of plant A} = \frac{52}{82} \times 100 = 63.4\%$$

$$\text{Coverage of plant B} = \frac{39}{63} \times 100 = 61.9\%$$

Therefore plant A covers the greater area.

Rates

When we say that someone's pulse rate is 72 beats per minute, we mean that it will beat 72 times in a minute. The frequency with which something happens in a certain time is a rate. If a pulse beats 33 times in half a minute (30 seconds), what is the rate per minute?

In 30 seconds the pulse beats 33 times

In 1 minute (60 seconds) the pulse beats

$$\frac{33 \times 60}{30} = 66 \text{ times}$$

Therefore the pulse rate is 66 beats per minute.

Suppose that in a respiration experiment some seeds used 1.8 cm^3 of oxygen in 23 minutes. What is the rate per minute at which they used oxygen?

In 23 minutes the oxygen used by the seeds is 1.8 cm^3

In 1 minute the oxygen used by the seeds is $\frac{1.8}{23} = 0.08 \text{ cm}^3$

Therefore the seeds used oxygen at the rate of 0.08 cm^3 per minute.

Many rates are calculated in biology, for example: bubbles of oxygen per minute; volume of oxygen per minute; heartbeats per minute; breaths per minute; increase in length per day; increase in mass per year.

Ratios

If a class contains 14 male and 16 female students, we can say the ratio of males to females is 14 to 16 or 7 to 8 and we can write it as 7:8. The ratio does not tell us how many males and females there are. It tells us only the numerical relation between them: we know that for every seven males there are eight females. If moth traps in a wood contain 125 dark peppered moths and 95 light peppered moths the ratio of dark to light peppered moths in the wood is 125:95 or 25:19 or 1.32:1. To express a ratio in the smallest possible whole numbers, divide the figures by any common factors: the only factor common to 125 and 95 was 5. To express a ratio to 1, divide the larger figure by the smaller. In the study of genetics, ratios approximating to 1:1 and 3:1 are common.

Tables

In a table figures can be listed both vertically, with a description at the top, and horizontally, with a description on the left. A table can therefore give you two descriptions of every figure, one above it and one to the left of it. Remember that, if figures are measurements, the table must give the units of measurement such as grams (g) and metres (m). Whenever possible, give the units of measurement in the descriptions at the top or on the left. When this is not possible, add the units of measurement to the individual figures in the table. All the units of measurement are given at the top in Table 21.2 and on the left in Table 21.1. In Table 29.1 all the units of measurement are added to the figures.

Graphs

A graph is a diagram showing numerical values. Though it is not as accurate as a table, it makes it easier to grasp the relation between different figures. It often makes it easier to see a trend. For example, if you graph temperature against time of day, you can see at a glance that temperature rises until the afternoon and then falls.

The only kind of graph you are likely to be asked to draw yourself in the GCSE examination is a jagged-line graph. You must also be able to interpret bar charts, histograms and pie charts.

Jagged-line graphs are the kind in Figures 12.4 and A.1. In a jagged-line graph the horizontal axis is the x-axis. This is the line on which you space the **independent variable**, the one you have been able to decide on for yourself. For example, in Experiment 5.5 you decided to measure the volume of dough mixture every five minutes. You decided on the five-minute intervals, whereas you could not know what the volume of dough mixture would be whenever you measured it. Time is therefore the independent variable and goes on the x-axis. Because you have decided it for yourself, it is in regular whole numbers without fractions: 5 min, 10 min, 15 min, etc.

The vertical axis is the y-axis. On this you space the **dependent variable**, the one you have not been able to decide for yourself. In Experiment 5.5 it is the volume of dough mixture. Because you have not decided it for yourself, it is in irregular numbers containing fractions: $22.5 cm^3$, $28.2 cm^3$, $31.8 cm^3$, etc.

In the GCSE examination you are usually told which set of figures should go on which axis.

When you draw a graph, make sure that the scales you choose for your vertical and horizontal axes cover all the figures you have to plot. You can use your graph paper either way round so that either the vertical or horizontal axis is on the longer side: choose the way round that allows you to use more of the sheet of paper.

After you have drawn your two axes, label each one with a description of what you will plot along it and of the unit in which it is measured unless it simply records numbers: for example, in a graph of the number of organisms in a habitat at different times of day, the unit of measurement on the x-axis will be hours and there will be no unit of measurement on the y-axis. If the figures plotted on an axis are percentages, you should put (%) after the description. A unit of measurement is usually given in brackets after the description of what is plotted on the axis. You should always draw both the axes and the graph in pencil, but you may label the axes in ink. Write the numerical values of the scales you have chosen against the darker lines on the graph paper. The point where the x-axis and y-axis meet need not equal 0. If the values you must plot on the x-axis range from 6.00 hours to 18.00 hours, you can make the point where the two axes meet equal to 6.00 hours on the x-axis. If the number of organisms in the habitat varies from 23 to 47, you can make the point where the two axes meet equal to 20 on the y-axis.

Suppose you have to draw the graph in Figure A.1 from the data in Table A.1. To plot each point, first find the vertical alignment that corresponds to the independent variable, i.e. the year. Then find the horizontal alignment that corresponds to the dependent variable, i.e. the percentage of adult males who smoke cigarettes. Mark the point where the two alignments meet. There are various ways of marking the points of a graph. The best two ways are a dot in a circle (⊙) and a plus sign (+). A plus sign is better than a multiplication sign (×), which is more likely to be hidden when you join up the crosses.

When you have plotted all the points of your graph, join them in order by drawing straight lines between them with a ruler. This is what is meant by drawing a **jagged-line graph**. A jagged-line graph is easier to draw neatly than a curved-line graph of 'best fit'.

Table A.1 Adult male cigarette smokers

	Males aged 16 and over					
	1972 %	1974 %	1976 %	1978 %	1980 %	1982 %
Cigarette smokers	52	51	46	45	42	38

Source: Office of Population Censuses and Surveys

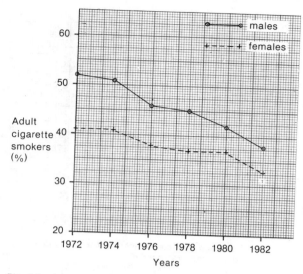

Fig. A.2 Adult cigarette smokers

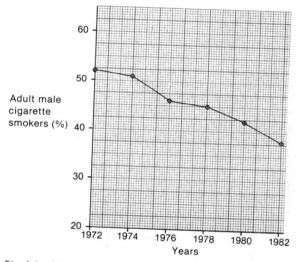

Fig. A.1 Adult male cigarette smokers

Suppose you have to draw the graph in Figure A.2 from the data in Table A.2. Use different kinds of points to plot the two sets of figures: this will prevent you from joining up the wrong points. Use different kinds of lines to join up the two sets of points and give a key to them: this makes it easier to understand the graph.

Table A.2 Adult male and female cigarette smokers

Year	Percentage cigarette smokers	
	Males aged 16 and over	Females aged 16 and over
1972	52	41
1974	51	41
1976	46	38
1978	45	37
1980	42	37
1982	38	33

Source: Office of Population Censuses and Surveys

Bar charts are a form of graph suitable when only one of the variables is numerical. Figure A.3 is a bar chart showing the proportions of fat, starch, sugar and protein in the diet of people in Britain in 1850 and today. Because the figures are percentages, the bar chart does not tell you whether people eat more or less today than in the past: it tells you only about changes in the kinds of things people eat. The key tells you what type of food the different bars (or blocks) represent; the numerical vertical axis tells you the percentage of each type of food in the total diet.

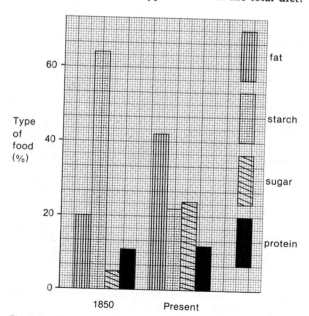

Fig. A.3 Diet content: 1850 and present

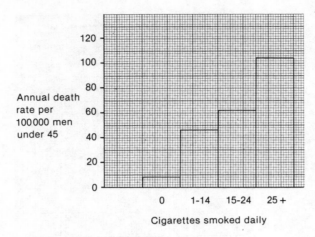

Fig. A.4 Death rate from coronary heart disease: male smokers and non-smokers under 45
(Source: *Report of the Royal College of Physicians, 1977*)

Histograms are used when it is convenient to present figures in continuous groups. Figure A.4 shows the annual death rate from coronary heart disease of male smokers and non-smokers in the under-45 age group. It shows how many men die each year of 100 000 who smoke an average each day of 0, 1–14, 15–24 and 25 or more cigarettes.

Pie charts are a form of graph that show proportions of a whole: because we are familiar with pies and cakes cut from the middle, we can grasp quickly what proportion of the whole is in each segment. Figure A.5 shows the estimated percentage use of different contraceptive methods, as well as other data on sexual activity, of women of child-bearing age in the United Kingdom in 1979.

Physical background

Energy is the capacity to do work. Energy is needed for activities as different as making protoplasm and sending a missile into space. In living organisms the important forms of energy are:

chemical energy the energy stored in chemical compounds;
radiant energy the energy from rays of light;
heat energy the energy in heat.

Vapour is gas which a liquid or solid gives off below the temperature at which it has to become gas (at normal pressure). When we say that 100°C is the boiling point of water, we mean this is the temperature at which all water will change from liquid to gas: keep a saucepan of water at 100°C for long enough and it will all become steam. This does not mean that water cannot become gas below 100°C. On the contrary there is always some gas given off by water (and by every other liquid and solid). We call the gas given off by water below 100°C **water vapour** to distinguish it from steam which is water gas with a temperature of 100°C or higher. The change of a substance to gas below the temperature at which it has to become gas is **evaporation**.

Solubility is the maximum amount of a substance that will dissolve in a certain mass or volume of another substance at a certain temperature. In short, it is a measure of how easily one substance dissolves in another. A very soluble substance is one that dissolves easily. A liquid that has another substance dissolved in it is a **solution**. A liquid that has another substance dispersed in it without being dissolved is a **suspension**.

Pressure is the force acting on a unit area. **Atmospheric pressure** is the pressure exerted by the air on a unit area: it is measured as kilopascals.

Chemical background

An **element** is a substance that cannot be divided by chemical methods.

A **compound** consists of two or more elements chemically combined.

An **atom** is the smallest stable part of an element. That an atom is stable does not mean it can exist by itself. Many atoms cannot exist by themselves.

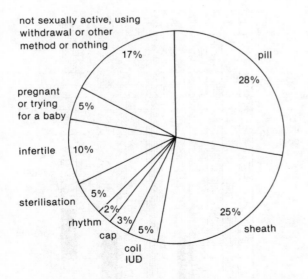

Fig. A.5 Contraception and sexual activity: percentages of UK women of child-bearing age in 1979
(Source: *Family Planning Association*)

A **molecule** is the smallest part of a substance that can exist by itself. The substance may be either an element or a compound. A molecule of carbon, which is an element, contains only one atom (C), i.e. a single atom of carbon can exist on its own. Molecules of hydrogen and oxygen, which are also elements, both contain two atoms (H_2 and O_2). A molecule of water, which is a compound, contains two atoms of hydrogen and one of oxygen (H_2O). Though the formula for oxygen (O_2) shows that a single atom of oxygen cannot exist on its own, the formula for water (H_2O) shows that a single atom of oxygen is nonetheless stable.

Other familiar biological compounds include carbon dioxide, which contains one atom of carbon and two of oxygen (CO_2), and glucose, which contains six atoms of carbon, twelve atoms of hydrogen and six atoms of oxygen ($C_6H_{12}O_6$). Like any other compounds, carbon dioxide and glucose can be broken down chemically into their elements but no further. The elements of carbon dioxide are carbon (C) and oxygen (O_2); the elements of glucose are carbon (C), hydrogen (H_2) and oxygen (O_2).

In aerobic respiration, on which nearly all life depends, one molecule of glucose is broken down with six molecules of oxygen to form six molecules of carbon dioxide and six molecules of water while energy is released:

$$C_6H_{12}O_6 + 6O_2 \rightarrow 6CO_2 + 6H_2O + \text{energy}$$

An **ion** is an electrically charged atom or group of atoms. Simple inorganic substances often move into and through living organisms as ions. As ions the parts of a compound such as sodium chloride can move independently: for example, the chloride ions may enter a cell while the sodium ions remain outside it.

pH is a measure of acidity and alkalinity related to hydrogen-ion concentration. The range is from pH 1 (very acid) to pH 14 (very alkaline); pH 7 is neutral.

Appendix B How to draw; how to answer examination questions; how to revise

How to draw

Artistic ability is not important but accuracy is. Biological drawings are not pictures: do not shade them to give them a three-dimensional look. Draw the outlines of structures to show their position and shape. Do not feel you must fill up every space: for example, it is better to draw a few cells accurately and leave the rest undrawn than to draw them all inaccurately. Always draw in pencil because you will have to rub out to achieve accuracy. It is easier to achieve accuracy in a large drawing: each finished drawing should cover at least a quarter of a page. Writing in the names of the parts is called labelling a drawing, and the names of the parts are labels. A line from a label to the drawing to indicate to which part it refers is a labelling line. To avoid confusion, never cross two labelling lines. Print your labels horizontally so that they are easy to read. Print them in pencil so that, if you decide they are wrong, you can rub them out.

Remember these rules:
- use an HB pencil with a sharp point;
- be accurate, not artistic;
- draw large;
- use a ruler to draw labelling lines;
- do not cross labelling lines;
- print labels horizontally.

How to answer examination questions

Allocating your time

In any examination it is very important to do as you are instructed. The examination papers will tell you how many marks are awarded for each question. Suppose that you are told to answer all the questions in a paper and the last two questions carry 6 marks and 7 marks respectively. If you spend too long on the earlier questions and do not tackle the last two, you are bound to lose 13 marks. Moreover it is usual for each question to carry a few easy marks as well as a few difficult ones: if you fail to tackle the last two questions, you are probably losing something between 5 and 7 easy marks.

Suppose that a paper lasts an hour and contains twelve questions, all compulsory. It may seem sensible to divide your time equally between the questions by allowing yourself five minutes for each. But there is no reason why the questions should be of equal difficulty and require equal time. The guide to their difficulty and to the time you should spend on them is the number of marks allowed for each. Suppose that a one-hour (60-minute) paper carries a total of 70 marks. That is an average of 6/7 of a minute per mark. You can afford to allow yourself three minutes for a question with 3–5 marks, six minutes for a question with 6–8 marks and nine minutes for a question with 9–11 marks and so on.

Suppose that a one-and-a-half hour paper contains six compulsory questions each carrying 15 marks (a total of 90 marks). Since each question carries the same number of marks, you can divide your time equally between them by allowing fifteen minutes for each, or one minute per mark. This makes it easy to allocate your time between the sections of questions: if a section carries 5 marks, you can spend five minutes on it, if it carries 10 marks, ten minutes, and so on.

If you allow yourself, for example, six minutes for a question or section and you have not finished it after six minutes, you must leave it and go to the next question or section. If you finish the last question before the time is up, you can go back and do more work on the questions you left unfinished. If you finish all the questions before the time is up, you can re-read your answers to see if there is anything you have missed out or if there is any other way in which you can improve them.

Even if you know the room where you will sit the examination has a clock, it is sensible to take your own watch so that you can see the time without looking up.

Answering a question

The most important thing is to read the question carefully. If you fail to do this and grasp only that, for example, a question is about respiration, you may write lots of things you do not need to write about respiration and yet not write what is required. You must answer the question exactly as it is set: no marks are given for irrelevant information. On the other hand you do not lose marks for giving wrong information: if you do not know the answer, you should guess at it.

Diagrams

A labelled diagram is often an easy way to get marks. If you give information in a diagram, there is no need to repeat it in your written answer.

How much should you write?

You should avoid writing more than you need to because it wastes time. In a short-answer question you may be given a space in which to write your answer. You can be sure that the space provided is the most that can be needed by someone with large handwriting. There is therefore no need to fill the space for the sake of doing so. Suppose that the question is:

What is the function of a chloroplast? . . . . . . (1 mark)

It is enough to write the one word 'Photosynthesis'. It is a waste of time to write: 'The function of a chloroplast is photosynthesis'.

Just as the marks allowed for each question or section of a question give you a guide to how long to spend on it, so they give you a guide to how much to write. Where only one mark is available, one word or a few words will be enough. If 4 marks are allowed, you are obviously expected to write more. Suppose the question is:

What happens to the iris of the eye when you come out of a dark cinema into sunlight? (4 marks)

Suppose you write: 'The iris gets bigger and the pupil gets smaller.' This is excellent but it makes only two points and will not earn you more than two marks. You must think of something else. You can say that the circular muscles of the iris contract while the radial muscles of the iris relax. Or you can draw the iris in the cinema and in sunlight and add notes to the diagram, as in Figure B.1, explaining that the change is brought about by contraction of the circular muscles and relaxation of the radial muscles.

Scientific terms

Use scientific terms if you know them but explain things any way you can if you do not. It is better to refer to an insect's antennae than to its feelers, but better to refer to its feelers than not to describe them at all.

Things you have not learnt about

Do not expect all the questions to be about things you have studied. The GCSE examination is more concerned to test what you understand than what you know. Some of the questions therefore tell you things you are not expected to know and ask you to explain them. You can always do so by applying the simple principles and techniques explained in this book. But you must read such questions very carefully to ensure that you understand them and make use of the information given in them. Do not be surprised if you need to read them two or three times: they will make more sense each time you read them.

Describing experiments

Your descriptions of experiments must give details. It is not enough to say: 'Keep the other conditions

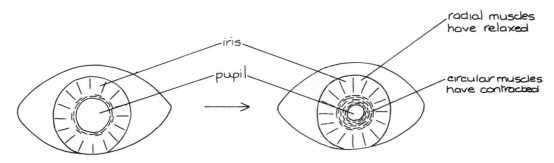

Fig. B.1

constant'. You must give details such as: 'Keep the temperature the same; keep the light intensity constant; add the same amount of fertiliser; add the same amount of water'. Give numerical values whenever you can. 'Sow a large number of seeds' is too vague. Say 'Sow 50 seeds' or however many are needed. 'Take many readings' is also too vague. Say 'Take a reading every five minutes for half an hour', or 'Take a reading once a day for two weeks'. A labelled drawing of an experiment will often be a simple way of giving a lot of necessary information. Instead of saying 'Sow 30 seeds, in five rows of six, in sterile sand in a seed tray', you can say 'Sow seeds as shown in the diagram' and draw the diagram in Figure B.2.

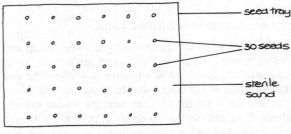

Fig. B.2

Past examination papers

Many of the questions at the end of the units are like those in the GCSE examination. To gain more practice you should answer questions from previous years' examination papers.

Examiner's reports

The Southern Group may publish reports on GCSE examinations. You can learn a lot from these reports and the past examination papers to which they refer. The reports explain what answers were required to the questions and what common mistakes were made.

How to revise

When you revise for the examination, remember that you learn best by making yourself think. Do not just read through this book or your notes. Look at the title of each unit in turn. Ask yourself what you know about each subject mentioned in it: perhaps make a few rough notes and drawings to jog your memory. Only after you have done all that should you go through the unit itself to check that what you have remembered is correct and to see what you have missed out. In this way you will discover the subjects and bits of subjects you really need to revise. Make a list of them all. When you have gone through all the units in this way, work your way once more through the list of subjects you needed to revise, seeing what you can remember of each one and only consulting this book after you have done so.

Revision is not something that need wait until the examination is near. You can revise the units you have already studied at any time. The better you understand and remember the units you have already studied, the easier and more interesting will be those you study in future.

Appendix C for teachers   Practical work

The experiments in this book assume: a laboratory with electricity, gas, mains water, hand-washing facilities and sufficient work space for a class; access to a refrigerator, to a dark cupboard at room temperature and to storage space where students can leave experimental material for up to six weeks; an autoclave or high-dome pressure cooker to sterilise glassware and reagents for class use; and the equipment, chemicals and biological material listed. Distilled water is required in quantities that justify a still.

The lists of equipment, chemicals and biological material indicate the experiment or experiments in which each item is required. They constitute what is desirable rather than what is essential: jars can be substituted for beakers, watches for stopclocks, empty yoghurt cartons for plant pots; one microscope, balance or Bunsen burner can often be shared by a number of students, though this will make experiments last longer. Except in the few experiments requiring them to work in pairs, students should perform practical work individually, even if it means sharing equipment, so that their skills can be adequately assessed.

In ordering mammalian parts teachers must take account of their students' religions: parts from pigs would offend Jews and Muslims; parts from cattle would offend Hindus. Sheep's parts are acceptable to virtually all religious communities represented in British schools and colleges.

There should be no economising in safety equipment: safety spectacles should be worn wherever indicated and solutions should be pipetted with a pipette filler, not by mouth.

Students can perform some of the simpler experiments at home.

General laboratory equipment

Numbers indicate the experiments in which each item is required

adhesive tape, waterproof 3.2, 15.1, 42.2
bag, small polythene 42.2
bag, large polythene 3.1, 3.2, 40.1
balance, minimum weight 0.05 g 5.4, 5.5, 5.6, 15.2, 16.1, 16.2, 38.1, 48.1
beaker, 100 cm^3 5.6, 15.2, 22.1
beaker, 250 cm^3 3.3, 5.5, 5.6, 7.1, 15.2, 16.1, 16.2, 17.1, 17.2, 18.1, 21.1, 22.1, 23.1, 23.2, 31.2, 32.3, 38.1, 40.1, 47.1
beaker, 500 cm^3 5.4, 7.2
beaker, clear spoutless, 500 cm^3, with glass cover 28.2
blood-grouping card and mixing stick 26.1
bottle, minimum volume 3 dm^3 28.3
bottle, weighing 5.4, 5.5, 5.6, 48.1
bowl (e.g. washing-up style) 16.1, 27.2, 28.2, 28.3, 28.5, 36.1, 40.1, 47.1
box, cardboard 31.1
brush, fine paint 48.1
bucket 35.1
bung, rubber, plain 27.1, 28.5
bung, rubber, one-hole 5.4
bung, rubber, two-hole 5.6, 28.1, 28.5
Bunsen burner 5.1, 5.2, 7.1, 17.1, 17.2, 18.1, 21.1, 22.1, 22.2, 23.1, 38.1
candle, night-light 5.3, 28.2
clamp, stand and boss head 7.2, 28.5
cling film 10.1, 15.2, 16.2
clip, screw-type 7.2
comb for use with quadrat frame 3.1
compass, plotting-type 3.2
cork borer 7.1
cotton wool 5.3, 5.6, 9.1, 15.1, 20.1, 25.1, 26.1, 28.5, 31.1, 39.1, 42.1
cup, plastic 3.2
cylinder, measuring, 100 cm^3 3.2, 5.5, 16.2
cylinder, measuring, 100 cm^3 with glass cover 15.1
dibber 47.1
dish, dissection 24.1, 25.1, 27.2, 29.1
dividers, geometrical 39.1
elastic bands 22.1
filter paper 7.1
filter paper, fluted 17.2
flask, conical, 100 cm^3 28.5
flask, conical, 250 cm^3 28.1
foil, aluminium cooking-type 5.2, 20.1, 27.1
forceps 5.2, 7.1, 9.1, 13.2, 15.2, 23.1, 24.1, 27.1, 28.5, 29.1, 37.1, 41.1, 42.2, 48.1
funnel, filter 7.1, 17.2

glass rod 5.5, 15.1, 32.3
gloves, disposable 24.1, 26.1, 27.2, 29.1
hand-lens (pocket magnifier) 23.2, 27.2, 41.1
heat-resistant mat (bench mat) 5.2, 38.1
inoculating loop 22.2
jar with tightly fitting lid 5.2, 5.3
label, plastic horticultural 36.1, 47.1
labstix for protein and reducing sugar 30.2
lamp, bench 7.2, 27.1, 31.1
lancet, sterile disposable 25.1, 26.1
line, washing-type, 21 m long 3.1
loam, garden 40.1, 47.1
microburette 7.2
microscope with at least a × 10 objective 13.1, 13.2, 14.1, 30.1, 32.2, 37.1
microscope with at least a × 40 objective 25.1
modelling clay 40.2, 46.1
moss peat 40.1, 47.1
needle, mounted dissection 31.2, 37.1
paper, absorbent 13.1, 13.2, 14.1, 16.1, 16.2, 25.1
peat pellet, compressed (Jiffy 7s) 36.1, 40.1
peat pot 40.1
peg, wooden or metal tent 3.1
pestle and mortar 7.1, 17.2
petri dish 3.3, 22.2, 23.1, 31.1, 42.1
pin, stainless dissecting 40.1
pipette, dropper (Pasteur-type) 3.3, 5.4, 13.1, 13.2, 14.1, 17.1, 17.2, 18.1, 18.2, 21.1, 22.1, 24.1, 25.1, 26.1, 30.2, 37.1
pipette, 1 cm^3 15.1
pipette, 2 cm^3 27.1
pipette, 5 cm^3 3.3, 16.1, 17.1, 18.1, 18.2
pipette, graduated, 5 cm^3 21.1, 22.1
pipette, 10 cm^3 17.2
pipette, graduated, 10 cm^3 5.4, 5.5, 7.1
pipette filler all experiments using pipettes
poppit beads (poppet beads) 40.2, 46.1
pot, plant 39.2, 47.1
potting compost 39.2
pump, vaccum-type 28.5
quadrat frame, 50 cm × 50 cm 3.1
quadrat frame, plastic, 10 cm × 10 cm 3.2
ruler graduated in mm 7.2, 15.2, 16.1, 23.2, 31.1, 31.2, 32.1, 36.1, 38.1, 39.1, 39.2, 40.1
sand, clean fine 7.1, 17.2
sand, sharp 40.1, 47.1
scalpel 7.2, 13.2, 17.2, 23.1, 23.2, 24.1, 27.2, 29.1, 36.1, 40.1
scissors, dissection-type 7.1, 16.1, 24.1, 29.1, 31.1
slide and coverslip 13.1, 13.2, 14.1, 25.1, 37.1
spatula 5.4, 5.5, 5.6
spectacles, safety 15.1, 18.2, 21.1, 27.1, 38.1
spirit marker 3.3, 5.3, 5.4, 5.6, 7.1, 9.1, 10.1, 15.2, 16.2, 17.1, 18.1, 18.2, 20.1, 21.1, 22.1, 22.2, 23.1, 27.1, 28.1, 31.1, 36.1, 38.1, 39.2, 42.1, 47.1, 48.1
spirit marker, fine 31.2
stopclock 5.2, 5.3, 5.5, 5.6, 7.1, 7.2, 15.1, 16.1, 16.2, 17.1, 17.2, 18.1, 18.2, 21.1, 22.1, 25.1, 28.2, 28.4, 35.1
string, 2 m long 3.2
syringe, plastic disposable, 2 cm^3 5.6, 7.2
tap, plastic three-way 5.6
test-tube (16 mm diameter) 3.3, 5.1, 5.4, 7.1, 17.1, 17.2, 18.1, 18.2, 20.1, 21.1, 22.1, 27.1, 30.2
test-tube rack 3.3, 5.1, 7.1, 17.1, 17.2, 18.1, 18.2, 20.1, 21.1, 22.1, 27.1, 30.2
thermometer to measure up to 60°C 5.1, 5.4, 10.1, 18.1, 42.1
tile, dimple (spotting) 17.1, 17.2, 18.1, 18.2, 22.1
tile, plain white 7.1, 7.2, 27.1, 41.1
tongs 5.1, 38.1
torch 32.3
towel, disposable 25.1, 26.1
tray, plastic seed 36.1, 40.1
tripod stand and gauze 5.6, 7.1, 17.1, 17.2, 18.1, 21.1, 22.1, 23.1
tube, U-type absorption with side arms 28.5
tube, boiling (25 mm diameter) 5.4, 5.6, 9.1, 15.2
tube, boiling-, rack (large test-tube rack) 5.4, 5.6, 9.1
tube, flat-bottomed specimen 28.5, 48.1
tubing, 1 mm capillary 5.6
tubing, dialysis (Visking) 16.1
tubing, delivery, 5–6 mm 5.4, 5.6, 28.1, 28.5
tubing, rubber 5.6, 7.2, 28.1, 28.2, 28.3, 28.5
wire, galvanised or copper 5.1, 5.2, 5.3, 28.2, 32.1

Chemicals

agar powder 9.1, 15.2, 22.2, 23.1, 48.1
albumen 21.1
ammonia solution 15.1
ammonium sulphate 48.1
amylase 17.1, 18.1, 18.2
ascorbic acid (vitamin C) 21.1
Benedict's qualitative solution 7.1, 17.1, 21.1, 22.1
calcium hydroxide 5.2, 5.4, 28.1
calcium phosphate 9.1
calcium sulphate-2-water 9.1, 20.1
calcium tetrahydrogenphosphate 9.1, 20.1, 48.1
cobalt-chloride paper 5.2
copper(II) sulphate 21.1
cresol red 27.1, 28.1
2,6-dichlorophenolindophenol (DCPIP) *see* phenol-indo-2,6-dichlorophenol
disinfectant 27.2

eosin (or methylene blue) 23.2
ethanol 25.1, 26.1, 27.1, 37.1
flour (plain) 5.5
gelatin 15.2
D(+) glucose 5.4, 5.5, 21.1
glucose-1-phosphate 17.2
hydrochloric acid 18.2, 27.1, 37.1, 38.1
hydrogencarbonate indicator solution 27.1, 28.1
iodine solution (1% in potassium iodide) 13.2, 17.1, 17.2, 18.1, 18.2, 21.1, 22.1, 22.2, 23.1
iron(III) chloride-6-water 20.1, 48.1
iron(II) phosphate 9.1
Leishman's stain 25.1
lime water *see* calcium hydroxide
litmus paper (red) 15.1, 38.1
magnesium sulphate-7-water 9.1, 20.1, 48.1
methylbenzene 30.2
methylene blue 13.1, 15.2, 23.2
Nipagin solution 15.2
oil (any vegetable) 5.4, 21.1
pH paper *see* Universal indicator paper
phenol-indo-2,6-dichlorophenol (PIPDC) 21.1
phloroglucinol 37.1
potassium chloride 9.1, 20.1
di-potassium hydrogenphosphate(V) 9.1, 48.1
potassium nitrate 20.1
propan-2-ol 21.1
rooting powder 36.1, 40.1
soda lime 5.6
sodium carbonate 18.2
sodium chlorate(I) 5.3, 9.1, 25.1, 26.1
sodium chloride 20.1, 48.1
sodium hydrogencarbonate 7.2, 27.1, 28.1
sodium hypochlorite *see* sodium chlorate(I)
sodium hydroxide 21.1, 27.1
starch 16.1, 17.1, 18.1, 18.2, 21.1, 22.1, 22.2, 23.1
talc (French chalk) 36.1
thymol blue 27.1, 28.1
Universal indicator paper and colour identification chart 15.1, 18.2
Universal indicator solution and colour identification chart 28.5
Wright's blood stain 25.1

Biological material

Artemia (brine shrimp) 3.3
Begonia 40.1
Bryophyllum (air plant) 40.1, 47.1
bread (or water biscuit) 5.2
celery (or *Impatiens*) 23.2
chicken ribs from a cooked carcass 38.1
Chlorophytum (spider plant) 40.1
cress seeds 31.1, 48.1
Daphnia (water flea) (or *Artemia*) 3.3
Elodea (Canadian pond weed) 7.2
French bean
  flower 41.1
  plant 42.2
  root with nodules 9.1
  seed (likely to lose viability if stored longer than a few months) 5.6, 9.1, 23.1, 42.1
  seedling 39.2
fruit (any soft-skinned) 10.1
fungus (any hyphal) 22.2
geranium *see Pelargonium*
Impatiens (Busy Lizzy) (or celery) 23.2
Iris 7.1
ivy shoot with leaves (or privet) 39.1
maize cob showing genetic segregation 45.1
mammalian
  heart 24.1
  kidney, prepared V.S. 30.1
  kidney, whole 29.1
  limb bone T.S. 37.1
  lungs and trachea 27.2
  skin, prepared V.S., human 32.2
moss leaf (any) 14.1
nitrogen-fixing bacteria 9.1
onion bulb 13.2
pea seed 5.3
peanut 5.1
Pelargonium (geranium)
  prepared T.S. of stem 37.1
  whole plant 36.1, 40.1
potato stem tuber 17.2, 23.1
privet shoot with leaves (or ivy) 39.1
stick-insect nymph (*Carausius morosus*) 39.1
sultana, dried 16.2
wheat seedling 20.1
yeast (dried baker's) 5.4, 5.5

Index

Bold type indicates relative importance. The index does not refer to Questions, Practical Work or Appendices.

abdominal cavity 100–2
ABO system **124–6**, 224–5
absorption 47–8, 100, 102–3
acid rain 55, 136
active site **79–80**, 82–3
active transport **69**, 90–1, 100, 103, 106–9, 121, 147
addiction 162–3
additive 96
adrenaline **165–6**, 169
afforestation 54
air 20, 134, 174
air sac 129, 131, 135–6
alcohol 23, 148, **162**, 169
allele 218–25, 231–2
allergy 124
alveolus 129, 131, 135–6
amino acid 78, 90, 100, 103, 106–7, 109, 111, 121, **141–2**, 169, 184
amnion 215–16
amylase 78, 100, 102, 106
anaemia 96
anther 198, **204–5**
antibody 120–1, **123–5**
antigen 123–6
aorta 113–16
artery 111–16
arthropod 2–7, 9–11, 174, **176**, 185
artificial selection 172, **235**
ascorbic acid 96–7
ATP **23**, 87
atrium 113–15
autotrophic 86, 103

bacterium 1, 4, 7, 30, 43–5, **47–9**, 57, 59, 65, 74, 83, 97, **103**, 120–1, **191**, 231–2
  denitrifying 43, 45
  nitrifying 43, 45
  nitrogen-fixing 44–5, 55, 92
baking 23
beer 23
bile 102–3, 169
binomial system 7
biodegradable 56
biological control 56
biomass **38–9**, 92
birth 216
  control 53, 212–13
bladder 142–3
blanching 83
bleeding 96, **116**, 210
blood **119–21**
  cell 60, **65**, **73**, 96, 111, **119–21**, 123–6, 169
  circulation 111–16
  clotting 96, 116, 120–1
  group **124–6**, 224–5
  sugar 111, 164–6

transfusion 124–6
bone 96, **179–81**
Bowman's capsule 146–7
brain 121, 157, **161–2**
bread 23
breeding 235
brewing 23
bronchiole 129–31
bronchitis 136
bronchus 129–30
burning 22, 41–3

calcium 91–2, 96–8
cancer 136
capillary 112–13, 116, 131–2, 146–7, 167–8, 215–16
capture–recapture technique 15
carbohydrase 100, 102–3
carbohydrate **62**, **95**, 100, 102
carbon 62, 90, 95
  cycle **41–2**, 49
carbon dioxide 41–2 *see also* blood, burning, decomposition, excretion, gas exchange, photosynthesis, respiration
carbon monoxide 136
carnivore **30–1**, 34
carpel 198, **204–5**
cartilage 179–80
cell 6, **59–62**, **64–6**
  division *see* mitosis, reduction division
  membrane **59–61**, 68–9, 72–4, 95, 100, 211
  sap 61
  size 59
  specialisation 64–6
  wall 7, 60–2, 68, 74, 90–1, 174–5
cellulose 62, 90–1, 95, 103
cerebellum 161
cerebral cortex 161
cerebrum 161
cervix **209**, 211–12
chemical control 56
chlorophyll **32**, 86–8, 91
chloroplast **32**, **60–1**, 65, 68, 86–7
chromatid 189–91
chromosome **189–91**, **195–9**, 218, 226, 229–31
circulation system **111–16**, 131–2
climax community 51
clone 191–2, 232, 234
co-dominant allele 219, 224
collecting duct 146–7
colon 101–3
colonisation 50
community 50–1
competition **51**, 235
concentration gradient *see* active transport, diffusion, gas exchange
consumer 4, **38–9**, 86
continuous growth 185
contraception 212–13
contraceptive pill 172, 212

contractile vacuole 73
coronary artery 116
cotyledon 201–2
cramp **23**, 116, 135, 182
cuticle 65, 174, 176
cutting 172, 192, 234
cytoplasm **59–62**, 65–6, 77–8, 107, 119–20

DDT 56
deamination **142**, 169
decomposer 4, 30, 38–9
decomposition 43–5, **47–9**
deficiency disease 95–6
deforestation 54
dehydration 83, 91
denaturing 82–3
dental decay 97
dermis 155–6
diabetes mellitus 164–5
dialysis 148–9
diaphragm 101, **134**
dichotomous key **8–11**
diet 39, **94–8**
diffusion **67–9**, 71–4, 90–1, 100, 103, 106–9, 121, **128–32**, 147–8
digestion 78, 100, 102–3, 106, 201
  extracellular **47–8**, 78, 103
diploid **195–8**, 218, 229–30
discontinuous growth 185
diversity of organisms 1–5
DNA **62**, 77–8, 90, 189–92, 195, 199, **218**, **231–2**
dominant allele 219
drug 162–3
drying 65, 83, 91, 174
dung 55, 91

ecology 37
ecosystem **37**, 41–4
effector 151
egg **59**, 65, 166, **195–8**, 209–12, 217, 226, 231
elbow 181–2
electron microscope 61
embryo 195, 198, 201, 205, 211–12, 217
emulsification 103
endoskeleton 179–80
energy
  capture 32–3, 86–7
  flow 37–9
  food 22–3, 59–60, 94–5, 97
  loss 33–4
  release 22–3
  transfer 29–31
  use 23, 29, 59–60, 69, 90
environment 18, 234
  influence on 54–7
environmental factor 52
  measurement of 18–20
enzyme
  activity 82–4
  function 47, **77–80**, 100, 102–3, 106, 201

names 78
 specificity 77, 79
 structure 62, 78–80
epidermis 64–6, 155–6
epithelial tissue 66
erosion 54
evaporation 108, 141, 167, 174
evolution 234
excretion 141–4, 146–9
exercise 116, 135
exoskeleton 174, **176**, 185
eye 151–2

faeces 33–4, 43, 48, 92, 103
Fallopian tube 209–11
farming technique 54–5, 91, 172
fat 22, **95**, 155–6, 168
fatigue **23**, 116, 135, 182
fatty acid 100, 103, 106, 111, 119, 121
fermentation 23
fertilisation 174, **195–9, 210–11**, 217, 220, 222, **230**
fertiliser 91
fibre 95, 103, 175
filtration 144, **146–7**
flower 197, 202–5
fluoride 97
food 94–8, 100, 102–3, 111, 121
 chain 30–1
 excess 97
 preservation 83–4
 reserve 22, 59, 62, 95, 106, 156, 201
 web 30–1
fossil fuel **41–2**, 55
freezing 83
French bean
 artificial selection 235
 germination **201–2**, 206
 growth **184–5**, 202, 206
 leaf 86, 202–3
 reproduction 197–8, **201–6**
 root nodule 44
fruit 92, 172, **205**
fungicide 56, 172
fungus 4, 7, 30, 43, **47–9**, 50, 74, 83, **103**, 191

gall bladder 102–3, 169
gamete *see* egg and sperm
gas exchange 128–32, 134–5
gene 218, 231
genetic engineering 172, **231–2**
genetic variation 191, 229–32
genetics 218–26
genotype 219–21, 223–5
genus 7–8
German measles 124
germination 184, **201–2**, 206
gland
 adrenal 142–3, **165**
 ductless (endocrine) 164
 grease (sebaceous) 155
 prostate 208–9
 salivary 100
 sweat 155
glomerulus 146–7
glucose 22, 78, 80, 86, 164–6
glucose (starch) phosphorylase 78, 80, 87
glucose-1-phosphate 78, 80
glue-sniffing 163
glycerol 100, 103, 106, 111
glycogen 22, 62, 95, 103, 164–6, 169
growth 90–1, 106, **184–6**, 189, 202
 curve 206

guard cell **64–5**, 67, 87
gum disease 97
gut 100–3, 111–13

habitat 4, **37**, 50–1
 method of study 13–15
haemoglobin 65, 96, **119**, 131
hair 1, 155–6, 168
haploid **195–8**, 229–30
hay fever 124
heart 66, 111–16
 attack 116
heat loss and gain 23, 33–4, 121, 156, **167–9**
hepatic portal vein 111–13
herbicide 56, **172**
herbivore **29**, 31, 33
heterotrophic 86, 103
hinge joint 182
homeostasis 147, **169**
Homo erectus 7, 94–5
Homo sapiens 7, 94–5
homozygous 219–25
hormone 96, 121, 151, **164–6**, 171–2, 186
 artificial use 172, 212
horticultural technique 91, 172, 192
host 48
humidity 19, 52, 108
humus 107
hydrochloric acid 100
hydrogen 62, 87, 90, 95
hypha 47
hypodermis 155
hypothesis 13, 78, 94

identification 8
immunisation 124
implantation 211–12
increase in complexity 6–7
independent assortment 230
inheritance 186, **218–26**
inhibitor **79–80**, 83
insect 3–5, 7, 10, 174
insecticide *see* pesticide
insemination 210–12
insulin 111, **164–6**, 169
intestine 100–3
involuntary action 152, 157
ion 20, 32, 43–5, 52, 90–2, 96, 103, 109, 121, 141, 147, 169
 ammonium 43–5, 90, 109
 nitrate 43–5, 90, 109
iris **152**, 157
iron 91, 96–8, 169
irritability 151

joint 176, 180–2

key 8–11
kidney 73, 112–13, **141–4**, 146–9
 failure 148
 machine 148–9
 transplant 149

lactic acid **23**, 116, 135, 182
land organism 4–5, 174
land use 54–5
leaching 92
lead 148
leaf **64–7**, **86–7**, 90–2, 108, 128–9, 202–3
lever 182
lichen 50
life cycle **195–9**, 201–6

ligament 181
light measurement 19
lightning 44
lignin 175
limestone 91
lipase 78, 100, 102–3, 106
lipid 62, 78, **95**, 97, 100, 102–3, 106, 111, 119, 121, 169
liver 95–6, 101–2, 111–13, 121, 142, **169**
longitudinal section (L.S.) 61
lung 112–14, **129–31**, **134–6**

Malpighian layer 155–6
mammal 1, 5, 7, 11, 216
mammary gland 1, 216
mapping 13
maturity 206
medulla 161–2
meiosis *see* reduction division
melanin 156
membrane
 cell **59–61**, 68–9, 72–4, 95, 100, 211
 impermeable 69, 71
 nuclear 59–61
 permeable 68, 71
 selectively permeable 68, **71–3**
menstrual cycle 166, 210–11
mesophyll 65–7, 86, 108
milk 216
mineral 90–2, 96
mitosis **189–92**, 218
monoculture 54
monohybrid cross 221–2
mould *see* fungus
moulting 176, 185
mouth cavity **100**, 152
mouth-to-mouth ventilation 135
movement 180–2
mucus 100, 129, 131, 136, 210–11
muscle 95–6, 100, 112–13, 115–16, 134, 157, 180–2, 216
 antagonistic 152, 181
 cell 66
 cramp **23**, 116, 135, 182
mutation 231
mutualism 44, 48
myoglobin 96
myxomatosis 56, 234–5

natural selection 234–5
nephron 146–7
nerve 157
 cell 66
 impulse 96, 151–2, **157–8**
nipple 1, 209, 216
nitrogen 62, **90–1**, 95
 cycle 43–5
non-renewable resource 55
nucleus **59–62**, 78, 120, 189–91, 195–8, 211, 218, 229–30

oesophagus 100–2
omnivore **30–1**, 34, 39
optimum 82
oral hygiene 97
organ 66
osmoregulation 147, 169
osmosis **71–4**, 90–1, 107–9, 147
ovary 197, **204**, **209–10**, 212
ovulation **209–10**, 212
ovule 197–8, 204–5
ovum *see* egg
oxidation 22

oxygen **42**, 62, 90, 95, 121, 174
  debt **23**, 116, 135, 182
  for respiration **22**–**3**, 67, **128**–**32**, 134–5
  from photosynthesis **32**–**3**, 67, **86**–**8** 141

pain killer 162–3
pancreas 100–2, 164
parasite 30–1, **48**
penis 208–10, 212
percentage cover 13
pest 54
pesticide 56
petal 203
pH 20, 52, 82–4, 100–2
phenotype 219–21, 223–5
phloem 87, 90, 106–7, 109
photosynthesis 32–3, 61, 64–5, 68, 86–8, 128–9, 141
pickling 84
placenta 215–16
plaque 97
plasma 69, 111, 119–21
plasmolysis 74
platelet 119–21
point frame 14
pollen 124, 198, 203–5
pollination 198, 204–5
pollution 55–7, 136–7
population **50**–**3**, 217
  control **51**–**3**, 212–13
predator 30
pregnancy 136, 163, **215**–**16**
prey 30
probability 221
producer 4, **38**–**9**, 86
protease 78, 100, 102–3, 106
protein 62, 77–8, 82–3, 90, **95**, 97–8, 100, 102, 106, 123, 141, 149, 176, 179, 184, 201
protoplasm 6, **59**, 62, 95, 184–5
puberty 166
pulmonary artery and vein **112**–**15**, 131
pulse 116
Punnett square 219–21
pupil 152
pyramid of biomass 39

quadrat 13
  frame and comb 13

receptor **151**–**2**, 155–8
recessive allele 219
recombinant DNA 231–2
rectum 102–3
recycling 56–7
reduction division 196, 209, 220, **229**–**30**
reflex action 152, **157**–**8**
refrigeration 83
renal artery and vein **142**–**4**, 146–7
reproduction
  asexual **191**–**2**, 195
  sexual 174, 191, **195**–**9**, **208**–**11**, **215**–**17**, 230
respiration
  aerobic **22**–**3**, 128–9
  anaerobic **22**–**3**, 116, 135, 182
response **151**–**2**, 157
retina 151–2, 157
root hair 90–1, **107**–**9**
root nodule **44**, 55
root pressure 108
roughage 95

saliva 100
salting 83–4
sampling 13–14
saprophyte 30–1, **48**–**9**, 57, 78, 83, 103, 191
scavenger 29, 31, 48
scrotum 208–9
scurvy 95–6
secondary sexual characteristic 166
sedative 162
seed 106, 184, **201**–**2**, **204**–**6**
  dispersal 205–6
selective reabsorption 144, 146–7
semen 209–13
seminal vesicle 208–9
senescence 205–6
sense organ 151–2, 156–7
sensitivity 151–2
sepal 203
sewage 55–7
sex chromosome 226
sex hormone **166**, 172, 210, 215
shape 176–7
shivering 169
sieve tube 106–7
size 176–7, 180
skeleton 174, 176, **179**–**80**
skin 141, 152, **155**–**6**, **167**–**8**
smoking 135–6
soil 52, **90**–**2**, 107, 109
  habitat 50
  measurement 20
species 7–8
sperm 59, **65**, 166, **195**–**6**, 208–13, 217, 226, 231
spinal cord 157–8, 161
spore 4, **191**, 195
stamen 198, **204**–**5**
starch 22, 61–2, 78, 80, 87, 95, 100, 106
STD 214
stigma 197–8, **204**–**5**
stimulant 162
stimulus 151–2, 157–8
stoma **64**–**5**, 67, 87
stomach 100–1
style 197–8, **204**–**5**
substrate 78–9, 83
succession 50–1
sucrose 106–7
sugar **22**–**3**, **32**–**3**, 71–3, 86–7, 94–5, 97–8, 100, 103, 106, 111, 164–6, 169
sulphur dioxide 55, 136–7
support 174–6, 179–80
surface area 176–7
sweat 141, 147, 156–7, **167**–**8**
synapse 157–8
syruping 83

target organ 164
temperature 52, 82–3
  measurement 18–19
  regulation 156, **167**–**9**
tendon 115, 180–1
testis 166, **208**–**9**
testosterone 166
test-tube baby 212
thrombosis 116, **120**–**1**
tissue 66
  culture 192
  fluid 121
  matching 149
  rejection 124, 149
tongue-rolling 223–4

tooth decay 97
toxic substance 148
transect 14
transpiration 108–9
transport 106–9, 111–16, 121
transverse section (T.S.) 61–2
trap 15
trophic level 38
tropism 151
trout 51, **217**, 234
turgidity **74**, 174–5
twin 124, 149, 192, **229**–**31**, 234

ultrafiltration 147
umbilical cord 215–16
urea 43, 121, **142**–**4**, 147–8, 169
ureter 142–3
urethra **142**–**3**, 208–9
urine 43, 92, **141**–**4**, 147
uterus 209–12, 215–16

vaccination 123–4
vacuole 61, 73–4, 174
vagina 209–11, 216
valve 100, 103, 113, 115–16
variation
  environmental 229, **234**
  genetic 191, **229**–**32**
vascular bundle 65, **106**–**9**
vasoconstriction 167–8
vasodilation 167–8
vegetative propagation 192
vein 65, 86–7, 90, **106**–**9**, **111**–**16**
vena cava 113–15
ventilation 67, 128–30, **134**–**5**
ventricle 113–15
villus 102–3, 215
virus 120–1, 123–4
vital capacity 134
vitamin 95–7
voluntary action 157

water 47–8, **67**–**8**, **71**–**4**, 83, 90–1, **96**, 100, 103, 119, 167, 174, 217
  excretion **141**–**4**, 147
  for photosynthesis 32, 86–8
  osmoregulation 147
  transport **107**–**9**, 121
weedkiller 56, **172**
windpipe 129
withdrawal reflex 157

xylem 90, **108**–**9**, 175

yeast 23

zygote **195**, 198, 205, 211, 217, 226, 230